SCENES AND LEGENDS

OF

THE NORTH OF SCOTLAND

OR

The Traditional History of Cromarty.

BY HUGH MILLER,

AUTHOR OF 'THE OLD RED SANDSTONE,' ETC. ETC.

SEVENTH EDITION.

EDINBURGH:

WILLIAM P. NIMMO.

1869.

SCENES AND LEGENDS

OF

THE NORTH OF SCOTLAND.

SCENES AND LEGENDS

OF

THE NORTH OF SCOTLAND

OR

The Traditional History of Cromarty

HUGH MILLER

DEDICATION

TO

SIR THOMAS DICK LAUDER

OF GRANGE AND FOUNTAINHALL, BARONET.

HONOURED SIR,

I am not much acquainted with what Goldsmith has termed the ceremonies of a dedication. I know, however, that like other ceremonies, they are sometimes a little tedious, and often more than a little insincere. But it is well that, though dulness be involuntary, no one need deceive unless he wills it. There are comparatively few who seem born to think vigorously, or to express themselves well ; but since all men may be honest, though all cannot be original or elegant, every one, surely, may express only what he feels. In dedicating this little volume to you, I obey the dictates of a real, though perhaps barren, gratitude ; nor can I think of the kind interest which you have taken in my amusements as a writer, and my fortunes as a man, without feeling that, though I may be dull, I cannot be insincere.

There are other motives which have led to this address. He who dedicates, more than expresses his gratitude. By his choice of a Patron, he intimates also, as if by specimen, the class which he would fain select as his readers ; or, as I should perhaps rather express myself, he specifies the peculiar cast of intellect and range of acquirement from which he anticipates the justest appreciation of his labours, and the deepest interest in the subject of them. Need I say that I regard you, Sir

Thomas, as a representative of the class whom it is most my ambition to please ?   My stories, arranged as nearly as possible in the chronological order, form a long vista into the *past* of Scotland, with all its obsolete practices and all its exploded beliefs.   And where shall I find one better qualified to decide regarding the truth of the scenery, the justness of the perspective, or the proportions and costume of the figures, than he whom contemporary genius has so happily designated as the "Poet and Painter of the great Morayshire Floods ?"   I can form no higher wish than that my work may prove worthy of so discerning a critic, or that you, Sir, may be as fortunate in your *protégé* as I in my patron.

I am, I trust, no *hypocrite* in literature, but a right-hearted devotee to whom composition is quite its own reward.   If my little volume succeed, I shall be gratified by reflecting that the pleasure derived from it has not been confined to myself ; if it fail, there will be some comfort in the thought that it has proved, to at least one mind, a copious source of entertainment.   Besides, I am pretty sure, I shall be sanguine enough to transfer to some production of the future, the few hopes which, in the past, I had founded on it.   And when thinking of it as the "poor deceased," I reflect that, at worst, it was rather dull than wicked, and that it rather failed in performance than erred in intention ; I shall not judge the less tenderly regarding it, when I further remember that it procured for me the honour of your notice, and furnished me with this opportunity of subscribing myself,

Honoured Sir,

With sincere respect,

Your humble friend, and obedient Servant,

THE AUTHOR.

CROMARTY, 1834.

NOTE TO THE SECOND EDITION.

THE present edition contains about one-third more matter than the first. The added chapters, however, like those which previously composed the work, were almost all written about twenty years ago,[1] in leisure hours snatched from a laborious employment, or during the storms of winter, when the worker in the open air has to seek shelter at home. But it is always less disadvantageous to a traditionary work, that it should have been written early than late. Of the materials wrought up into the present volume, the greater part was gathered about from fifteen to twenty years earlier still ; and though some thirty-five or forty years may not seem a very lengthened period, such has been the change that has taken place during the lapse of the generation which has in that time disappeared from the earth, that perhaps scarce a tithe of the same matter could be collected now. We live in an age unfavourable to tradition, in which the written has superseded the oral. As the sun rose in his strength, the manna wasted away like hoar-frost from off the ground.

In preparing my volume a second time for the press, I have felt rather gratified than otherwise, that, at least, much of what it contains should have been preserved. The reader will here and there find snatches of dissertation, which would perhaps not be missed if away—which, at all events, had they not been written before, would have remained unwritten now ;

---

[1] Chiefly between the years 1829 and 1832, inclusive. A few of the paragraphs were, however, introduced at a later time.

but which I have spared, partly for the sake of the associations connected with them, and partly under the impression that the other portions of the work would have less of character if they were wanting. Some of these dissertative fragments I have, however, considerably abridged, and there were others of a similar kind in the first edition which have been wholly suppressed. In my longer stories I have, I find, exercised the same sort of liberty in filling up the outlines as that taken by the ancient historians in their earlier chapters. Livy in the the times of the Empire could write speeches for Romulus and Junius Brutus, and introduce them into his narrative as authentic; and Tacitus details as minutely, in his Life of Agricola, the deliberations of the warlike Caledonians as if he had formed one of their councils. Even the sober Hume puts arguments for and against toleration into the mouths of Cardinal Pole and his opponents which belonged to neither the men nor the age. But though I have, in some cases, given shade and colour to the original lines, in no case have I altered the character of the drawing. I have only to state further, that the reader, when he finds reference made, in the indefinite style of the traditionary historian, to the years which have elapsed since the events related took place, must add in every instance twenty additional twelvemonths to the number; the some thirty bygone years of my narratives have stretched out into half a century, and the half century into the threescore years and ten.

CONTENTS.

SCENES AND LEGENDS.

THE TRADITIONAL HISTORY OF CROMARTY.

CHAPTER I.

" Tradition is a meteor, which, if once it falls, cannot be rekindled."—Johnson.

EXTREMES may meet in the intellectual as certainly as in the moral world. I find, in tracing to its first beginnings the slowly accumulated magazine of facts and inferences which forms the stock in trade on which my mind carries on its work of speculation and exchange, that my greatest benefactors have been the philosophic Bacon and an ignorant old woman, who, of all the books ever written, was acquainted with only the Bible.

When a little fellow of about ten or twelve years of age, I was much addicted to reading, but found it no easy matter to gratify the propensity ; until, having made myself acquainted with some people in the neighbourhood who were possessed of a few volumes, I was permitted to ransack their shelves, to the no small annoyance of the bookworm and the spider. I read incessantly ; and as the appetite for reading, like every other kind of appetite, becomes stronger the more it is indulged, I felt, when I had consumed the whole, a still keener craving than before. I was quite in the predicament of the shipwrecked

A

sailor, who expends his last morsel when on the open sea, and, like him too, I set myself to prey on my neighbours. Old greyheaded men, and especially old women, became my books; persons whose minds, not having been preoccupied by that artificial kind of learning which is the result of education, had gradually filled, as they passed through life, with the knowledge of what was occurring around them, and with the information derived from people of a similar cast with themselves, who had been born half an age earlier. And it was not long before I at least *thought* I discovered that their narratives had only to be translated into the language of books, to render them as interesting as even the better kind of written stories. They abounded with what I deemed as true delineations of character, as pleasing exhibitions of passion, and as striking instances of the vicissitudes of human affairs—with the vagaries of imaginations as vigorous, and the beliefs of superstitions as wild. Alas! the epitaph of the famous American printer may now be written over the greater part of the volumes of this my second library; and so unfavourable is the present age to the production of more, that even that wise provision of nature which implants curiosity in the young, while it renders the old communicative, seems abridged of one-half its usefulness. For though the young must still learn, the old need not teach; the press having proved such a supplanter of the past-world schoolmaster, Tradition, as the spinning-wheel proved in the last age to the distaff and spindle. I cannot look back on much more than twenty years of the past; and yet in that comparatively brief space, I see the stream of tradition rapidly lessening as it flows onward, and displaying, like those rivers of Africa which lose themselves in the burning sands of the desert, a broader and more powerful volume as I trace it towards its source.

It has often been a subject of regret to me, that this oral knowledge of the past, which I deem so interesting, should be thus suffered to be lost. The meteor, says my motto, if it once fall, cannot be rekindled. Perhaps had I been as conversant, some five or ten years ago, with the art of the writer as with the narratives of my early monitors, no one at this time of day would have to entertain a similar feeling ; but I was not so conversant with it, nor am I yet, and the occasion still remains. The Sibyline tomes of tradition are disappearing in this part of the country one by one ; and I find, like Selkirk in his island when the rich fruits of autumn were dropping around him, that if I myself do not preserve them they must perish. I therefore set myself to the task of storing them up as I best may, and urge as my only apology the emergency of the case. Not merely do I regard them as the produce of centuries, and like the blossoms of the Aloe, interesting on this account alone, but also as a species of produce which the harvests of future centuries may fail to supply. True it is, that superstition is a weed indigenous to the human mind, and will spring up in the half-cultivated corners of society in every coming generation ; but then the superstitions of the future may have little in common with those of the past. True it is, that human nature is intrinsically the same in all ages and all countries ; but then it is not so with its ever-varying garb of custom and opinion, and never again may it wear this garb in the curious obsolete fashion of a century ago.——Geologists tell us that the earth produced its plants and animals at a time when the very stones of our oldest ruins existed only as mud or sand ; but they were certainly not the plants and animals of Linnæus or Buffon.

The traditions of this part of the country, and of perhaps every other, may be divided into three great classes. Those of the first and simplest class are strictly local ; they record real events, and owe their chief interest to their delineations of character. Those of the second are pure inventions. They are formed mostly after a set of models furnished perhaps by the later bards, and are common—though varying in different places according to the taste of the several imitators who first introduced them, or the chance alterations which they afterwards received—to almost every district of Scotland. The traditions of the third and most complex class are combinations of the two others, with in some instances a dash of original invention, and in others a mixture of that superstitious credulity which can misconceive as ingeniously as the creative faculty can invent. The value of stories of the first class is generally in proportion to their truth, and there is a simple test by which we may ascertain the degree of credit proper to be attached to them. There is a habit of minute attention almost peculiar to the common people (in no class, at least, is it more perfect than in the commonest), which leads them to take a kind of microscopic survey of every object suited to interest them; and hence their narratives of events which have really occurred are as strikingly faithful in all the minor details as Dutch paintings. Not a trait of character, not a shade of circumstance, is suffered to escape. Nay more, the *dramatis personæ* of their little histories are almost invariably introduced to tell their own stories in their own language. And though this be the easiest and lowest style of narrative, yet to invent in this style is so far from being either low or easy, that with the exception of Shakspere, and one or two more, I know not any who have excelled in it. Nothing more common than those faithful memories which can record whole conversations, and every attendant circumstance, however minute; nothing less so than that just conception of character and vigour of imagination, which can alone construct a natural dialogue, or depict, with the nice pencil of truth, a scene wholly fictitious. And thus though any one, even the weakest, can mix up falsehoods with the truths related in this way, not one of a million can make them amalgamate. The iron and clay, to use Bacon's illustration, retain their separate natures, as in the feet of the image, and can as easily be distinguished.

The traditions of the second class, being in most instances

only imperfect copies of extravagant and ill-conceived originals, are much less interesting than those of the first ; and such of them as are formed on the commoner models, or have already, in some shape or other, been laid before the public, I shall take the liberty of rejecting. A very few of them, however, are of a superior and more local cast, and these I shall preserve. Their merit, such as it is, consists principally in their structure as stories—a merit, I am disposed to think, which, when even at the best, is of no high order. I have observed that there is more of plot and counter-plot in our commonest novels and lowest kind of plays, than in the tales and dramas of our best writers ; and what can be more simple than the fables of the Iliad and the Paradise Lost !—From the third class of traditions I trust to derive some of my choicest materials. Like those of the first, they are rich in character and incident, and to what is natural in them and based on fact, there is added, as in Epic poetry, a kind of machinery, supplied either by invention or superstition, or borrowed from the fictions of the bards, or from the old classics. In one or two instances I have met with little strokes of fiction in them, of a similar character with some of even the finest strokes in the latter, but which seem to be rather coincidences of invention, if I may so express myself, than imitations.—There occurs to me a story of this class which may serve to illustrate my meaning.

In the upper part of the parish of Cromarty there is a singularly curious spring, termed Sludach, which suddenly dries up every year early in summer, and breaks out again at the close of autumn. It gushes from the bank with an undiminished volume until within a few hours before it ceases to flow for the season, and bursts forth on its return in a full stream. And it acquired this peculiar character, says tradition, some time in the seventeenth century. On a very warm day of summer, two farmers employed in the adjacent fields were approaching the spring in opposite directions to quench their thirst. One of them was tacksman of the farm on which the spring rises, the other tenanted a neighbouring farm. They had lived for some time previous on no very friendly terms. The tacksman, a coarse, rude man, reached the spring first, and taking a hasty draught, he gathered up a handful of mud, and just as his neighbour came up, flung it into the water. " Now," said he, turning away as he spoke, " you may drink your fill." Scarcely had he uttered the words, however, when the offended stream began to boil like a caldron, and after bubbling a while among the grass and rushes, sunk into the ground. Next day at noon the heap of grey sand which had been incessantly rising and falling within it, in a little conical jet, for years before, had become as dry as the dust of the fields ; and the strip of white flowering cresses which skirted either side of the runnel that had issued from it, lay withering in the sun. What rendered the matter still more extraordinary, it was found that a powerful spring had burst out on the opposite side of the firth, which at this place is nearly five miles in breadth, a few hours after the Cromarty one had disappeared. The story spread ; the tacksman, rude and coarse as he was, was made unhappy by the forebodings of his neighbours, who seemed to regard him as one resting under a curse ; and going to an elderly person in an adjoining parish, much celebrated for his knowledge of the supernatural, he craved his advice. " Repair," said the seer, " to the old hollow of the fountain, and as nearly as you can guess, at the hour in which you insulted the water, and after clearing it out with a clean linen towel lay yourself down beside it and abide the result." He did so, and waited on the bank above the hollow from noon until near sunset, when the water came rushing up with a noise like the roar of the sea, scattering the sand for several yards around ; and then, subsiding to its common level, it flowed on as formerly between the double row of cresses. The spring on the opposite side of the firth withdrew its waters about the time of the rite of the cleansing, and they have not since re-appeared ; while those of Sludach, from that day to the present, are presented, as if in scorn, during the moister seasons, when no one regards them as valuable, and withheld in the seasons of drought, when they would be prized. We recognise in this singular tradition a kind of soul or Naiad of the spring, susceptible of offence, and conscious of the attentions paid to it ; and the passage of the waters beneath the sea reminds us of the river Alpheus sinking at Peloponnesus to rise in Sicily.

Next in degree to the pleasure I have enjoyed in collecting these traditions, is the satisfaction which I have felt in contemplating the various cabinets, if I may so speak, in which I found them stored up according to their classes. For I soon discovered that the different sorts of stories were not lodged indiscriminately in every sort of mind—the people who cherished the narratives of one particular class frequently rejecting those of another. I found, for instance, that the traditions of the third class, with all their machinery of wraiths and witches, were most congenial to the female mind ; and I think I can now perceive that this was quite in character. Women, taken in the collective, are more poetical, more timid, more credulous than men. If we but add to these general traits one or two that are less so, and a few very common circumstances ; if we but add a judgment not naturally vigorous, an imagination more than commonly active, an ignorance of books and of the world, a long-cherished belief in the supernatural, a melancholy old age, and a solitary fireside—we have compounded the elements of that terrible poetry which revels among skulls, and coffins, and enchantments, as certainly as Nature did when she moulded the brain of a Shakspere. The stories of the second class I have almost never found in communion with those of the third ; and never heard well told—except as jokes. To tell a story avowedly untrue, and to tell it as a piece of humour, requires a very different cast of mind from that which character-

ized the melancholy people who were the grand depositories of
the darker traditions : they entertained these only because they
deemed them mysterious and very awful truths, while they
regarded open fictions as worse than foolish.   Nor were their
own stories better received by a third sort of persons, from
whom I have drawn some of my best traditions of the first class,
and who were mostly shrewd, sagacious men, who, having
acquired such a tinge of scepticism as made them ashamed of
the beliefs of their weaker neighbours, were yet not so deeply
imbued with it as to deem these beliefs mere matters of amusement.   They did battle with them both in themselves and the
people around them, and found the contest too serious an affair
to be laughed at.   Now, however (and the circumstance is
characteristic), the successors of this order of people venture
readily enough on telling a good ghost story, when they but
get one to tell.   Superstition, so long as it was living superstition, they deemed, like the live tiger in his native woods, a
formidable, mischievous thing, fit only to be destroyed ; but
now that it has perished, they possess themselves of its skin and
its claws, and store them up in their cabinets.

I have thus given a general character of the contents of my
departed library, and the materials of my proposed work.   My
stories form a kind of history of the district of country to which
they belong—hence the title I have chosen for them ; and, to
fill up some of those interstices which must always be occurring
in a piece of history purely traditional, I shall avail myself of
all the little auxiliary facts with which books may supply me.
The reader, however, need be under no apprehension of meeting much he was previously acquainted with ; and, should I
succeed in accomplishing what I have purposed, the local aspect
of my work may not militate against its interest.   Human
nature is not exclusively displayed in the histories of only great
countries, or in the actions of only celebrated men; and human
nature may be suffered to assert its claim on the attention of

the beings who partake of it, even though the specimens exhibited be furnished by the traditions of an obscure village. Much, however, depends on the manner in which a story is told ; and thus far I may vouch for the writer. I have seriously resolved not to be tedious, unless I cannot help it ; and so, if I do not prove amusing, it will be only because I am unfortunate enough to be dull. I shall have the merit of doing my best—and what writer ever did more? I pray the reader, however, not to form any very harsh opinion of me for at least the first four chapters, and to be not more than moderately critical on the two or three that follow. There is an obscurity which hangs over the beginnings of all history—a kind of impalpable fog—which the writer can hardly avoid transferring from the first openings of his subject to the first pages of his book. He sees through this haze the men of an early period "like trees walking;" and, even should he believe them to be beings of the same race with himself, and of nearly the same shape and size—a belief not always entertained—it is impossible for him, from the atmosphere which surrounds them, to catch those finer traits of form and feature by which he could best identify them with the species. And hence a necessary lack of interest.

CHAPTER II.

"Consider it warilie; read aftiner than anis."—GAVIN DOUGLAS.

THE histories of single districts of country rarely ascend into
so remote an antiquity as to be lost like those of nations in the
ages of fable.   It so happens, however, whether fortunately or
otherwise, for the writer, that in this respect the old shire of
Cromarty differs from every other in the kingdom.   Sir Thomas
Urquhart, an ingenious native of the district, who flourished
about the middle of the seventeenth century, has done for it all
that the chroniclers and senachiés of England and Ireland have
done for their respective countries; and as he united to a
vigorous imagination a knowledge of what is excellent in character, instead of peopling it with the caco-demons of the one
kingdom, or the resuscitated antediluvians of the other, he has
bestowed upon it a longer line of heroes and demigods than can
be exhibited by the annals of either.   I avail myself of his
writings on the strength of that argument which O'Flaherty uses
in his Ogygia as an apology for the story of the three fishermen who were driven by tempest into a haven of Ireland fifteen
days before the universal deluge.   " Where there is no room,"
says this historian, " for just disquisition, and no proper field of
inquiry, we must rely on the common suffrages of the writers
of our country; to whose opinions I voluntarily subscribe."

Alypos, the forty-third in a direct line from Japhet, was the
first, says Sir Thomas, who discovered that part of Scotland
which has since been known by the name of Cromarty.   He
was contemporary with Rehoboam, the fourth king of Israel,

and a very extraordinary personage, independent of his merits as a navigator. For we must regard him as constituting a link which divides into ancestors and descendants—a chain that depends unbroken from the creation of Adam to the present times; and which either includes in itself, or serves to connect by its windings and involutions, some of the most famous people of every age of the world. His grandmother was a daughter of Calcido the Tyrian, who founded Carthage, and who must have lived several ages before the Dido of Virgil; his mother travelled from a remote eastern country to profit by the wisdom of Solomon, and is supposed by many, says Sir Thomas, to have been the queen of Sheba. Nor were his ancestors a whit less happy in their friends than in their consorts. There was one of them intimately acquainted with Nimrod, the founder of the Assyrian Empire, and the builder of Babel; another sat with Abraham in the door of his tent, sharing with him his feelings of sorrow and horror when the fire of destruction was falling on the cities of the plain; a third, after accompanying Bacchus in his expedition to the Indies, and receiving from him in marriage the hand of Thymelica his daughter, was presented with a rich jewel when passing through Syria, by Deborah, the judge and prophetess of Israel. The gem might have been still in the family had not one of his descendants given it to Penthesilea, that queen of the Amazons who assisted the Trojans against Agamemnon. Buchanan has expressed his astonishment that the chroniclers of Britain, instead of appropriating to themselves honourable ancestors out of the works of the poets, should rather, through a strange perversity, derive their lineage from the very refuse of nations: Sir Thomas seems to have determined not to furnish a similar occasion of surprise to any future historian. There were princes of his family who reigned with honour over Achaia and Spain, and a long line of monarchs who flourished in Ireland before the expedition of Fergus I.

The era of Alypos was one of the most important in the

history of Britain. It was that in which the inhabitants first began to build cities, and to distinguish their several provinces by different names. It witnessed the erection of the city of York by one Elborak, a brother-in-law of Alypos, and saw the castle of Edinburgh founded by a contemporary chieftain of Scotland, who had not the happiness of being connected to him, and whose name has therefore been lost. The historian assigns, too, to the same age the first use of the term Olbion as a name for the northern division of the island—a term which afterwards, " by an Eolic dialect," came to be pronounced Albion, or Albyn; and the first application of the name Sutors, from the Greek σωτηρες, *preservers*, to those lofty promontories which guard the entrance of the bay of Cromarty—a fact which Aikman the historian recommends, with becoming gravity, to the consideration of Gaelic etymologists. Much of a similar character, as appears from Sir Thomas, could have been brought under their notice in the reign of Charles I., when, as he states in one of his treatises, the names of all places in the shire of Cromarty, whether promontories, fountains, rivers, or lakes, were of pure and perfect Greek. Since that time, however, many of these names have been converted into choice trophies of the learning and research of those very etymologists;—even the derivation of the term Sutors has been disputed, but by the partisans of languages less ancient than either Greek or Gaelic. The one party write the contested dissyllable Suitors, the other Soutars, and defend their different modes of spelling each by a different legend—a species of argument practised at one time with much ingenuity and success by the contending Orders of St. Dominic and Loyola.

The promontories which bear this name are nearly equal in height, but when viewed from the west they differ considerably in appearance. The one, easy of access, crowned with a thick wood of pine, divided into corn-fields, and skirted at the base by a broad line of ash and elm, seems feminine in its character;

while the other, abrupt, stern, broken into precipices, and tufted with furze, is of a cast as decidedly masculine. Two lovers of some remote age, had met by appointment in a field of Cromarty which commands a full view of the promontories in the aspect described. The young man urged his suit with the characteristic warmth of his sex—his mistress was timid and bashful. He accused her of indifference; and with all the fervour of a passion which converts even common men into poets, he exclaimed, pointing to the promontories, "See, Ada! they too are lovers—they are hastening to embrace; and stern and rugged as that carle-hill of the north may seem to others, he is not reckoned so by his lady-hill of the south;—see how, with all her woods and her furrows, she advances to meet him."— "And think you," rejoined the maiden, entering into the poetry of the feeling, "that these tongueless *suitors* cannot express their mutual regards without the aid of language; or that that carle of the north, rude as he is, would once think of questioning the faith and affection of his advancing mistress, merely because she advances in silence?" Her reply, say the people who contend for the English derivation of the word, furnished the promontories with a name; and as those alchemists of mind who can transmute etymology into poetry have not been produced everywhere, few names have anecdotes equally pleasing connected with their origin. The other legend is of a different character, and has a merit peculiar to itself, to be amenable to any known law of criticism.

In some age of the world more remote than even that of Alypos, the whole of Britain was peopled by giants—a fact amply supported by early English historians and the traditions of the north of Scotland. Diocletian, king of Syria, say the historians, had thirty-three daughters, who, like the daughters of Danaus, killed their husbands on their wedding night. The king, their father, in abhorrence of the crime, crowded them all into a ship, which he abandoned to the mercy of the waves, and which was drifted by tides and winds till it arrived on the coast of Britain, then an uninhabited island. There they lived solitary, subsisting on roots and berries, the natural produce of the soil, until an order of demons, becoming enamoured of them, took them for their wives; and a tribe of giants, who must be regarded as the true aborigines of the country, if indeed the demons have not a prior claim, were the fruit of these marriages. Less fortunate, however, than even their prototypes the Cyclops, the whole tribe was extirpated a few ages after by Brutus the parricide, who, with a valour to which mere bulk could render no effectual resistance, overthrew Gog-Magog, and Termagol, and a whole host of others, with names equally terrible. Tradition is less explicit than the historians in what relates to the origin and extinction of the race, but its narratives of their prowess are more minute. There is a large and very ponderous stone in the parish of Edderton, which a giantess of the tribe is said to have flung from the point of a spindle across the Dornoch Firth; and another within a few miles of Dingwall, still larger and more ponderous, which was thrown from a neighbouring eminence by a person of the same family, and which still bears the marks of a gigantic finger and thumb impressed on two of its sides. The most wonderful, however, of all their achievements was that of a lady, distinguished even among the tribe as the *Cailliach-more,* or great woman, who, from a pannier filled with earth and stones, which she carried on her back, formed almost all the hills of Ross-shire. When standing on the site of the huge Ben-Vaichard, the bottom of the pannier is said to have given way, and the contents falling through the opening, produced the hill, which owes its great height and vast extent of base to the accident. Prior to the invasion of Brutus, the promontories of Cromarty served as work-stools to two giants of this tribe, who supplied their brethren with shoes and buskins. They wrought together; for, being furnished with only one set of implements, they could not carry on their trade apart; and these, when needed, they used to fling to each other across the opening of the firth, where the promontories are only about two miles asunder. In process of time the name Soutar, a shoemaker, was transferred by a common metonymy from the craftsmen to their stools—the two promontories; and by this name they have ever since been distinguished. Such are the etymological legends of the Sutors, opposed each to the other, and both to the scholarlike derivation of Sir Thomas; which must be confessed, however, to have been at one time a piece of mere commonplace, though it has since become learning.

I have seen in the museum of the Northern Institution a very complete collection of stone battle-axes, some of which were formed little earlier than the last age by the rude natives of America and the South Sea Islands; while others, which had been dug out of the cairns and tumuli of our own country, witnessed to the unrecorded feuds and forgotten battle-fields of twenty centuries ago. I was a good deal struck by the resemblance which they bore to each other—a resemblance so complete, that the most practised eye could hardly distinguish between the weapons of the old Scot and those of the New Zealander. Both seemed to have selected the same rude materials, employed the same imperfect implements, and wrought after the same uncouth model. But man in a savage state is the same animal everywhere, and his constructive powers, whether employed in the formation of a legendary story or of a battle-axe, seem to expatiate almost everywhere in the same rugged track of invention. For even the traditions of this first stage may be identified, like its weapons of war, all the world over. Mariner, in his account of the Tonga Islands, tells us that the natives pointed out to him a perforated rock, in the hollow of which, they said, one of their gods, when employed in fishing, entangled his hook, and that, pulling lustily to disengage it, he pulled up the whole island (one of the largest of the group) from the bottom of the sea. Do not this singular story, and the wild legend of Ben-Vaichard, though the product of ages and countries so widely separated, belong obviously to the same rude stage of invention?

There may be some little interest in tracing the footprints of what I may term the more savage traditions of a country in the earlier pages of its history, and in marking how they blend with its imperfect narratives of real but ill-remembered events, on the one hand, and its mutilated imitations of the master-pieces of a classical literature, on the other. The fabulous pages of English history furnish, when regarded in this point of view, a not uninteresting field to the legendary critic. They are suited to remind him of those huts of the wild Arab, com-posed of the fragments of ruined grandeur which the traveller finds amid the ruins of Palmyra or Balbec, and in which, as he prosecutes his researches, he sees the capital totter over the architrave, the base overtop the capital, masses of turf heaped round the delicate volute, which emulated in granite the curled tresses of a beautiful female, and the marble foliage of the acanthus crushed by the rude joist which bends under a roof of clay and rushes. Perhaps the reader may indulge me in a few brief remarks on this rather curious subject.

Diocletian, the Syrian king of the English legend, is, as Buchanan justly remarks, a second Danaus, and owes his exist-ence to the story of his prototype; but the story of the mar-riages of his daughters with an order of demons, which, accord-ing to that historian, the English have invented through a pride of emulating the Gauls and Germans, who derive their lineage from Pluto, does not appear to me to be so legitimately traced to its original. The oldest of all the traditions of Britain seem to be those which describe it as peopled at some remote era by giants;—they are the broken vestiges, it is possible, of those incidents of Mosaic history which are supposed to be shadowed out in the fables of the giants of Grecian mythology, or they are perhaps mutilated remains of the fables themselves. It seems more probable, however, that they should have originated in that belief, common to the vulgar of all countries, that the race of men is degenerating in size and prowess with every succeeding generation, and that at some early period their bulk and strength must have been gigantic. Judging of them from their appearance, they must have been known in a very early age—an age as early perhaps as that of the stone battle-axe; and what more probable than that they should have attracted the notice of the chroniclers, who would naturally consult tradition for the materials of their first pages? But tradition, though it records the achievements of the giants, is silent respecting their origin. A first link would therefore be wanting, which could only be supplied by imagination; and as, like every other class of writers, the chroniclers would find it easier to imitate than to invent, it is not difficult to conceive how, after having learned in their cloisters that in an early age of the world the sons of God had contracted marriages with the daughters of men, and that heroes and giants were the fruit of the connexion—they should blend a legend imitative of the event with the stories of the giants of Britain. Their next employment, for it would be too bold an attempt to link so terrible a tribe to the people of their own times, would be to show how this tribe became extinct, and the manner in which the country was first peopled with men like themselves.

There is but one way in which anything probable can be acquired concerning the origin of a people who have no early history; but the process is both difficult and laborious. There is another sufficiently easy, which barely reaches the possible, and which the historians of eight hundred years ago would have deemed the more eligible of the two. Instead of setting themselves to ascertain those circumstances by which the several families of men are distinguished, or to compare the language, character, and superstitions of the people of their own country

B

with those of the various tribes of the Continent, they would apply for such assistance as the imitator derives from his copy, to the histories of other kingdoms. From their connexion with the Latin Church they would be conversant with Roman literature, and acquainted with the story of Æneas as related by the historians, and amplified and adorned by Virgil. And thus, what may be termed the third link of their history, has come to bear a discernible resemblance to the early history of Rome. The occasion of the wanderings of Brutus resembles that of the expatriation of Tydeus, or rather that of the madness of Œdipus, but he is the Æneas of England notwithstanding. His history is a kind of national epic. Cornæus is his Achates. He finds hostile Rutulians, headed by a Turnus, in the giants and their leader; and Britain is both his Italy and his Trinacria, though, instead of fleeing from the Cyclops, he conquers them.

The legend of Scotland may also be regarded as a national epic. It is formed on the same model with the story of Brutus, but it has the merit of being a somewhat more skilful imitation, and there is nothing outrageously improbable in any of its circumstances. Galethus, its hero, is the Æneas of Scotland. He was the son of Cecrops, the founder of Athens, and, like Romulus, made himself famous as a captain of robbers before he became the founder of a nation. Having repeatedly invaded Macedonia and the neighbouring provinces of Greece, he was in imminent danger of being overpowered by a confederacy of the states he had injured, when, assembling his friends and followers, he retreated into Egypt, at a time when that kingdom was ravaged from its southern boundary to the gates of Memphis by an army of Ethiopians. Assuming on the sudden a new character, he joined his forces to those of Pharaoh, gave battle to the invaders, routed them with much slaughter, pursued them into Ethiopia, and after a succession of brilliant victories over them, compelled them to sue for peace. On his return he was presented by the king with the hand of his daughter Scota, and made general in chief of all the forces of the kingdom. Disgusted, however, by the cruelties practised on the Israelites, and warned by Moses and an oracle of the judgments by which these cruelties were to be punished, he fitted out a fleet, and, accompanied by great numbers of Greeks and Egyptians, set sail from the river Nile with the intention of forming a settlement on the shores of the Mediterranean. After a tedious voyage he arrived at a port of Numidia, where no better success awaited him than was met with by Æneas in the scene of his first colony. Again putting to sea, he passed the Pillars of Hercules, and after having experienced in the navigation of the straits dangers similar to those which appalled Ulysses when passing through the Straits of Messina, he landed in that part of Spain which has ever since been known by the name of Portugal. He found in this country a second Tiber in the river Munda, and a fierce army of Rutulians in the inhabitants. But his good fortune did not desert him. He vanquished his enemies in one decisive battle, dispossessed them of their fairest provinces, built cities, instituted laws, conquered and colonized Ireland, and, dying after a long and prosperous reign, left his kingdom to his children. Prior to his decease, his subjects, both Greeks and Egyptians, were termed Scots, from their having sunk their original designations in that name, out of courtesy to their Queen Scota—a name afterwards transferred to Albyn by a colony from Ireland, who took possession of it a few ages subsequent to the age of Galethus. Such is the fable of what may be regarded both as the historic epic of Scotland, and as the most classical of all the imitations of the Æneid which were fabricated during the middle ages.

Sir Thomas has recorded nothing further of his ancestor Alypos, than that he followed up his discovery of Cromarty by planting it with a colony of his countrymen, who, though some of his ancestors had settled in Portugal several ages before, seem to have been Greeks. Of sixteen of his immediate descendants, it is only known that they were born, and that they married—some of them finding honourable consorts in Ireland, some in Greece, and one in Italy. The wife of that one was a sister of Marcus Coriolanus—a daughter of Agesilaus the Spartan, a daughter of Simeon Breck, the first crowned king of the Irish Scots, a daughter of Alcibiades, the friend and pupil of Socrates, and a niece of Lycurgus the lawgiver, were wives to some of the others. Never was there a family that owed more to its marriages.

Nomaster, the son-in-law of Alcibiades, disgusted by the treatment which that great but ambitious statesman had received from his country, took leave of Greece, and, " after many dangerous voyages both by sea and land, he arrived at the harbour Ochoner, now called Cromarty." It owed its more ancient name to Bestius Ochoner, one of the sixteen immediate descendants of Alypos, and the father, says the genealogist, of the Irish O'Connors; the name which it now bears is derived by Gaelic etymologists from the windings and indentations of its shores.[1] Nomaster, immediately on his landing, was recognised by the colonists as their legitimate prince, and he reigned over them till his death, when he was succeeded by his son Astorimon, a valiant and accomplished warrior, in whom the genius and heroism of his grandfather seem to have been revived. And the events of his time were suited to find employment. For in this age an immense body of Scythians, after voyaging along the shores of Europe in quest of a settlement, were incited by the great natural riches of the country to make choice of Scotland; and, pouring in upon its western coasts, they dispossessed the natives of some of their fairest provinces. But the little territory of Astorimon, though one of the invaded, was not one of the conquered provinces. The Scythians, under Ethus their general, intrenched upon an extensive moor, which now forms the upper boundary of the parish of Cromarty; and the grandson of Alcibiades drew out his forces to oppose them. A battle ensued, in which the Scythian general was killed in single combat

---

[1] *Cromba, i.e.,* crooked bay.

by Astorimon ; and his followers, dispirited by his death, and
unable to contend with an army trained to every evolution of
Greek and Roman discipline, were routed with immense slaughter.
The Scythians afterwards became famous as the Picts of Scottish
history ; and *Ethus*, their leader, is reckoned their first king. Sir
Thomas, to the details of this battle, which he terms the great
battle of *Farna*,[1] has added, that " the trenches, head-quarters,
and castrametation" of the invading army can still be traced on
a moor of Cromarty.

This moor, which formed a few years ago an unappropriated
common, but which was lately divided among the proprietors
whose lands border on it, has evidently at some remote period
been a field of battle. It is sprinkled over with tumuli and
little heathy ridges resembling the graves of a churchyard. The
southern shore of the Cromarty Firth runs almost parallel to it
for nearly fourteen miles ; and upon a hill in the parish of
Resolis, which rises between it and the firth, and which is
separated from it by a deep valley, there are the vestiges of
Danish encampments. And there is perhaps scarcely an eminence in Scotland on which in the early ages an invading army
could have encamped with more advantage than on this hill, or
a moor upon which the invaders could have been met with on
more equal terms than on the moor adjacent. The eminence is
detached on the one side from the other rising grounds of the
country by a valley, the bottom of which is occupied by a bog,
and it commands on the other an extensive bay, in which whole
fleets may ride with safety ; while the neighbouring moor is of
great extent, and has few inequalities of surface. Towards its
eastern boundary, about six miles from the town of Cromarty,
there is a huge heap of stones, which from time immemorial has
been known to the people of the place as *The Grey Cairn*, a
name equally descriptive of other lesser cairns in its vicinity,

[1] Two ancient farms in the neighbourhood bear the names of Meikle and Little *Farness*,
and a third that of *Eathie*.

but which with the aid of the definite article serves to distinguish it. Not more than thirty years ago the stones of a similar cairn of the moor were carried away for building by a farmer of the parish. There were found on their removal human bones of a gigantic size, among the rest a skull sufficiently capacious, according to the description of a labourer employed by the farmer, to contain " two lippies of beer."

About fifteen years ago, a Cromarty fisherman was returning from Inverness by a road which for several miles skirts the upper edge of the moor, and passes within a few yards of the cairn. Night overtook him ere he had half completed his journey; but, after an interval of darkness, the moon, nearly at full, rose over the eminence on his right, and restored to him the face of the country—the hills which he had passed before evening, but which, faint and distant, were sinking as he advanced, the wood which, bordering his road on the one hand, almost reached him with its shadow, and the bleak, unvaried, interminable waste, which, stretching away on the other, seemed lost in the horizon. After he had entered on the moor, the stillness which, at an earlier stage of his journey, had occasionally been broken by the distant lowing of cattle, or the bark of a shepherd's dog, was interrupted by only his own footsteps, which, from the nature of the soil, sounded hollow as if he trod over a range of vaults, and by the low monotonous murmur of the neighbouring wood. As he approached the cairn, however, a noise of a different kind began to mingle with the other two; it was one with which his profession had made him well acquainted—that of waves breaking against a rock. The nearest shore was fully three miles distant, the nearest cliff more than five, and yet he could hear wave after wave striking as if against a precipice, then dashing upwards, and anon descending, as distinctly as he had ever done when passing in his boat beneath the promontories of Cromarty. On coming up to the cairn, his astonishment was converted into terror.—Instead of the brown heath, with here and there a fir seedling springing out of it he saw a wide tempestuous sea stretching before him, with the large pile of stones frowning over it, like one of the Hebrides during the gales of the Equinox. The pile appeared as if half enveloped in cloud and spray, and two large vessels, with all their sheets spread to the wind, were sailing round it.

The writer of these chapters had the good fortune to witness at this cairn a scene which, without owing anything to the supernatural, almost equalled the one described. He was, like the fisherman, returning from Inverness to Cromarty in a clear frosty night in December. There was no moon, but the whole sky towards the north was glowing with the Aurora Borealis, which, shooting from the horizon to the central heavens, in flames tinged with all the hues of the rainbow, threw so strong a light, that he could have counted every tree of the wood, and every tumulus of the moor. There is a long hollow morass which runs parallel to the road for nearly a mile;—it was covered this evening by a dense fleece of vapour raised by the frost, and which, without ascending, was rolling over the moor before a light breeze. It had reached the cairn, and the detached clump of seedlings which springs up at its base.—The seedlings rising out of the vapour appeared like a fleet of ships, with their sails dropping against their masts, on a sea where there were neither tides nor winds;—the cairn, grey with the moss and lichens of forgotten ages, towered over it like an island of that sea.

But I daresay I have imparted to the reader more of the fabulous history of Cromarty than he will well know how to be grateful for. One other remark, however, in better language, and a more vigorous style of thinking than my own, and I shall have done;—it may show that Sir Thomas, however unique as a man, forms, as a historian, only one of a class.

" The last century," says the philosophic Gibbon, " abounded with antiquarians of profound learning and easy faith, who, by the dim light of legends and traditions, of conjectures and etymologies, conducted the great-grandchildren of Noah from the tower of Babel to the extremities of the globe. Of these judicious critics," continues the historian, "one of the most entertaining was Olaus Rudbeck, professor in the university of Upsal. Whatever is celebrated either in history or fable, this zealous patriot ascribes to his country. From Sweden, the Greeks themselves derived their alphabetical characters, their astronomy, and their religion. Of that delightful region (for so it appeared to the eyes of a native), the Atlantis of Plato, the country of the Hyperboreans, the Garden of the Hesperides, the Fortunate Islands, and even the Elysian fields, were all but faint and imperfect transcripts. A clime so profusely favoured by nature could not long remain desert after the flood. The learned Rudbeck allows the family of Noah a few years to multiply from eight to about twenty thousand persons. He then disperses them into small colonies to replenish the earth and to propagate the human species. The Swedish detachment (which marched, if I am not mistaken, under the command of Askenos, the son of Gomer, the son of Japhet), distinguished itself by more than common diligence in the prosecution of this great work. The northern hive cast its swarms over the greater part of Europe, Africa, and Asia; and (to use the author's metaphor), the blood circulated from the extremities to the heart."

CHAPTER III.

"The wild sea, baited by the fierce north-east,
So roar'd, so madly raged, so proudly swell'd,
As it would thunder full into our streets."—ARMSTRONG.

THE Bay of Cromarty was deemed one of the finest in the world at a time when the world was very little known ; and modern discovery has done nothing to lower its standing or character. We find it described by Buchanan in very elegant Latin as " formed by the waters of the German Ocean, opening a way through the stupendous cliffs of the most lofty precipices, and expanding within into a spacious basin, affording certain refuge against every tempest." The old poet could scarce have described it better had he sat on the loftiest pinnacle of the southern Sutor during a winter storm from the north-east, and seen vessel after vessel pressing towards the opening through spray and tempest ;—like the inhabitants of an invaded country hurrying to the gateway of some impregnable fortress, their speed quickened by the wild shouts of the enemy, and pursued by the smoke of burning villages.

Viewed from the Moray Firth in a clear morning of summer, the entrance of the bay presents one of the most pleasing scenes I have ever seen. The foreground is occupied by a gigantic wall of brown precipices, beetling for many miles over the edge of the Firth, and crested by dark thickets of furze and pine. A multitude of shapeless crags lie scattered along the base, and we hear the noise of the waves breaking against them, and see the reflected gleam of the foam flashing at intervals into the darker recesses of the rock. The waters of the bay find entrance, as described by the historian, through a natural postern scooped out of the middle of this immense wall. The huge projection of cliff on either hand, with their alternate masses of light and shadow, remind us of the out-jets and buttresses of an ancient fortress ; and the two Sutors, towering over the opening, of turrets built to command a gateway. The scenery within is of a softer and more gentle character. We see hanging woods, sloping promontories, a little quiet town, and an undulating line of blue mountains, swelling as they retire into a bolder outline and a loftier altitude, until they terminate, some twenty miles away, in the snow-streaked, cloud-capped Ben Wevis. When I last gazed on this scene, and contrasted the wild sublimity of the foreground with the calm beauty of the interior, I was led to compare it, I scarcely knew how, to the exquisite masterpiece of his art which the Saxon sculptor Nahl placed over the grave of a lady who had died in the full bloom of youth and loveliness. It represents the ruins of a tomb shattered as if by the last trumpet ; but the chisel has not been employed on it in merely imitating the uncouth ravages of accident and decay ; for through the yawning rifts and fissures there is a beautiful female, as if starting into life, and rising in all the ecstasy of unmingled happiness to enjoy the beatitudes of heaven.

There rises within the bay, to the height of nearly a hundred feet over the sea level, a green sloping bank, in some places covered with wood, in others laid out into gardens and fields. We may trace it at a glance all along the shores of the firth, from where it merges into the southern Sutor, till where it sinks at the upper extremity of the bay of Udoll ; and, fronting it on the opposite side, we may see a similar escarpment, winding along the various curves and indentations of the coast —now retiring far into the country, along the edge of the bay of Nigg—now abutting into the firth, near the village of Invergordon. The Moray and Dornoch firths are commanded by re-

sembling ramparts of bank of a nearly corresponding elevation,
and a thorough identity of character ; and, as in the Firth of
Cromarty, the space between their bases and the shore is occupied by a strip of level country, which in some places' encroaches
on the sea in the form of long low promontories, and is hollowed out in others to nearly the base of the escarpment.
Wherever we examine, we find data to conclude, that in some
remote era this continuous bank formed the line of coast, and
that the plain at its base was everywhere covered by the waters
of the sea. We see headlands, rounded as if by the waves, advancing the one beyond the other, into the waving fields and
richly-swarded meadows of this lower terrace ; and receding
bays with their grassy unbeaten shores comparatively abrupt at
the entrance, and reclining in a flatter angle within. We may
find, too, everywhere under the vegetable soil of the terrace,
alternate layers of sand and water-worn pebbles, and occasionally,
though of rarer occurrence, beds of shells of the existing species,
and the bones of fish. In the valley of Munlochy, the remains
of oyster-beds, which could not have been formed in less than
two fathoms of water, have been discovered a full half mile
from the sea ; beds of cockles still more extensive, and the
bones of a porpoise, have been dug up among the fields which
border on the bay of Nigg ; similar appearances occur in the
vicinity of Tain ; and in digging a well about thirty years ago,
in the western part of the town of Cromarty, there was found
in the gravel a large fir-tree, which, from the rounded appearance of the trunk and branches, seems to have been at one time
exposed to the action of the waves. In a burying-ground of
the town, which lies embosomed in an angle of the bank, the
sexton sometimes finds the dilapidated spoils of our commoner
shell-fish mingling with the ruins of a nobler animal ; and in
another inflection of the bank, which lies a short half mile to
the east of the town, there is a vast accumulation of drift peat,
many feet in thickness, and the remains of huge trees.

The era of this old coast line we find it impossible to fix ; but there are grounds enough on which to conclude that it must have been remote—so remote, perhaps, as to lie beyond the beginnings of our more authentic histories. We see, in the vicinity of Tain, one of the oldest ruins of the province situated far below the base of the escarpment ; and meet in the neighbourhood of Kessock, at a still lower level, with old Celtic cairns and tumuli. It is a well-established fact, too, that for at least the last three hundred years the sea, instead of receding, has been gradually encroaching on the shores of the Bay of Cromarty; and that the place formerly occupied by the old burgh, is now covered every tide by nearly two fathoms of water.

The last vestige of this ancient town disappeared about eighteen years ago, when a row of large stones, which had evidently formed the foundation line of a fence, was carried away by some workmen employed in erecting a bulwark. But the few traditions connected with it are not yet entirely effaced. A fisherman of the last century is said to have found among the title-deeds of his cottage a very old piece of parchment, with a profusion of tufts of wool bristling on one of its sides, and bearing in rude antique characters on the other a detail of the measurement and boundaries of a garden which had occupied the identical spot on which he usually anchored his skiff. I am old enough to have conversed with men who remembered to have seen a piece of corn land, and a belt of planting below two properties in the eastern part of the town, that are now bounded by the sea. I reckon among my acquaintance an elderly person, who, when sailing along the shore about half a century ago in the company of a very old man, heard the latter remark, that he was now guiding the helm where, sixty years before, he had guided the plough. Of Elspat Hood, a native of Cromarty, who died in the year 1701, it is said that she attained to the extraordinary age of 120 years, and that in her recollection, which embraced the latter part of the sixteenth

century, the *Clach Malacha*, a large stone covered with sea-
weed, whose base only partially dries during the ebb of Spring
and Lammas tides, and which lies a full quarter of a mile from
the shore, was surrounded by corn fields and clumps of wood.
And it is a not less curious circumstance than any of these,
that about ninety years ago, after a violent night storm from
the north-east, the beach below the town was found in the
morning strewed over with human bones, which, with several
blocks of hewn stone, had been washed by the surf out of what
had been formerly a burying-place.   The bones were carried to
the churchyard, and buried beneath the eastern gable of the
church ; and one of the stones—the corner stone of a ponderous
cornice—is still to be seen on the shore.   In the firths of
Beauly and Dornoch the sea seems to have encroached to fully
as great an extent as in the bay of Cromarty.   Below the town
of Tain a strip of land, once frequented by the militia of the
county for drill and parade, has been swept away within the
recollection of some of the older inhabitants ; and there may
be traced at low water (says Carey in his notes to Craig Phad-
rig), on the range of shore that stretches from the ferry of
Kessock to nearly Redcastle, the remains of sepulchral cairns,
which must have been raised before the places they occupy
were invaded by the sea, and which, when laid open, have been
found to contain beams of wood, urns, and human bones.—But
it is full time that man, the proper inhabitant of the country,
should be more thoroughly introduced into this portion of its
history.   We feel comparatively little interest in the hurricane
or the earthquake which ravages only a desert, where there is
no intelligent mind to be moved by the majesty of power, or
the sublimity of danger ; while on the other hand, there is no
event, however trivial in itself, which may not be deemed of
importance if it operate influentially on human character and
human passion.

It is not much more than twenty years since a series of violent storms from the hostile north-east, which came on at almost re-gular intervals for five successive winters, seemed to threaten the modern town of Cromarty with the fate of the ancient. The tides rose higher than tides had ever been known to rise before; and as the soil exposed to the action of the waves was gradually disappearing, instead of the gentle slope with which the land formerly merged into the beach, its boundaries were marked out by a dark abrupt line resembling a turf wall. Some of the people whose houses bordered on the sea looked exceedingly grave, and affirmed there was no danger whatever; those who lived higher up thought differently, and pitied their poor neigh-bours from the bottom of their hearts. The consternation was heightened too by a prophecy of Thomas the Rhymer, handed down for centuries, but little thought of before. It was pre-dicted, it is said, by the old wizard, that Cromarty should be twice destroyed by the sea, and that fish should be caught in abundance on the Castle-hill—a rounded projection of the es-carpment which rises behind the houses, and forms the ancient coast line.

Man owes much of his ingenuity to his misfortunes; and who does not know that, were he less weak and less exposed as an animal, he would be less powerful as a rational creature? On a principle so obvious, these storms had the effect of converting not a few of the townsfolk into builders and architects. In the eastern suburb of the town, where the land presents a low yet projecting front to the waves, the shore is hemmed in by walls and bulwarks, which might be mistaken by a stranger approach-ing the place by sea for a chain of little forts. They were erected during the wars of the five winters by the proprietors of the gardens and houses behind ; and the enemy against whom they had to maintain them, was the sea. At first the contest seemed well-nigh hopeless ;—week after week was spent in throwing up a single bulwark, and an assault of a few hours demolished the whole line. But skill and perseverance prevailed at last ;—the storms are all blown over, but the gardens and houses still remain. Of the many who built and planned during this war, the most indefatigable, the most skilful, the most successful, was Donald Miller.

Donald was a true Scotchman. He was bred a shoemaker; and painfully did he toil late and early for about twenty-five years with one solitary object in view, which, during all that time, he had never lost sight of—no, not for a single moment. And what was that one?—independence—a competency sufficient to set him above the necessity of further toil; and this he at length achieved, without doing aught for which the severest censor could accuse him of meanness. The amount of his savings did not exceed four hundred pounds; but, rightly deeming himself wealthy, for he had not learned to love money for its own sake, he shut up his shop. His father dying soon after, he succeeded to one of the snuggest, though most perilously-situated little properties within the three corners of Cromarty—the sea bounding it on the one side, and a stream, small and scanty during the droughts of summer, but sometimes more than sufficiently formidable in winter, sweeping past it on the other. The series of storms came on, and Donald found he had gained nothing by shutting up his shop.

He had built a bulwark in the old, lumbering, Cromarty style of the last century, and confined the wanderings of the stream by two straight walls. Across the walls he had just thrown a wooden bridge, and crowned the bulwark with a parapet, when on came the first of the storms—a night of sleet and hurricane—and lo! in the morning, the bulwark lay utterly overthrown, and the bridge, as if it had marched to its assistance, lay beside it, half buried in sea-wrack. " Ah," exclaimed the neighbours, " it would be well for us to be as sure of our summer's employment as Donald Miller, honest man!" Summer came; the bridge strided over the stream as before; the bulwark was built anew, and with such neatness and apparent strength, that no bulwark on the beach could compare with it. Again came winter; and the second bulwark, with its proud parapet, and rock-like strength, shared the fate of the first. Donald fairly took to his bed. He rose, however, with renewed vigour; and a third bulwark, more thoroughly finished than even the second, stretched ere the beginning of autumn between his property and the sea. Throughout the whole of that summer, from grey morning to grey evening, there might be seen on the shore of Cromarty a decent-looking, elderly man, armed with lever and mattock, rolling stones, or raising them from their beds in the sand, or fixing them together in a sloping wall—toiling as never labourer toiled, and ever and anon, as a neighbour sauntered the way, straightening his weary back, and tendering the ready snuff-box. That decent-looking, elderly man, was Donald Miller. But his toil was all in vain. Again came winter and the storms; again had he betaken himself to his bed, for his third bulwark had gone the way of the two others. With a resolution truly indomitable, he rose yet again, and erected a fourth bulwark, which has now presented an unbroken front to the storms of twenty years.

Though Donald had never studied mathematics as taught in books or the schools, he was a profound mathematician notwithstanding. Experience had taught him the superiority of the sloping to the perpendicular wall in resisting the waves; and he set himself to discover that particular angle which, without being inconveniently low, resists them best. Every new bulwark was a new experiment made on principles which he had discovered in the long nights of winter, when, hanging over the fire, he converted the hearth-stone into a tablet, and, with a pencil of charcoal, scribbled it over with diagrams. But he could never get the sea to join issue with him by changing in the line of his angles; for, however deep he sunk his foundations, his insidious enemy contrived to get under them by washing away the beach; and then the whole wall tumbled into the cavity. Now, however, he had discovered a remedy. First he laid a row of large flat stones on their edges in the line of the foundation and paved the whole of the beach below until it presented the appearance of a sloping street—taking care that his pavement, by running in a steeper angle than the shore, should, at its lower edge, lose itself in the sand. Then, from the flat stones which formed the upper boundary of the pavement, he built a ponderous wall, which, ascending in the proper angle, rose to the level of the garden; and a neat firm parapet surmounted the whole.— Winter came, and the storms came; but though the waves broke against the bulwark with as little remorse as against the Sutors, not a stone moved out of its place. Donald had at length fairly triumphed over the sea.

The progress of character is fully as interesting a study as the progress of art; and both are curiously exemplified in the history of Donald Miller. Now that he had conquered his enemy, and might realize his long-cherished dream of unbroken leisure, he found that constant employment had, through the force of habit, become essential to his comfort. His garden was the very paragon of gardens; and a single glance was sufficient to distinguish his furrow of potatoes from every other furrow in the field; but, now that his main occupation was gone, much time hung on his hands, notwithstanding his attentions to both. First, he set himself to build a wall quite round his property; and a very neat one he did build; but unfortunately, when once erected, there was nothing to knock it down again. Then he whitewashed his house, and built a new sty for his pig, the walls of which he also whitewashed. Then he enclosed two little patches on the side of the stream, to serve as bleaching-greens. Then he covered the upper part of his bulwark with a layer of soil, and sowed it with grass. Then he repaired a well, the common property of the town. Then he constructed a path for foot-passengers on the side of a road, which, passing his garden on the south, leads to Cromarty House. His labours for the good of the public were wretchedly recompensed, by, at least, his more immediate neighbours. They would dip their dirty pails into the well which he had repaired, and tell him, when he hinted at the propriety of washing them, that they were no dirtier than they used to be.   Their pigs would break into his bleaching-greens, and furrow up the sward with their snouts : and when he threatened to pound them, he would be told " how unthriving a thing it was to keep the puir brutes aye in the fauld," and how impossible a thing "to watch them ilka time they gae'd out."   Herd-boys would gallop their horses and drive their cattle along the path which he had formed for foot-passengers exclusively : and when he stormed at the little fellows, they would canter past, and shout out, from what they deemed a safe distance, that their " horses and kye had as good a right to the road as himsel'."   Worse than all the rest, when he had finished whitening the walls of his pig-sty, and gone in for a few minutes to the house, a mischievous urchin, who had watched his opportunity, sallied across the bridge, and, seizing on the brush, whitewashed the roof also. Independent of the insult, nothing could be in worse taste ; and yet, when the poor man preferred his complaint to the father of the urchin, the boor only deigned to mutter in reply, that " folk would hae nae peace till three Lammas tides, joined intil ane, would come and roll up the *Clach Malacha*" (it weighs about twenty tons) "frae its place i' the sea till flood water-mark."   The fellow, rude as he was, had sagacity enough to infer that a tide potent enough to roll up the *Clach Malacha*, would demolish the bulwark, and concentrate the energies of Donald for at least another season.

But Donald found employment, and the neighbours were left undisturbed to live the life of their fathers without the intervention of the three Lammas tides.   Some of the gentlemen farmers of the parish who reared fields of potatoes, which they sold out to the inhabitants in square portions of a hundred yards, besought Donald to superintend the measurement and the sale. The office was one of no emolument whatever, but he accepted it with thankfulness ; and though, when he had potatoes of his own to dispose of, he never failed to lower the market for the benefit of the poor, every one now, except the farmers, pronounced him rigid and narrow to a fault. On a dissolution of Parliament, Cromarty became the scene of an election, and the honourable member-apparent deeming it proper, as the thing had become customary, to whitewash the dingier houses of the town, and cover its dirtier lanes with gravel, Donald was requested to direct and superintend the improvements. Proudly did he comply ; and never before did the same sum of *election*-money whiten so many houses, and gravel so many lanes. Employment flowed in upon him from every quarter. If any of his acquaintance had a house to build, Donald was appointed inspector. If they had to be enfeoffed in their properties Donald acted as bailie, and tendered the earth and stone with the gravity of a judge. He surveyed fields, suggested improvements, and grew old without either feeling or regretting it. Towards the close of his last, and almost only illness, he called for one of his friends, a carpenter, and gave orders for his coffin ; he named the seamstress who was to be employed in making his shroud ; he prescribed the manner in which his lyke-wake should be kept, and both the order of his funeral and the streets through which it was to pass. He was particular in his injunctions to the sexton, that the bones of his father and mother should be placed directly above his coffin ; and professing himself to be alike happy that he had lived, and that he was going to die, he turned him to the wall, and ceased to breathe a few hours after. With all his rage for improvement, he was a good old man of the good old school. Often has he stroked my head, and spoken to me of my father, a friend and namesake, though not a relative ; and when, at an after period, he had learned that I set a value on whatever was antique and curious, he presented me with the fragment of a large black-letter Bible which had once belonged to the Urquharts of Cromarty.

CHAPTER IV.

"All hail, Macbeth! thou shalt be king hereafter."—SHAKSPERE.

IT is, perhaps, not quite unworthy of remark, that not only is Cromarty the sole district of the kingdom whose annals ascend into the obscure ages of fable, but that the first passage of even its real history derives its chief interest, not from its importance as a fact, but from what may be termed its chance union with a sublime fiction of poetry. Few, I daresay, have so much as dreamed of connecting either its name or scenery with the genius of Shakspere, and yet they are linked to one of the most powerful of his achievements as a poet, by the bonds of a natural association. The very first incident of its true history would have constituted, had the details been minutely preserved, the early biography of the celebrated Macbeth; who, according to our black-letter historians, makes his first appearance in public life as Thane of Cromarty, and Maormor, or great man of Ross. But I am aware I do not derive from the circumstance any right to become his biographer. For though his character was probably formed at a time when he may be regarded as the legitimate property of the provincial annalist, no sooner is it exhibited in action than he is consigned over to the chroniclers of the kingdom.

For the earlier facts of our history the evidence is rather circumstantial than direct. We see it stamped on the face of the country, or inscribed on our older obelisks, or sometimes disinterred from out of hillocks of sand, or accumulations of moss; but very rarely do we find it deposited in our archives. Let us examine it, however, wherever it presents itself, and strive, should it seem at all intelligible, to determine regarding its pur· port and amount.   Not more than sixty years ago a bank of blown sand, directly under the northern Sutor, which had been heaped over the soil ages before, was laid open by the winds of a stormy winter, when it was discovered that the nucleus on which the hillock had originally formed, was composed of the bones of various animals of the chase, and the horns of deer. It is not much more than twelve years since there were dug up in the same sandy tract two earthen urns, the one filled with ashes and fragments of half-burned bones, the other with bits of a black bituminous-looking stone, somewhat resembling jet, which had been fashioned into beads, and little flat parallelo-grams, perforated edgewise, with four holes apiece.   Nothing could be ruder than the workmanship : the urns were clumsily modelled by the hand, unassisted by a lathe ; the ornaments, rough and unpolished, and still bearing the marks of the tool, resembled nothing of modern production, except, perhaps, the toys which herd-boys sometimes amuse their leisure in forming with the knife.   We find remains such as these fraught with a more faithful evidence regarding the early state of our country than the black-letter pages of our chroniclers.   They testify of a period when the chase formed, perhaps, the sole employment of the few scattered inhabitants ; and of the practice, so preva-lent among savages, of burying with their dead friends whatever they most loved when alive.   It may be further remarked as a curious fact, and one from which we may infer that trinkets wrought in so uncouth a style could have belonged to only the first stage of society, that man's inventive powers receive their earliest impulses rather from his admiration of the beautiful, than his sense of the useful.   He displays a taste in ornament, and has learned to dye his skin, and to tatoo it with rude figures of the sun and moon, before he has become ingenious enough to discover that he stands in need of a covering.

There is a tradition of this part of the country which seems not a great deal more modern than the urns or their ornaments, and which bears the character of the savage nearly as distinctly impressed on it. On the summit of Knock-Ferril, a steep hill which rises a few miles to the west of Dingwall, there are the remains of one of those vitrified forts which so puzzle and interest the antiquary; and which was originally constructed, says tradition, by a gigantic tribe of *Fions*, for the protection of their wives and children, when they themselves were engaged in hunting. It chanced in one of their excursions that a mean-spirited little fellow of the party, not much more than fifteen feet in height, was so distanced by his more active brethren, that, leaving them to follow out the chase, he returned home, and throwing himself down, much fatigued, on the side of the eminence, fell fast asleep. Garry, for so the unlucky hunter was called, was no favourite with the women of the tribe;—he was spiritless and diminutive, and ill-tempered; and as they could make little else of him that they cared for, they converted him into the butt of many a teasing little joke, and the sport of many a capricious humour. On seeing that he had fallen asleep, they stole out to where he lay, and after fastening his long hair with pegs to the grass, awakened him with their shouts and laughter. He strove to extricate himself, but in vain; until at length, infuriated by their gibes and the pain of his own exertions, he wrenched up his head, leaving half his locks behind him, and, hurrying after them, set fire to the stronghold into which they had rushed for shelter. The flames rose till they mounted over the roof, and broke out at every slit and opening; but Garry, unmoved by the shrieks and groans of the sufferers within, held fast the door until all was silent; when he fled into the remote Highlands, towards the west. The males of the tribe, who had, meanwhile, been engaged in hunting on that part of the northern Sutor which bears the name of the hill of Nigg, alarmed by the vast column of smoke which they saw ascending from their dwelling, came pressing on to the Firth of Cromarty, and leaping across on their hunting-spears, they hurried home. But they arrived to find only a huge pile of embers, fanned by the breeze, and amid which the very stones of the building were sputtering and bubbling with the intense heat, like the contents of a boiling caldron. Wild with rage and astonishment, and yet collected enough to conclude that none but Garry could be the author of a deed so barbarous, they tracked him into a nameless Highland glen, which has ever since been known as Glen-*Garry*, and there tore him to pieces. And as all the women of the tribe perished in the flames, there was an end, when this forlorn and widowed generation had passed away, to the whole race of the *Fions*. The next incident of our history bears no other connexion to this story, than that it belongs to a very early age, that of the Vikingr and Sea-King, and that we owe our data regarding it, not to written records, but to an interesting class of ancient remains, and to a doubtful and imperfect tradition.

In this age, says the tradition, the Maormor of Ross was married to a daughter of the king of Denmark, and proved so barbarous a husband, that her father, to whom she at length found the means of escape, fitted out a fleet and army to avenge on him the cruelties inflicted on her. Three of her brothers accompanied the expedition; but, on nearing the Scottish coast, a terrible storm arose, in which almost all the vessels of the fleet either foundered or were driven ashore, and the three princes were drowned. The ledge of rock at which this latter disaster is said to have taken place, still bears the name of the King's Sons; a magnificent cave which opens among the cliffs of the neighbouring shore is still known as the King's Cave; and a path that winds to the summits of the precipices beside it, as the King's Path. The bodies of the princes, says the tradition, were interred, one at Shandwick, one at Hilton, and one at Nigg; and the sculptured obelisks of these places, three very

curious pieces of antiquity, are said to be monuments erected to
their memory by their father.   In no part of Scotland do stones
of this class so abound as on the shores of the Moray Firth.
And they have often attracted the notice and employed the
ingenuity of the antiquary ; but it still appears somewhat doubtful whether we are to regard them as of Celtic or of Scandinavian origin.   It may be remarked, however, that though their
style of sculpture resembles, in its general features, that exhibited in the ancient crosses of Wales, which are unquestionably British, and though they are described in a tradition current
on the southern shore of the Moray Firth, as monuments raised
by the inhabitants on the expulsion of the Danes, the amount
of evidence seems to preponderate in the opposite direction ;
when we consider that they are invariably found bordering on
the sea; that their design and workmanship display a degree
of taste and mechanical ability which the Celtæ of North Britain
seem never to have possessed; that the eastern shores of the
German Ocean abound in similar monuments, which, to a com-
plexity of ornament not more decidedly Runic, add the Runic
inscription ; and that the tradition just related—which, wild as
it may appear, can hardly be deemed less authentic than the one
opposed to it, seeing that it belongs to a district still peopled
by the old inhabitants of the country, whereas the other seems
restricted to the lowlands of Moray—assigns their erection not
to the natives, but to their rapacious and unwelcome visitors,
the Danes themselves.   The reader may perhaps indulge me in
a few descriptive notices of the three stones connected with the
tradition ; they all lie within six miles of Cromarty, and their
weathered and mossy planes, roughened with complicated tracery
and doubtful hieroglyphics, may be regarded as pages of provincial history—as pages, however, which we must copy rather
than translate.   May I not urge, besides, that men who have
visited Egypt to examine monuments not much more curious,
have written folios on their return ?

The obelisk at Hilton, though perhaps the most elegant of its class in Scotland, is less known than any of the other two, and it has fared more hardly. For, about two centuries ago, it was taken down by some barbarous mason of Ross, who converted it into a tombstone, and, erasing the neat mysterious hieroglyphics of one of the sides, engraved on the place which they had occupied a rude shield and label, and the following laughable inscription; no bad specimen, by the bye, of the taste and judgment which could destroy so interesting a monument, and of that fortuitous species of wit which lies within the reach of accident, and of accident alone.

HE ' THAT ' LIVES ' WEIL ' DYES ' WEIL ' SAYS ' SOLOMON ' THE ' WISE.

HEIR ' LYES ' ALEXANDER ' DVFF ' AND ' HIS ' THRIE ' WIVES.

The side of the obelisk which the chisel has spared is surrounded by a broad border, embossed in a style of ornament that would hardly disgrace the frieze of an Athenian portico; —the centre is thickly occupied by the figures of men, some on horseback, some afoot—of wild and tame animals, musical instruments, and weapons of war and of the chase. The stone of Shandwick is still standing,[1] and bears on the side which corresponds to the obliterated surface of the other, the figure of a large cross, composed of circular knobs wrought into an involved and intricate species of fretwork, which seems formed by the twisting of myriads of snakes. In the spaces on the sides of the shaft there are two huge, clumsy-looking animals, the one resembling an elephant, and the other a lion; over each of these a St. Andrew seems leaning forward from his cross; and on the reverse of the obelisk the sculpture represents processions, hunting-scenes, and combats. These, however, are but meagre notices; the obelisk at Nigg I shall describe more minutely as an average specimen of the class to which it belongs.

It stands in the parish burying-ground, beside the eastern

---

[1] Since, however, blown down during a storm, and broken into three pieces.

gable of the church ; and bears on one of its sides, like the
stone at Shandwick, a large cross, which, it may be remarked,
rather resembles that of the Greek than of the Romish Church,
and on the other a richly embossed frame, enclosing, like the
border of the obelisk at Hilton, the figures of a crowded assemblage of men and animals.   Beneath the arms of the cross the
surface is divided into four oblong compartments, and there are
three above—one on each side, which form complete squares,
and one a-top, which, like the pediment of a portico, is of a
triangular shape.   In the lower angle of this upper compart-
ment, two priest-like figures, attired in long garments, and
furnished each with a book, incline forwards in the attitude of
prayer ; and in the centre between them there is a circular cake
or wafer, which a dove, descending from above, holds in its bill.
Two dogs seem starting towards the wafer from either side ;
and directly under it there is a figure so much weathered, that
it may be deemed to represent, as fancy may determine, either
a little circular table, or the sacramental cup.   A pictorial
record cannot be other than a doubtful one ; and it is difficult
to decide whether the hieroglyphic of this department denotes
the ghostly influence of the priest in delivering the soul from
the evils of an intermediate state ; for, at a slight expense of
conjectural analogy, we may premise that the mysterious dove
descends in answer to the prayer of the two kneeling figures, to
deliver the little emblematical cake from the " power of the
dog ;"—or, whether it may not represent a treaty of peace between rival chiefs whose previous hostility may be symbolized
by the two fierce animals below, and their pacific intentions by
the bird above, and who ratify the contract by an oath, solem-
nized over the book, the cup, and the wafer.   A very few such
explanations might tempt one to quote the well-known story of
the Professor of signs and the Aberdeen butcher ; the weight
of the evidence, however, rests apparently with those who adopt
the last.   We see the locks of the kneeling figures curling upon

their shoulders in unclerical profusion, unbroken by the tonsure; while the presence of the two books, with the absence of any written inscription, seems characteristic of the mutual memorial of tribes, who, though not wholly illiterate, possess no common language save the very doubtful language of symbol. If we hold further that the stone is of Scandinavian origin—and it seems a rather difficult matter to arrive at a different conclusion—we can hardly suppose that the natives should have left unmutilated the monument of a people so little beloved had they had no part in what it records, or no interest in its preservation.

We pass to the other compartments;—some of these and the plane of the cross are occupied by a species of fretwork exceedingly involved and complicated, but formed, notwithstanding, on regular mathematical figures. There are others which contain squares of elegantly arrayed tracery, designed in a style which we can almost identify with that of the border illuminations of our older manuscripts, or of the ornaments, imitative of these, which occur in works printed during the reigns of Elizabeth and James. But what seem the more curious compartments of the stone are embossed into rows of circular knobs, covered over, as if by basket-work, with the intricate foldings of myriads of snakes; and which may be either deemed to allude to the serpent and apple of the Fall—thus placed in no inapt neighbourhood to the cross; or to symbolize (for even the knobs may be supposed to consist wholly of serpents) that of which the serpent has ever been held emblematic, and which we cannot regard as less appositely introduced—a complex wisdom, or an incomprehensible eternity.

The hieroglyphics of the opposite side are in lower relief, and though the various fretwork of the border is executed in a style of much elegance, the whole seems to owe less to the care of the sculptor. The centre is occupied by what, from its size, we must deem the chief figure of the group—that of a man attired in long garments, caressing a fawn; and directly fronting him, there are the figures of a lamb and a harp. The whole is, perhaps, emblematical of peace, and may be supposed to tell the same story with the upper hieroglyphic of the reverse. In the space beneath there is the figure of a man furnished with cymbals, which he seems clashing with much glee, and that of a horse and its rider, surrounded by animals of the chase; while in the upper part of the stone there are dogs, deer, an armed huntsman, and, surmounting the whole, an eagle or raven. It may not be deemed unworthy of remark, that the style of the more complex ornaments of this stone very much resembles that which obtains in the sculptures and tatooings of the New-Zealander. We see exhibited in both the same intricate regularity of pattern, and almost similar combinations of the same waving lines. And we are led to infer, that though the rude Scandinavian of perhaps nine centuries ago had travelled a long stage in advance of the New-Zealander of our own times, he had yet his ideas of the beautiful cast in nearly the same mould. Is it not a curious fact, that man, in his advances towards the just and graceful in desgin, proceeds not from the simple to the complex, but from the complex to the simple?

The slope of the northern Sutor which fronts the town of Cromarty, terminates about a hundred and fifty feet above the level of the shore in a precipitous declivity surmounted by a little green knoll, which for the last six centuries has borne the name of Dunskaith (*i.e.* the fort of mischief). And in its immediate vicinity there is a high-lying farm, known all over the country as the farm of Castle-Craig. The prospect from the edge of the eminence is one of the finest in the kingdom. We may survey the entire Firth of Cromarty spread out before us as in a map; the town, though on the opposite shore, seems so completely under our view that we think of looking down into its streets; and yet the distance is sufficient to conceal all but what is pleasing in it. The eye, in travelling over the country beyond, ascends delighted through the various regions of corn, and wood, and moor, and then expatiates unfatigued amid a wilderness of blue-peaked hills. And where the land terminates towards the east, we may see the dark abrupt cliffs of the southern Sutor flinging their shadows half-way across the opening, and distinguish among the lofty crags, which rise to oppose them, the jagged and serrated shelves of the Diamond-rock, a tall beetling precipice which once bore, if we may trust to tradition, a wondrous gem in its forehead. Often, says the legend, has the benighted boatman gazed from amid the darkness, as he came rowing along the shore, on its clear beacon-like flame, which, streaming from the rock, threw a long fiery strip athwart the water ; and the mariners of other countries have inquired whether the light which they saw shining so high among the cliffs, right over their mast, did not proceed from the shrine of some saint, or the cell of some hermit. But like the carbuncle of the Ward-hill of Hoy, of which the author of Waverley makes so poetical a use, " though it gleamed ruddy as a furnace to them who viewed it from beneath, it ever became invisible to him whose daring foot had scaled the precipices from whence it darted its splendour." I have been oftener than once interrogated on the western coast of Scotland regarding the " Diamond-rock of Cromarty;" and an old campaigner who fought under Abercromby has told me that he has listened to the familiar story of its diamond amid the sand wastes of Egypt. But the jewel has long since disappeared, and we see only the rock. It used never to be seen, it is said, by day, nor could the exact point which it occupied be ascertained ; and on a certain luckless occasion an ingenious ship-captain, determined on marking its place, brought with him from England a few balls of chalk, and, charging with this novel species of shot, took aim at it in the night-time with one of his great guns. Ere he had fired, however, it vanished, as if suddenly withdrawn by some guardian hand ; and its place on the rock has ever since remained as undistinguishable as the scaurs and cliffs around it.   And now the eye, after completing its circuit, rests on the eminence of Dunskaith;—the site of a royal fortress erected by William the Lion, to repress, says Lord Hailes in his Annals of Scotland, the oft-recurring rebellions and disorders of Ross-shire.   We can still trace the moat of the citadel, and part of an outwork which rises towards the hill ; but the walls have sunk into low grassy mounds, and the line of the outer moat has long since been effaced by the plough.   The disorders of Ross-shire seem to have outlived, by many ages, the fortress raised to suppress them.   I need hardly advert to a story so well known as that of the robber of this province who nailed horse-shoes to the feet of the poor widow who had threatened him with the vengeance of James I., and who, with twelve of his followers, was brought to Edinburgh by that monarch, to be horse-shoed in turn.   Even so late as the reign of James VI. the clans of Ross are classed among the peculiarly obnoxious, in an Act for the punishment of theft, rief, and oppression.

Between the times of Macbeth and an age comparatively recent, there occurs a wide chasm in the history of Cromarty. The Thane, magnified by the atmosphere of poetry which surrounds him, towers like a giant over the remoter brink of the gap, while, in apparent opposition to every law of perspective, the people on its nearer edge seem diminished into pigmies. And yet the Urquharts of Cromarty—though Sir Thomas, in his zeal for their honour, has dealt by them as the poets of ancient Greece did by the early history of their country—were a race of ancient standing and of no little consideration.   The editor of the second edition of Sir Thomas's *Jewel*, which was not published until the first had been more than a hundred years out of print, states in his advertisement that he had compared the genealogy of his author with another genealogy of the family in possession of the Lord Lyon of Scotland, and that from the reign of Alexander II. to that of Charles I. he had found them perfectly to agree. The lands of the family extended from the furthest point of the southern Sutor to the hill of *Kinbeakie* (*i.e.* end of the living), a tract which includes the parishes of Cromarty, Kirkmichael, and Cullicuden; and, prior to the imprisonment and exile of Sir Thomas, he was vested with the patronage of the churches of these parishes, and the admiralty of the eastern coast of Scotland, from Caithness to Inverness.

The first of his ancestors, whose story receives some shadow of confirmation from tradition, was a contemporary of Wallace and the Bruce. When ejected from his castle, he is said to have regained it from the English by a stratagem, and to have held it out with only forty men for about seven years. " During that time," says Sir Thomas, " his lands were wasted and his woods burnt; and having nothing he could properly call his own but the moat-hill of Cromarty, which he maintained in defiance of all the efforts of the enemy, he was agnamed *Gulielmus de monte alto.* At length," continues the genealogist, " he was relieved by Sir William Wallace, who raised the siege after defeating the English in a little den or hollow about two miles from the town." Tradition, though silent respecting the siege, is more explicit than Sir Thomas in her details of the battle.

Somewhat more than four miles to the south of Cromarty, and about the middle of the mountainous ridge which, stretching from the Sutors to the village of Rosemarkie, overhangs at the one edge the shores of the Moray Firth, and sinks on the other into a broken moor, there is a little wooded eminence. Like the ridge which it overtops, it sweeps gradually towards the east until it terminates in an abrupt precipice that overhangs the sea, and slopes upon the west into a marshy hollow, known to the elderly people of the last age and a very few of the present as *Wallace-slack*—*i.e.*, ravine. The direct line of communication with the southern districts, to travellers who cross the Firth at the narrow strait of Ardersier, passes within

a few yards of the hollow.   And when, some time during the
wars of Edward, a strong body of English troops were marching by this route to join another strong body encamped in the
peninsula of Easter Ross, this circumstance is said to have
pointed it out to Wallace as a fit place for forming an ambus-
cade.   From the eminence which overtops it, the spectator can
look down on a wide tract of country, while the ravine itself is
concealed by a flat tubercle of the moor, which to the traveller
approaching from the south or west, seems the base of the
eminence.   The stratagem succeeded; the English, surprised and
panic-struck, were defeated with much slaughter, six hundred
being left dead in the scene of the attack; and the survivors,
closely pursued and wholly unacquainted with the country, fled
towards the north along the ridge of hill which terminates at the
bay of Cromarty.   From the top of the ridge the two Sutors
seem piled the one over the other, and so shut up the opening,
that the bay within assumes the appearance of a lake; and the
English deeming it such, pressed onward, in the hope that a
continued tract of land stretched between them and their countrymen on the opposite shore.   They were only undeceived when,
on climbing the southern Sutor, where it rises behind the town,
they saw an arm of the sea more than a mile in width, and
skirted by abrupt and dizzy precipices, opening before them.
The spot is still pointed out where they made their final stand;
and a few shapeless hillocks, that may still be seen among the
trees, are said to have been raised above the bodies of those
who fell; while the fugitives, for they were soon beaten from
this position, were either driven over the neighbouring precipices, or perished amid the waves of the Firth.   Wallace, on
another occasion, is said to have fled for refuge to a cave of the
Sutors; and his metrical historian, Blind Harry, after narrating
his exploits at St. Johnstone's, Dunotter, and Aberdeen, describes
him as

" Raiding throw the North-land into playne,
Till at *Crummade* fell Inglismen he'd slayne."

Hamilton, in his modernized edition of the "Achievements," renders the *Crummade* here Cromarty; and as shown by an ancient custom-house seal or cocket (supposed to belong to the reign of Robert II.), now in the Inverness Museum, the place was certainly designated of old by a word of resembling sound— *Chrombhte.*

Of all the humbler poets of Scotland—and where is there a country with more?—there is hardly one who has not sung in praise of Wallace. His exploit, as recorded in the Jewel, connected with the tradition of the cave, has been narrated by the muse of a provincial poet, who published a volume of poems at Inverness about five years ago; and, in the lack of less questionable materials for this part of my history, I avail myself of his poem.

Thus ran the tale :—proud England's host
Lay 'trench'd on Croma's winding coast,
And rose the Urquhart's towers beneath
Fierce shouts of wars, deep groans of death.
The Wallace heard ;—from Moray's shore
One little bark his warriors bore.
But died the breeze, and rose the day,
Ere gained that bark the destined bay;
When, lo ! these rocks a quay supplied,
These yawning caves meet shades to hide.
Secure, where rank the nightshade grew,
And patter'd thick th' unwholesome dew,
Patient of cold and gloom they lay,
Till eve's last light had died away.
    It died away ;—in Croma's hall
No flame glanced on the trophied wall,
Nor sound of mirth nor revel free
Was heard where joy had wont to be.
With day had ceased the siege's din,
But still gaunt famine raged within.
    In chamber lone, on weary bed,
That castle's wounded lord was laid;
His woe-worn lady watch'd beside.
To pain devote, and grief, and gloom,
No taper cheer'd the darksome room ;
Yet to the wounded chieftain's sight
Strange shapes were there, and sheets of light
And oft he spoke, in jargon vain,
Of ruthless deed and tyrant reign,
For maddening fever fired his brain.

D

O hark! the warder's rousing call—
" Rise, warriors, rise, and man the wall!"
Starts up the chief, but rack'd with pain,
And weak, he backward sinks again:
" O Heaven, they come!" the lady cries,
" The Southrons come, and Urquhart dies!"
    Nay, 'tis not fever mocks his sight;
His broider'd couch is red with light;
In light his lady stands confest,
Her hand clasp'd on her heaving breast.
And hark; wild shouts assail the ear,
Loud and more loud, near and more near
They rise!—hark, frequent rings the blade,
On crested helm relentless laid;
Yells, groans, sharp sounds of smitten mail,
And war-cries load the midnight gale;
O hark! like Heaven's own thunder high,
Swells o'er the rest one ceaseless cry,
Racking the dull cold ear of night,
" The Wallace wight!—the Wallace wight!"
    Yes, gleams the sword of Wallace there,
Unused his country's foes to spare;
Roars the red camp like funeral pyre,
One wild, wide, wasteful sea of fire;
Glow red the low-brow'd clouds of night,
The wooded hill is bathed in light,
Gleams wave, and field, and turret height.
Death's vassals dog the spoiler's horde,
Burns in their front th' unsparing sword;
The fired camp casts its volumes o'er;
Behind spreads wide a skiffless shore;
Fire, flood, and sword, conspire to slay.
How sad shall rest morn's early ray
On blacken'd strand, and crimson'd main,
On floods of gore, and hills of slain;
But bright its cheering beams shall fall
Where mirth whoops in the Urquharts' Hall.

        *   *   *

There occurs in our narrative another wide chasm, which extends from the times of Wallace to the reign of James IV. Like the earlier gap, however, it might be filled up by a recital of events, which, though they belong properly to the history of the neighbouring districts, must have affected in no slight degree the interests and passions of the people of Cromarty. Among these we may reckon the descents on Ross by the Lords of the Isles, which terminated in the battles of Harlaw and Driemderfat, and that contest between the Macintoshes and Munros, which took place in the same century at the village of Clachnaherry. I might avail myself, too, on a similar principle, of the pilgrimage of James IV. to the neighbouring chapel of St. Dothus, near Tain. But as all these events have, like the story of Macbeth, been appropriated by the historians of the kingdom, they are already familiar to the general reader. In an after age, Cromarty, like Tain, was honoured by a visit from royalty. I find it stated by Calderwood, that in the year 1589, on the discovery of Huntly's conspiracy, and the discomfiture of his followers at the Bridge of Dee, James VI. rode to Aberdeen, ostensibly with the intention of holding justice-courts on the delinquents; but that, deputing the business of trial to certain judges whom he instructed to act with a lenity which the historian condemns, he set out on a hunting expedition to Cromarty, from which he returned after an absence of about twenty days.

We find not a great deal less of the savage in the records of these later times than in those of the darker periods which went before. Life and property seem to have been hardly more secure, especially in those hapless districts which, bordering on the Highlands, may be regarded as constituting the battle-fields on which needy barbarism, and the imperfectly-formed vanguard of a slowly advancing civilisation, contended for the mastery. Early in the reign of James IV. the lands of Cromarty were wasted by a combination of the neighbouring clans, headed by Hucheon Rose of Kilravock, Macintosh of Macintosh, and Fraser of Lovat; and so complete was the spoliation, that the entire property of the inhabitants, to their very household furniture, was carried away. Restitution was afterwards enforced by the Lords of Council. We find it decreed in the *Acta Dominorum Concilii* for 1492, that Hucheon Rose of Kilravock do restore, content, and pay to Mr. Alexander Urquhart, sheriff of Cromarty, and his tenants, the various items carried off by him and his accomplices; viz., six hundred cows, one hundred horses, one thousand sheep, four hundred goats, two hundred swine, and four hundred bolls of victual. Kilravock is said to have conciliated the justice-general on this occasion by resigning into his hands his grand-daughter, the heiress of Calder, then a child; and her lands the wily magistrate secured to his family by marrying her to one of his sons.

There lived in the succeeding reign a proprietor of Cromarty, who, from the number of his children, received, says the genealogist, the title, or agname, of Paterhemon. He had twenty-five sons who arrived at manhood, and eleven daughters who ripened into women, and were married. Seven of the sons lost their lives at the battle of Pinkie; and there were some of the survivors who, settling in England, became the founders of families which, in the days of the Commonwealth, were possessed of considerable property and influence in Devonshire and Cumberland. Tradition tells the story of Paterhemon somewhat differently. His children, whom it diminishes to twenty, are described as robust and very handsome men; and he is said to have lived in the reign of Mary. On the visit of that princess to Inverness, and when, according to Buchanan, the Frasers and Munros, two of the most warlike clans of the country, were raised by their respective chieftains to defend her against the designs of Huntly, the Urquhart is said also to have marched to her assistance with a strong body of his vassals, and accompanied by all his sons, mounted on white horses. At the moment of his arrival Mary was engaged in reviewing the clans, and surrounded by the chiefs and her officers. The venerable chieftain rode up to her, and, dismounting with all the ease of a galliard of five-and-twenty, presented to her, as his best gift, his little troup of children. There is yet a third edition of the story :—About the year 1652, one Richard Franck, a native of the sister kingdom, and as devoted an angler as Isaac Walton himself, made the tour of Scotland, and then published a book descriptive of what he had seen. His notice of Cromarty is mostly summed up in a curious little anecdote of the patriarch, which he probably derived from some tradition current at the time of his visit. Sir Thomas he describes as his eldest son ; and the number of his children who arrived at maturity he has increased to forty. " He had thirty sons and ten daughters," says the tourist, "standing at once before him, and not one natural child amongst them." Having attained the extreme verge of human life, he began to consider himself as already dead ; and in the exercise of an imagination, which the genealogist seems to have inherited with his lands, he derived comfort from the daily repetition of a kind of ceremony, ingenious enough to challenge comparison with any rite of the Romish Church. For every evening about sunset, being brought out in his couch to the base of a tower of the castle, he was raised by pulleys, slowly and gently, to the battlements ; and the ascent he deemed emblematical of the resurrection. Or to employ the graphic language of the tourist—" The declining age of this venerable laird of Urquhart, for he had now reached the utmost limit of life, invited him to contemplate mortality, and to cruciate himself by fancying his cradle his sepulchre; therein, therefore, was he lodged night after night, and hauled up by pulleys to the roof of his house, approaching, as near as the summits of its higher pinnacles would let him, to the beautiful battlements and suburbs of heaven."

I find I must devote one other chapter to the consideration of the interesting remains which form almost the sole materials of this earlier portion of my history. But the class of these to which I am now about to turn, are to be found, not on the face of the country, but locked up in the minds of the inhabitants. And they are falling much more rapidly into decay—mouldering away in their hidden recesses, like bodies of the dead ; while others, which more resemble the green mound and the monumental tablet, bid fair to abide the inquiry of coming generations. Those vestiges of ancient superstition, which are to be traced in the customs and manners of the common people, share in a polite age a very different fate from those impressions of it, if I may so express myself, which we find stamped upon matter. For when the just and liberal opinions which originate with philosophers and men of genius are diffused over a whole people, a modification of the same good sense which leads the scholar to treasure up old beliefs and usages, serves to emancipate the peasant from their influence or observance.

CHAPTER V.

"She darklins grapit for the bauks,
And in the blue clue throws then."—Burns.

Violence may anticipate by many centuries the natural pro-gress of decay. There are some of our Scottish cathedrals less entire than some of our old Picts' houses, though the latter have been deserted for more than a thousand years, and the former for not more than three hundred. And the remark is not less applicable to the beliefs and usages of other ages, than to their more material remains. It is a curious fact, that we meet among the Protestants of Scotland with more marked traces of the Paganism of their earlier, than of the Popery of their later ancestors. For while Christianity seems to have been introduced into the country by slow degrees, and to have travelled over it by almost imperceptible stages—leaving the less obnoxious practices of the mythology which it supplanted to the natural course of decay—it is matter of history that the doctrines of the Reformation overspread it in a single age, and that the observances of the old system were effaced, not by a gradual current of popular opinion, but by the hasty surges of popular resentment. The saint-days of the priest have in con-sequence been long since forgotten—the festivals of the Druid still survive.

There is little risk of our mistaking these latter; the rites of Hallowe'en, and the festivities of Beltane, possess well-authen-ticated genealogies. There are other usages, however, which, though they bear no less strongly the impress of Paganism, show a more uncertain lineage. And regarding these, we find it difficult to determine whether they have come down to us from the days of the old mythology, or have been produced in a later period by those sentiments of the human mind to which every false religion owes its origin. The subject, though a curious, is no very tangible one. But should I attempt throwing together a few simple thoughts respecting it, in that wandering desultory style which seems best to consort with its irregularity of outline, I trust I may calculate on the forbearance of the reader. I shall strive to be not very tedious, and to choose a not very beaten path.

Man was made for the world, and the world for man. Hence we find that every faculty of the human mind has in the things which lie without some definite object, or particular class of circumstances, on which to operate. There is a thorough adaptation of that which acts to that which is acted upon—of the moving power to the machine ; and woe be to him who deranges this admirable order, in the hope of rendering it more complete. It is prettily fabled by the Brahmins, that souls are moulded by pairs, and then sent to the earth to be linked together in wedlock, and that matches are unhappy merely in consequence of the parties disuniting by the way, and choosing for themselves other consorts. One might find more in this fable than any Brahmin ever found in it yet. There is a prospective connexion of a similar kind formed between the powers of the mind and the objects on which these are to be employed, and should they be subsequently united to objects other than the legitimate, a wretchedness quite as real as that which arises out of an ill-mated marriage is the infallible result.

Were I asked to illustrate my meaning by an example or two, I do not know where I could find instances better suited to my purpose than in the imaginative extravagancies of some of our wilder sectaries. There is no principle which so deals in unhappy marriages, and as unhappy divorces, as the fanatical ; or that so ceaselessly employs itself in separating what Heaven has joined, and in joining what Heaven has separated. Man, I have said, was made for the world he lives in ;—I should have added, that he was intended also for another world. Fanaticism makes a somewhat similar omission, only it is the other way. It forgets that he is as certainly a denizen of the present as an heir of the future ; that the same Being who has imparted to him the noble sentiment which leads him to anticipate an hereafter, has also bestowed upon him a thousand lesser faculties which must be employed now ; and that, if he prove untrue to even the minor end of his existence, and slight his proper though subordinate employments, the powers which he thus separates from their legitimate objects must, from the very activity of their nature, run riot in the cloisters in which they are shut up, and cast reproach by their excesses on the cause to which they are so unwisely dedicated. For it is one thing to condemn these to a life of celibacy, and quite another to keep them chaste. We may shut them up, like a sisterhood of nuns, from the objects to which they ought to have been united, but they will infallibly discover some less legitimate ones with which to connect themselves. Self-love, and the natural desire of distinction—proper enough sentiments in their own sphere—make but sad work in any other. The imagination, which was so bountifully given us to raise its ingenious theories as a kind of scaffolding to philosophical discovery, is active to worse purpose when revelling intoxicated amid the dim fields of prophecy, or behind the veil of the inner mysteries. Reason itself, though a monarch in its own proper territories, can exert only a doubtful authority in the provinces which lie beyond. Indeed, the whole history of fanaticism, from when St. Anthony retired into the deserts of Upper Egypt to burrow in a cell like a fox-earth, down to the times that witnessed some of the wilder heresiarchs of our own country, working what they had faith enough to deem miracles, is little else than a detail of the disorders occasioned by perversions of this nature.

There is an exhibition of phenomena equally curious when the religious sentiment, instead of thus swallowing up all the others, is deprived of even its own proper object. I once saw a solitary hen bullfinch, that retired one spring into a dark corner of her cage and laid an egg, over which she sat until it was addled. It is always thus when the devotional sentiment is left to form a religion for itself. Encaged like the poor bullfinch, it proves fruitful in just a similar way, and moping in its dark recesses, brings forth its pitiful abortions unassisted and alone. I have ever thought of the pantheons and mythological dictionaries of our libraries as a kind of museums, stored, like those of the anatomist, with embryos and abortions.

It must be remarked further, that the devotional sentiment operates in this way not only when its proper object is wanting, but even, should the mind be dark and uninformed, when that is present. Every false religion may be regarded as a wild irregular production, springing out of that basis of sentiment (one of the very foundations of our nature) which, when rendered the subject of a right course of culture, and sown with the good seed, proves the proper field of the true. But on this field, even when occupied the better way, there may be the weeds of a rank indigenous mythology shooting up below—a kind of subordinate superstition, which, in other circumstances, would have been not the underwood, but the forest. Hence our difficulty in fixing the genealogy of the Pagan-like usages to which I allude; there are two opposite sources, from either of which they may have sprung :—they may form a kind of undergrowth, thrown up at no very early period by a soil occupied by beliefs the most serious and rational, or they may constitute the ancient and broken vestiges of an obsolete and exploded mythology. I shall briefly describe a few of the more curious.

I. People acquainted with seafaring men, and who occasionally accompany them in their voyages, cannot miss seeing them, when the sails are drooping against the mast, and the vessel lagging in her course, earnestly invoking the wind in a shrill tremulous whistling—calling on it, in fact, in its own language; and scarcely less confident of being answered than if preferring a common request to one of their companions. I rarely sail in calm weather with my friends the Cromarty fishermen, without seeing them thus employed—their faces anxiously turned in the direction whence they expect the breeze; now pausing, for a light uncertain air has begun to ruffle the water, and now resuming the call still more solicitously than before, for it has died away. On thoughtlessly beginning to whistle one evening about twelve years ago, when our skiff was staggering under a closely-reefed foresail, I was instantly silenced by one of the fishermen with a "Whisht, whisht, boy, we have more than wind enough already;" and I remember being much struck for the first time by the singularity of the fact, that the winds should be as sincerely invoked by our Scottish seamen of the present day, as by the mariners of Themistocles. There was another such practice common among the Cromarty fishermen of the last age, but it is now obsolete. It was termed soothing the waves. When beating up in stormy weather along a lee-shore, it was customary for one of the men to take his place on the weather gunwale, and there continue waving his hand in a direction opposite to the sweep of the sea, in the belief that this species of appeal to it would induce it to lessen its force. We recognise in both these singular practices the workings of that religion natural to the heart, which, more vivid in its personifications than poetry itself, can address itself to every power of nature as to a sentient being endowed with a faculty of will, and able, as it inclines, either to aid or injure. The seaman's prayer to the winds, and the thirty thousand gods of the Greek, probably derive their origin from a similar source.

II. Viewed in the light of reason, an oath owes its sacredness, not to any virtue in itself, but to the Great Being to whom it is so direct an appeal, and to the good and rational belief that

He knows all things, and is the ultimate judge of all. But the same uninformed principle which can regard the winds and waves as possessed of a power independent of His, seems also to have conferred on the oath an influence and divinity exclusively its own. I have met with many among the more grossly superstitious, who deemed it a kind of ordeal, somewhat similar to the nine ploughshares of the dark ages, which distinguished between right and wrong, truth or falsehood, by some occult intrinsic virtue. The innocent person swears, and like the guiltless woman when she had drunk the waters of jealousy, thrives none the worse;—the guilty perjure themselves, and from that hour cease to prosper. I remember—by the way, a very early recollection—that when a Justice of Peace Court was sitting in my native town, many years ago, a dark cloud came suddenly over the sun; and that a man who had been lounging on the street below, ran into the Court-room to see who it was that, *by swearing a false oath*, had occasioned the obscuration. It is a rather singular coincidence, and one which might lead us to believe in the existence of something analogous to principle in even the extravagancies of human belief, that the only oath deemed binding on the gods of classical mythology—the oath by the river Styx—was one of merely intrinsic power and virtue. Bacon, indeed, in his "Wisdom of the Ancients" (a little book but a great work), has explained the fable as merely an ingenious allegory; but who does not know that the Father of modern philosophy found half the Novum Organum in superstitions which existed before the days of Orpheus?

III. There seems to have once obtained in this part of the country a belief that the natural sentiment of justice had its tutelary spirit, which, like the Astræa of the Greeks, existed for it, and for it alone; and which not only seconded the dictates of conscience, but even punished those by whom they were disregarded. The creed of superstition is, however, rarely a well defined or consistent one; and this belief seems to have partaken, as much as any of the others, of the incoherent obscurity in which it originated. The mysterious agent (the object of it) existed no one knew where, and effected its purposes no one knew how. But the traditions which illustrate it, narrate better than they define. Many years ago, says one of these, a woman of Tarbat was passing along the shores of Loch-Slin, with a large web of linen on her back. There was a market held that morning at Tain, and she was bringing the web there to be sold. In those times it was quite as customary for farmers to rear the flax which supplied them with clothing, as the corn which furnished them with food; and it was of course necessary, in some of the earlier processes of preparing the former, to leave it for weeks spread out on the fields, with little else to trust to for its protection than the honesty of neighbours. But to the neighbours of this woman the protection was, it would seem, incomplete; and the web she carried on this occasion was composed of stolen lint. She had nearly reached the western extremity of the lake, when, feeling fatigued, she seated herself by the water edge, and laid down the web beside her. But no sooner had it touched the earth than up it bounded three Scots ells into the air, and slowly unrolling fold after fold, until it had stretched itself out as when on the bleaching-green, it flew into the middle of the lake, and disappeared for ever. There are several other stories of the same class, but the one related may serve as a specimen of the whole.

IV. The evils which men dread, and the appearances which they cannot understand, are invariably appropriated by superstition: if her power extend not over the terrible and the mysterious, she is without power at all. And not only does she claim whatever is inexplicable in the great world, but also in some cases what seems mysterious in the little; some, for instance, of the more paradoxical phenomena of human nature. It has been represented to me as a mysterious, unaccountable fact, that persons who have been rescued from drowning regard their deliverers ever after with a dislike which borders almost on enmity. I have heard it affirmed, too, that when the crew of some boat or vessel have perished, with the exception of one individual, the relatives of the deceased invariably regard that one with a deep, irrepressible hatred; and in both cases the feelings described are said to originate in some occult and supernatural cause. Alas! neither envy nor ingratitude lie out of our ordinary every-day walk. There occurs to me a little anecdote illustrative of this kind of apotheosis of the envious principle. Some fifty years ago there was a Cromarty boat wrecked on the rough shores of Eathie. All the crew perished with the exception of one fisherman; and the poor man was so persecuted by the relatives of the drowned, who even threatened his life, that he was compelled, much against his inclination, to remove to Nairn. There, however, only a few years after, he was wrecked a second time, and, as in the first instance, proved the sole survivor of the crew. And so he was again subjected to a persecution similar to the one he had already endured; and compelled to quit Nairn as he had before quitted Cromarty. And in both cases the relatives of the deceased were deemed as entirely under the influence of a mysterious, irresistible impulse, which acted upon their minds from without, as the Orestes of the dramatist when pursued by the Furies.

One may question, as I have already remarked, whether one sees, in these several instances, polytheism in the act of forming, and but barely forming, in the human mind, or the mutilated remnants of a long-exploded mythology. The usages to which I have alluded as more certain in their lineage, are perhaps less suited to employ speculation. But they are curious; and the fact that they are fast sinking into an oblivion, out of which the diligence of no future excavator will be able to restore them, gives them of itself a kind of claim on our notice. I pass over Beltane; its fires in this part of the country have long since been extinguished; but to its half-surviving partner,

Halloween, I shall devote a few pages; and this the more readily, as it chances to be connected with a story of humble life which belongs to that period of my history at which I have now arrived. True, the festival itself has already sat for its picture, and so admirable was the skill of the artist, that its very name recalls to us rather the masterly strokes of the transcript than the features of the original. But, with all its truth and beauty, the portrait is not yet complete.

The Scottish Halloween, as held in the solitary farmhouse and described by Burns, differed considerably from the Halloween of our villages and smaller towns. In the farmhouse it was a night of prediction only; in our towns and villages there were added a multitude of wild mischievous games, which were tolerated at no other season—a circumstance that serves to identify the festival with those pauses of license peculiar to the nonage of civil government, in which men are set free from the laws they are just learning to respect;—partly, it would seem, as a reward for the deference which they have paid them, partly to serve them as a kind of breathing-spaces in which to recover from the unwonted fatigue of being obedient. After nightfall, the young fellows of the town formed themselves into parties of ten or a dozen, and breaking into the gardens of the graver inhabitants, stole the best and heaviest of their cabbages. Converting these into bludgeons, by stripping off the lower leaves, they next scoured the streets and lanes, thumping at every door as they passed, until their uncouth weapons were beaten to pieces. When disarmed in this way, all the parties united into one, and providing themselves with a cart, drove it before them with the rapidity of a chaise and four through the principal streets. Woe to the inadvertent female whom they encountered! She was instantly laid hold of, and placed aloft in the cart—brothers, and cousins, and even sons, it is said, not unfrequently assisting in the capture; and then dragged backwards and forwards over the rough stones, amid shouts, and screams, and roars of laughter. The younkers within doors were meanwhile engaged in a manner somewhat less annoying, but not a whit less whimsical. The bent of their ingenuity for weeks before, had been turned to the accumulating of little hoards of apples—all for this night; and now a large tub, filled with water, was placed in the middle of the floor of some out-house, carefully dressed up for the occasion; and into the tub every one of the party flung an apple. They then approached it by turns, and, placing their hands on the edges, plunged forward to fish for the fruit with their teeth. I remember the main chance of success was to thrust the head fearlessly into the tub, amid the booming of the water, taking especial care to press down one of the apples in a line with the mouth, and to seize it when jammed against the bottom. When the whole party, with their dripping locks and shining faces, would seem metamorphosed into so many mermaids, this sport usually gave place to another :—A small beam of wood was suspended from the ceiling by a cord, and when fairly balanced, an apple was fastened to the one end, and a lighted candle to the other. It was then whirled round, and the boys in turn, as before, leaped up and bit at the fruit; not unfrequently, however, merely to singe their faces and hair at the candle. Neither of these games were peculiar to the north of Scotland : we find it stated by Mr. Polewhele, in his " Historical Views of Devonshire," that the Irish peasants assembled on the eve of *La Samon* (the 2d November), to celebrate the festival of the sun, with many rites derived from Paganism, among which was the dipping for apples in a tub of water, and the catching at an apple stuck on the one end of a kind of hanging beam.

There belonged to the north of Scotland two Halloween rites of augury which have not been described by Burns : and one of these, an elegant and beautiful charm, is not yet entirely out of repute. An ale-glass is filled with pure water, and into the water is dropped the white of an egg. The female whose future fortunes are to be disclosed (for the charm seems appropriated. exclusively by the better sex) lays her hand on the glass's mouth, and holds it there for about the space of a minute. In that time the heavier parts of the white settle to the bottom, while the lighter shoot up into the water, from which they are distinguished by their opacity, into a variety of fantastic shapes, resembling towers and domes, towns, fleets, and forests; or, to speak more correctly, into forms not very unlike those icicles which one sees during a severe frost at the edge of a waterfall. A resemblance is next traced, which is termed reading the glass, between the images displayed in it and some objects of either art or nature; and these are regarded as constituting a hieroglyphic of the person's future fortunes. Thus, the ramparts of a fortress surmounted by streamers, a plain covered with armies, or the tents of an encampment, show that the female whose hand covered the glass is to be united to a soldier, and that her life is to be spent in camps and garrisons. A fleet of ships, a church or pulpit, a half-finished building, a field stripped into furrows, a garden, a forest—all these, and fifty other scenes, afford symbols equally unequivocal. And there are melancholy hieroglyphics, too, that speak of death when interrogated regarding marriage;—there are the solitary tomb, the fringed shroud, the coffin, and the skull and cross-bones. "Ah!" said a young girl, whom I overheard a few years ago regretting the loss of a deceased companion, "Ah! I knew when she first took ill that there was little to hope. Last Halloween we went together to Mrs. —— to break our eggs. Betsie's was first cast, and there rose under her hand an ugly skull. Mrs. —— said nothing, but reversed the glass, while poor Betsie laid her hand on it a second time, and then there rose a coffin. Mrs. —— called it a boat, and I said I saw the oars; but Mrs. —— well knew what it meant, and so did I."

The other north country charm, which, of Celtic origin, bears evidently the impress of the romance and melancholy so pro-

E

dominant in the Celtic character, is only known and practised (if, indeed, still practised anywhere) in a few places of the remote Highlands. The person who intends trying it must steal out unperceived to a field whose furrows lie due south and north, and, entering at the western side, must proceed slowly over eleven ridges, and stand in the centre of the twelfth, when he will hear either low sobs and faint mournful shrieks, which betoken his early death, or the sounds of music and dancing, which foretell his marriage. But the charm is accounted dangerous. About twelve years ago, I spent an autumn in the mid-Highlands of Ross-shire, where I passed my Halloween, with nearly a dozen young people, at a farmhouse. We burned nuts and ate apples; and when we had exhausted our stock of both, some of us proposed setting out for the steading of a neighbouring farm, and robbing the garden of its cabbages; but the motion was overruled by the female members of the party; for the night was pitch dark, and the way rough; and so we had recourse for amusement to story-telling. Naturally enough most of our stories were of Halloween rites and predictions; and much was spoken regarding the charm of the rig. I had never before heard of it; and, out of a frolic, I stole away to a field whose furrows lay in the proper direction, and after pacing steadily across the ridges until I had reached the middle of the twelfth, I stood and listened. But spirits were not abroad: —I heard only the wind groaning in the woods, and the deep sullen roar of the Conan. On my return I was greeted with exclamations of wonder and terror, and it was remarked that I looked deadly pale, and had certainly heard something very terrible. "But whatever you may have been threatened with," said the author of the remark, "you may congratulate yourself on being among us in your right mind; for there are instances of people returning from the twelfth rig raving mad; and of others who went to it as light of heart as you, who never returned at all.'

The Maccullochs of the parish of Cromarty, a family now extinct, were, for about two centuries, substantial respectable farmers. The first of this family, says tradition, was Alaster Macculloch, a native of the Highlands. When a boy he quitted the house of his widow mother, and wandered into the low country in quest of employment, which he at length succeeded in procuring in the parish of Cromarty, on the farm of an old wealthy tacksman. For the first few weeks he seemed to be one of the gloomiest little fellows ever bred among the solitudes of the hills;—all the social feelings of his nature had been frozen within him; but they began to flow apace; and it was soon discovered that neither reserve nor melancholy formed any part of his real character. A little of the pride of the Celt he still retained; when he attended chapel he wore a gemmy suit of tartan, and his father's dirk always depended from his belt; but, in every other respect, he seemed a true Lowland Scot, and not one of his companions equalled him in sly humour, or could play off a practical joke with half the effect.

His master was a widower, and the father of an only daughter, a laughing warm-hearted girl of nineteen. She had more lovers than half the girls of the parish put together; and when they avowed to her their very sincere attachment, she tendered them her very hearty thanks in return. But then one's affections are not in one's own power; and as certainly as they loved her just because they could not help it, so certainly was she indifferent to them from the same cause. Their number received one last accession in little Alaster the herd-boy. He shared in the kindness of his young mistress, and his cattle shared in it too, with every living thing connected with her father or his farm; but his soul-engrossing love lay silent within him, and not only without words, but, young and sanguine as he was, almost without hope. Not that he was unhappy. He had the knack of dreaming when broad awake, and of making his dreams as pleasant as he willed them; and so his passion rather increased than diminished the amount of his happiness. It taught him, too, the very best species of politeness—that of the heart; and young Lillias could not help wondering where it was that the manners of the red-cheeked Highland boy had received so exquisite a polish, and why it was that she herself was so much the object of his quiet unobtrusive attentions. When night released him from labour, he would take up his seat in some dark corner of the house, that commanded a full view of the fire, and there would he sit for whole hours gazing on the features of his mistress. A fine woman looks well by any light, even by that of a peat fire; and fine women, it is said, know it; but little thought the maiden of the farmhouse of the saint-like halo which, in the imagination of her silent worshipper, the red smoky flames shed around her. How could she even dream of it? The boy Alaster was fully five years younger than herself, and it surely could not be forgotten that he herded her father's cattle. The incident, however, which I am just going to relate, gave her sufficient cause to think of him as a lover.

The Halloween of the year 1560 was a very different thing in the parish of Cromarty from that of the year 1829. It is now as dark and opaque a night—unless it chance to be brightened by the moon—as any in the winter season; it was then clear as the glass of a magician;—people looked through it and saw the future. Late in October that year, Alaster overheard his mistress and one of her youthful companions—the daughter of a neighbouring farmer—talking over the rites of the coming night of frolic and prediction. "Will you really venture on throwing the clue?" asked her companion; "the kiln, you ken, is dark and lonely; and there's mony a story no true if folk havena often been frightened." "Throw it?—oh, surely!" replied the other; "who would think it worth while to harm the like o' me? and, besides, you can bide for me just a wee bittie aff. One would like, somehow, to know the name o'

one's gudeman, or whether one is to get a gudeman at all."
Alaster was a lover, and lovers are fertile in stratagem. In the
presence of his mistress he sought leave from the old man, her
father, with whom he was much a favourite, to spend his Halloween at a cottage on a neighbouring farm, where there were
several young people to meet; and his request was readily
granted. The long-expected evening came; and Alaster set
out for the cottage, without any intention of reaching it for at
least two hours. When he had proceeded a little way he
turned back, crept warily towards the kiln, climbed like a
wild-cat up the rough circular gable, entered by the chimney,
and in a few seconds was snugly seated amid the ashes of the
furnace. There he waited for a full hour, listening to the beatings of his own heart. At length a light footstep was heard
approaching; the key was applied to the lock, and as the door
opened, a square patch of moonshine fell upon the rude wall
of the kiln. A tall figure stepped timidly forward, and stood
in the stream of faint light. It was Alaster's young mistress.
She looked fearfully round her, and then producing a small clue
of yarn, she threw it towards Alaster, and immediately began
to wind.[1] He suffered it to turn round and round among the
ashes, and then cautiously laid hold of it. "Wha hauds?"
said his mistress in a low startled whisper, looking as she spoke,
over her shoulder towards the door; "Alaster Macculloch," was
the reply; and in a moment she had vanished like a spectre.
Soon after, the tread of two persons was heard approaching the
door. It was now Alaster's turn to tremble. "Ah!" he
thought, "I shall be discovered, and my stratagem come to
worse than nothing." "An' did ye hear onything when you
came out yon gate?" said one of the persons without. "Oh,
naething, lass, naething!" replied the other, in a voice whose
faintest echoes would have been recognised by the lover within;
"steek too the door an' lock it;—it's a foolish conceit." The

[1] See Burns's Halloween.

door was accordingly locked, and Alaster left to find his way out in the manner he had entered.

It was late that night before he returned from the cottage to which, after leaving the kiln, he had gone. Next day he saw his mistress. She by no means exhibited her most amiable phase of character, for she was cold and distant, and not a little cross. In short, it was evident she had a quarrel with destiny. This mood, however, soon changed for the one natural to her ; years passed away, and suitor after suitor was rejected by the maiden, until, in her twenty-fourth year, Alaster Macculloch paid her his addresses. He was not then a little herd-boy, but a tall, handsome, young man of nineteen, who, active and faithful, was intrusted by his master with the sole management of his farm. A belief in destiny often becomes a destiny of itself ; and it became such to Alaster's mistress. How could the predestined husband be other than a successful lover? In a few weeks they were married ; and when the old man was gathered to his fathers, his son-in-law succeeded to his well-stocked farm.

There are a few other traditions of this northern part of the country—some of them so greatly dilapidated by the waste of years, that they exist as mere fragments—which bear the palpable impress of a pagan or semi-pagan origin. I have heard imperfectly-preserved stories of a lady dressed in green, and bearing a goblin child in her arms, who used to wander in the night-time from cottage to cottage, when all the inhabitants were asleep. She would raise the latch, it is said, take up her place by the fire, fan the embers into a flame, and then wash her child in the blood of the youngest inmate of the cottage, who would be found dead next morning. There was another wandering green lady, her contemporary, of exquisite beauty and a majestic carriage, who was regarded as the Genius of the smallpox, and who, when the disease was to terminate fatally, would be seen in the grey of the morning, or as the evening was passing into night, sitting by the bedside of her victim. I have heard wild stories, too, of an unearthly, squalid-looking thing, somewhat in the form of a woman, that used to enter farmhouses during the day, when all the inmates, except perhaps a solitary female, were engaged in the fields.   More than a century ago, it is said to have entered, in the time of harvest, the house of a farmer of Navity, who had lost nearly all his cattle by disease a few weeks before.   The farmer's wife, the only inmate at the time, was engaged at the fireside in cooking for the reapers ; the goblin squatted itself beside her, and shivering, as if with cold, raised its dingy, dirty-looking vestments over its knees.   " Why, ye nasty thing," said the woman, " hae ye killed a' our cattle?"—"An' why," inquired the goblin in turn, " did the gudeman, when he last roosed them, forget to gie them his blessing?"

Immediately over the sea, the tract of table-land, which forms the greater part of the parish of Cromarty, terminates, as has been already said, in a green sloping bank, that for several miles sweeps along the edge of the bay.   In the vicinity of the town, a short half mile to the west, we find it traversed by a deep valley, which runs a few hundred yards into the interior ; 'tis a secluded, solitary place, the sides sprinkled over with the sea-hip, the sloe, and the bramble—the bottom occupied by a blind pathway, that, winding through the long grass like a snake, leads to the fields above.   It has borne, from the earliest recollections of tradition, the name of *Morial's Den*, a name which some, on the hint of Sir Thomas Urquhart, ingeniously derive from the Greek, and others, still more ingeniously, from the Hebrew ; and it has, for at least the last six generations, been a scene of bird-nesting and truant-playing during the day, and of witch and fairy meetings, it is said, during the night. Rather more than a century ago, it was the *locale*, says tradition, of an interesting rencounter with one of the unknown class of spectres. On a Sabbath noon a farmer of the parish was herding a flock of sheep in a secluded corner of the den.   He was an old grey-

haired man, who for many years had been affected by a deaf-
ness, which grew upon him as the seasons passed, shutting out
one variety of sounds after another, until at length he lived in
a world of unbroken silence.   Though secluded, however, from
all converse with his brother men, he kept better company than
ever, and became more thoroughly acquainted with his Bible,
and the fathers of the Reformation, than he would have been had
he retained his hearing, or than almost any other person in the
parish.   He had just despatched his herd-boy to church, for he
himself could no longer profit by his attendance there ; his flock
was scattered over the sides of the hollow ; and with his Bible
spread out before him on a hillock of thyme and moss, which
served him for a desk, and sheltered on either hand from the
sun and wind by a thicket of sweetbriar and sloethorn, he was
engaged in reading, when he was startled by a low rushing sound,
the first he had heard for many months.   He raised his eyes
from the book ; a strong breeze was eddying within the hollow, waving the ferns and the bushes ; and the portion of sea
which appeared through the opening was speckled with white ;
—but to the old man the waves broke and the shrubs waved
in silence.   He again turned to the book—the sound was again
repeated ; and on looking up a second time, he saw a beautiful,
sylph-looking female standing before him.   She was attired in
a long flowing mantle of green, which concealed her feet, but
her breast and arms, which were of exquisite beauty, were uncovered.   The old man laid his hand on the book, and raising
himself from his elbow, fixed his eyes on the face of the lady.
" Old man," said she, addressing him in a low sweet voice,
which found prompt entrance at the ears that had so long been
shut up to every other sound, " you are reading *the book ;* tell
me if there be any offer of salvation in it to *us.*"—" The gospel
of this book," said the man, " is addressed to the lost children
of Adam, but to the creatures of no other race."   The lady
shrieked as he spoke, and gliding away with the rapidity of a

swallow on the wing, disappeared amid the recesses of the hollow.

About a mile further to the west, in an inflection of the bank, there is the scene of a story, which, belonging to a still earlier period than the one related, and wholly unlike it in its details, may yet be deemed to resemble it in its mysterious, and, if I may use the term, unclassified character.

A shipmaster, who had moored his vessel in the upper roadstead of the bay, some time in the latter days of the first Charles, was one fine evening sitting alone on deck, awaiting the return of some of his seamen who had gone ashore, and amusing himself in watching the lights that twinkled from the scattered farmhouses, and in listening in the extreme stillness of the calm to the distant lowing of cattle, or the abrupt bark of the watch-dog. As the hour wore later, the sounds ceased, and the lights disappeared—all but one solitary taper, that twinkled from the window of a cottage situated about two miles west of the town. At length, however, it also disappeared, and all was dark around the shores of the bay as a belt of black velvet. Suddenly a hissing noise was heard overhead; the shipmaster looked up, and saw one of those meteors that are known as falling stars, slanting athwart the heavens in the direction of the cottage, and increasing in size and brilliancy as it neared the earth, until the wooded ridge and the shore could be seen as distinctly from the ship-deck as by day. A dog howled piteously from one of the outhouses, an owl whooped from the wood. The meteor descended until it almost touched the roof, when a cock crew from within. Its progress seemed instantly arrested; it stood still; rose about the height of a ship's mast, and then began again to descend. The cock crew a second time. It rose as before, and after mounting much higher, sunk yet again in the line of the cottage. It almost touched the roof, when a faint clap of wings was heard, as if whispered over the water, followed by a still louder note of defiance from the cock. The meteor rose with a bound, and continuing to ascend until it seemed lost among the stars, did not again appear. Next night, however, at the same hour, the same scene was repeated in all its circumstances—the meteor descended, the dog howled, the owl whooped, the cock crew. On the following morning the shipmaster visited the cottage, and, curious to ascertain how it would fare when the cock was away, he purchased the bird; and sailing from the bay before nightfall, did not return until about a month after.

On his voyage inwards he had no sooner doubled an intervening headland, than he stepped forward to the bows to take a peep at the cottage: it had vanished. As he approached the anchoring ground, he could discern a heap of blackened stones occupying the place where it had stood; and he was informed, on going ashore, that it had been burnt to the ground, no one knew how, on the very night he had quitted the bay. He had it rebuilt and furnished, says the story, deeming himself, what one of the old schoolmen would have perhaps termed, the *occasional* cause of the disaster. About fifteen years ago there was dug up, near the site of the cottage, a human skeleton, with the skull and the bones of the feet lying together, as if the body had been huddled up twofold into a hole; and this discovery led to that of the story, which, though at one time often repeated and extensively believed, had been suffered to sleep in the memories of a few elderly people for nearly sixty years.

CHAPTER VI.

" Subtill muldrie wrocht mony day agone."—GAVIN DOUGLASS.

As house after house in the old town of Cromarty was yielding its place to the sea, the inhabitants were engaged in building
new dwellings for themselves in the fields behind.   A second
town was thus formed, the greater part of which has since also
disappeared, though under the influence of causes less violent
than those which annihilated the first.   Shortly after the Union,
the trade of the place, which prior to that event had been pretty
considerable, fell into decay, and the town gradually dwindled
in size and importance until about the year 1750, when it had
sunk into an inconsiderable village.   After this period, however,
trade began to revive, and the town again to increase ; and as
the old site was deemed inconveniently distant from the harbour,
it was changed for the present.   The main street of this second
town, which is still used as a road, and bears the name of the
Old Causeway, is situated about two hundred yards to the east
of the houses, and is now bounded by the fences of gardens and
fields, with here and there an antique-looking, high-gabled domicile rising over it.   A row of large trees, which have sprung
up since the disappearance of the town, runs along one of the
fences.

About the beginning of the last century, the Old Causeway
presented an aspect which, though a little less rural than at
present, was still more picturesque.   An irregular line of houses
thrust forward their gables on either side, like two parties of
ill-trained cavalry drawn up for the charge ;—some jutted for-

ward, others slunk backward, some slanted sideways, as if meditating a retreat, others, as if more decided, seemed in the act of turning round. They varied in size and character, from the little sod-covered cottage, with round moor stones sticking out of its mud walls, like skulls in the famous pyramid of Malta, to the tall narrow house of three storeys, with its court and gateway. Between every two buildings there intervened a deep narrow close, bounded by the back of one tenement and the front of another, and terminating in a little oblong garden, fringed with a deep border of nettles, and bearing in the centre plots of cabbage and parsnips ;—the latter being a root much used before the introduction of the potato. Here and there a gigantic ash or elm sprung out of the fence, and shot its ponderous arms over the houses. A low door, somewhat under five feet, and a few stone steps which descended from the level of the soil to that of the floor (for the latter was invariably sunk from one to three feet beneath the former), gave access to each of the meaner class of buildings. One little window, with the sill scarcely raised above the pavement, fronted the street, another, still smaller and equally low, opened to the close: they admitted through their unbevelled apertures and diminutive panes of brownish-yellow, a sort of umbery twilight, which even the level sunbeams, as they fell at eve or morn in long rules athwart the motty atmosphere within, scarce served to dissipate. An immense chimney, designed for the drying of fish, which formed the staple food of the poorer inhabitants, stretched from the edge of the window in the gable to near the opposite wall ; and on the huge black lintel were inscribed, in rude characters, the name of the builder of the tenement, and that of his wife, with the date of the erection. The walls, naked and uneven, were hollowed in several places into little square recesses, termed *bowels* or *boles ;* and at a height of not more than six feet above the floor, which was formed of clay and stone, and marvellously uneven, were the bare rafters varnished over with smoke.

The larger houses were built in a style equally characteristic of the age and country. A taste for ornamental masonry was considerably more prevalent in our Scottish villages about the beginning of the seventeenth century than at present. Palladio began to be studied about that period by a few architects of the southern parts of the kingdom ; and some of our provincial builders had picked up from them an imperfect acquaintance with the old classical style of architecture : but as they could avail themselves of only a few of its forms, and knew nothing of its proportions, they became, all unwittingly, the founders of a kind of school of their own. And some of the houses of the old town were no bad specimens of this half Grecian half Gothic school. The high narrow gables, jagged like the teeth of a saw, the diminutive, heavily-framed windows, and chamfered rybats, remained unaltered ; but there were stuck round the low doors, which still retained their Gothic proportions, imitations of Palladio's simpler door-pieces ; and huge Grecian cornices, more than sufficiently massy for halls twenty feet in height, with circular pateras designed in the same taste, and roughened with vile imitations of the vine and laurel, adorned the better rooms within. The closes leading to buildings of this superior class were lintelled at the entrance, and over each lintel there was fixed a tablet of stone, bearing the arms and name of the proprietor. A large house of this kind, on the eastern side of the street, was haunted, it was said, by a *green lady*, one of the old Scottish spectres, who flourished before the introduction of shrouds and dead linens ; and another on the opposite side, by a capricious brownie, who disarranged the pieces of furniture and the platters every night the domestics set them in order, and set them in order every night they were left disarranged. Directly in the middle of the street stood the town's cross, over the low-browed entrance of a stone vault, furnished with seats, also of stone. The formidable *jougs* depended from one of the abutments. A little higher up was the jail, an antique ruinous structure, with stone floors, and a roof of ponderous grey slate. The manse, a mean-looking house of two low storeys, with very small windows, and bearing above the door the initials of the first Protestant minister of the parish, nearly fronted it : while the only shop of the place was situated so much lower down, that, like the houses of the earlier town, it was carried away by the sea during a violent storm from the north-east. There mingled with the other domiciles a due proportion of roofless tenements, with their red weather-wasted gables, and melancholy-looking unframed windows and doors ; and, as trade decayed, even the more entire began to fall to pieces, and to show, like so many mouldering carcasses, their bare ribs through the thatch. Such was the old town of Cromarty in the year 1720.

Directly behind the site of the old town, the ground, as described in a previous chapter, rises abruptly from the level to the height of nearly a hundred feet, after which it forms a kind of table-land of considerable extent, and then sweeps gently to the top of the hill. A deep ravine, with a little stream running through it, intersects the rising ground at nearly right angles with the front which it presents to the houses ; and on the eastern angle, towering over the ravine on the one side, and the edge of the bank on the other, stood the old castle of Cromarty. It was a massy, time-worn building, rising in some places to the height of six storeys, battlemented at the top, and roofed with grey stone. One immense turret jutted out from the corner, which occupied the extreme point of the angle ; and looking down from an altitude of at least one hundred and sixty feet on the little stream, and the struggling row of trees which sprung up at its edge, commanded both sides of the declivity, and the town below. Other turrets of smaller size, but pierced like the larger one with rows of little circular apertures, which in the earlier ages had given egress to the formidable bolt, and in the more recent, when the crossbow was thrown aside for the petronel, to the still more formidable bullet, were placed by pairs on the several projections that stood out from the main body of the building, and were connected by hanging bartisans. There is a tradition that some time in the seventeenth century a party of Highlanders, engaged in some predatory enterprise, approached so near the castle on this side, that their leader, when in the act of raising his arm to direct their march, was shot at from one of the turrets and killed, and that the party, wrapping up the body in their plaids, carried it away.

The front of the castle opened to the lawn, from which it was divided by a dry moat, nearly filled with rubbish, and a high wall indented with embrasures, and pierced by an arched gateway. Within was a small court, flagged with stone, and bounded on one of the sides by a projection from the main building, bartisaned and turreted like all the others, but only three storeys in height, and so completely fallen into decay that the roof and all the floors had disappeared. From the level of the court, a flight of stone steps led to the vaults below; another flight of greater breadth, and bordered on both sides by an antique balustrade, ascended to the entrance; and the architect, aware of the importance of this part of the building, had so contrived it, that a full score of loopholes in the several turrets and outjets which commanded the court, opened directly on the landing-place. Round the entrance itself there jutted a broad, grotesquely-proportioned moulding, somewhat resembling an old-fashioned picture-frame, and directly over it there was a square tablet of dark blue stone, bearing in high relief the arms of the old proprietors; but the storms of centuries had defaced all the nicer strokes of the chisel, and the lady with her palm and dagger, the boars' heads, and the greyhounds, were transformed into so many attenuated spectres of their former selves;— no unappropriate emblem of the altered fortunes of the house. The windows, small and narrow, and barred with iron, were thinly sprinkled over the front: and from the lintel of each there rose a triangular cap of stone, fretted at the edges, and terminating at the top in two knobs fashioned into the rude semblance of thistles.   Initials and dates were inscribed in raised characters on these triangular tablets.   The aspect of the whole pile was one of extreme antiquity.   Flocks of crows and jays, that had built their nests in the recesses of the huge tusked cornices which ran along the bartisans, wheeled ceaselessly around the gables and the turrets, awakening with their clamorous cries the echoes of the roof.   The walls, grey and weather-stained, were tapestried in some places with sheets of ivy ; and an ash sapling, which had struck its roots into the crevices of the outer wall, rose like a banner over the half-dilapidated gateway.

The castle, for several years before its demolition, was tenanted by only an old female domestic, and a little girl whom she had hired to sleep with her.   I have been told by the latter, who, at the time when I knew her, was turned of seventy, that two threshers could have plied their flails within the huge chimney of the kitchen ; and that in the great hall, an immense dark chamber lined with oak, a party of a hundred men had exercised at the pike.   The lower vaults she had never the temerity to explore ; they were dark and gousty, she said, and the slits which opened into them were nearly filled up with long rank grass.   Some of her stories of the castle associated well with the fantastic character of its architecture, and the ages of violence and superstition which had passed over it.   A female domestic who had lived in it before the woman she was acquainted with, and who was foolhardy enough to sleep in it alone, was frightened one night out of her wits, and never again so far recovered them as to be able to tell for what.   At times there would echo through the upper apartments a series of noises, as if a very weighty man was pacing the floors ; and " Oh," said my informant, " if you could but have heard the shrieks, and moans, and long whistlings, that used to come sounding in the stormy evenings of winter from the chimneys and the turrets. Often have I listened to them as I lay a-bed, with the clothes drawn over my face." Her companion was sitting one day in a little chamber at the foot of the great stair, when, hearing a tapping against the steps, she opened the door. The light was imperfect—it was always twilight in the old castle—but she saw, she said, as distinctly as ever she saw any thing, a small white animal resembling a rabbit, rolling from step to step, head over heels, and dissolving, as it bounded over the last step, into a wreath of smoke. On another occasion, a Cromarty shoemaker, when passing along the front of the building in a morning of summer, was horrified by the apparition of a very diminutive, greyheaded, greybearded old man, with a withered meagre face scarcely bigger than one's fist, that seemed seated at one of the windows. On returning by the same path about half an hour after, just as the sun was rising out of the Firth, he saw the same figure wringing its hands over a little cairn in a neighbouring thicket, but he had not courage enough to go up to it.

The scene of all these terrors has long since disappeared ; the plough and roller have passed over its foundations ; and all that it recorded of an ancient and interesting, though unfortunate family, with its silent though impressive narratives of the un-settled lives, rude manners, uncouth tastes, and warlike habits of our ancestors, has also perished. It was pulled down by a proprietor of Cromarty, who had purchased the property a few years before ; and, as he was engaged at the time in building a set of offices and a wall to his orchard, the materials it fur-nished proved a saving to him of several pounds. He was a man of taste, too, as well as of prudence, and by smoothing down the eminence on which the building had stood, and then sowing it with grass, he bestowed upon it, for its former wild aspect, so workmanlike an appearance, that one might almost suppose he had made the whole of it himself. Two curious

F

pieces of sculpture were, by some accident, preserved entire in the general wreck. In a vaulted passage which leads from the modern house to the road, there is a stone slab about five feet in length, and nearly two in breadth, which once served as a lintel to one of the two chimneys of the great hall. It bears, in low relief, the figures of hares and deer sorely beset by dogs, and surrounded by a thicket of grapes and tendrils. The hunts-man stands in the centre, attired in a sort of loose coat that reaches to his knees, with his horn in one hand, and his hunt-ing-spear in the other, and wearing the moustaches and peaked beard of the reign of Mary. The lintel of the second chimney, a still more interesting relic, is now in Kinbeakie Cottage, parish of Resolis : and a good lithographic print of it may be seen in the museum of the Northern Institution, Inverness ; but of it more anon. All the other sculptures of the castle, including several rude pieces of Gothic statuary, were destroyed by the workmen. An old stone dial which had stood in front of the gate, was dug up by the writer, out of a corner of the lawn, about twelve years ago, and is now in his possession. When entire, it indicated the hour in no fewer than nineteen different places, and though sorely mutilated and divested of all its gnomons, it is still entire enough to show that the mathematical ability of the artist must have been of no ordinary kind. It was probably cut under the inspection of Sir Thomas, who, among his other accomplishments, was a skilful geometrician.

"The old castle of Cromarty," says the statistical account of the parish (Sir John Sinclair's), " was pulled down in the year 1772. Several urns, composed of earthenware, were dug out of the bank immediately around the building, with several coffins of stone. The urns were placed in square recesses formed of flags, and when touched by the labourers instantly mouldered away, nor was it possible to get up one of them entire. They were filled with ashes mixed with fragments of half-burned bones. The coffins contained human skeletons, some of which wanted the head ; while among the others which were entire, there was one of a very uncommon size, measuring seven feet in length."

The old proprietors of the castle, among the other privileges derived to them as the chiefs of a wide district of country, and the system of government which obtained during the ages in which they flourished, were hereditary Sheriffs of Cromarty, and vested with the power of pit and gallows. The highest knoll of the southern Sutor is still termed the Gallow-hill, from its having been a place of execution ; and a low cairn nearly hidden by a thicket of furze, which still occupies its summit, retains the name of the gallows. It is said that the person last sentenced to die at this place was a poor Highlander who had insulted the Sheriff, and that when in the act of mounting the ladder, he was pardoned at the request of the Sheriff's lady. At a remoter period the usual scene of execution was a little eminence in the western part of the town, directly above the harbour, where there is a small circular hollow still known to the children of the place as the *Witch's Hole ;* and in which, says tradition, a woman accused of witchcraft was burnt for her alleged crime some time in the reign of Charles II. The Court-hill, an artificial mound of earth, on which, at least in the earlier ages, the cases of the sheriffdom were tried and decided, was situated several hundred yards nearer the old town. Some of the sentences passed at this place are said to have been flagrantly unjust. There is one Sheriff in particular, whom tradition describes as a cruel, oppressive man, alike regardless of the rights and lives of his poor vassals ; and there are two brief anecdotes of him which still survive. A man named Macculloch, a tenant on the Cromarty estate (probably the same person introduced to the reader in the foregoing chapter), was deprived of a cow through the injustice of one of the laird's retainers, and going directly to the castle, disposed rather to be energetic than polite, he made his complaint more in the tone of one who had a right to demand, than in the usual style of submission. The laird, after hearing him patiently, called for the key of the dungeon, and going out, beckoned on Macculloch to follow. He did so ; they descended a flight of stone steps together, and came to a massy oak door, which the laird opened ; when suddenly, and without uttering a syllable, he laid hold of his tenant with the intention of thrusting him in. But he had mistaken his man ; the grasp was returned by one of more than equal firmness, and a struggle ensued, in which Macculloch, a bold, powerful Highlander, had so decidedly the advantage, that he forced the laird into his own dungeon, and then locking the door, carried away the key in his pocket. The other anecdote is of a sterner cast :—A poor vassal had been condemned on the Court-hill under circumstances more than usually unjust ; and the laird, after sentence had been executed on the eminence at the *Witch's Hole*, was returning homewards through the town, surrounded by his retainers, when he was accosted in a tone of prophecy by an old man, one of the Hossacks of Cromarty, who, though bed-ridden for years before, had crawled to a seat by the wayside to wait his coming up. Tradition has preserved the words which follow as those in which he concluded his prediction ; but they stand no less in need of a commentary than the obscurest prophecies of Merlin or Thomas the Rhymer :—" Laird, laird, what mayna skaith i' the brock, maun skaith i' the stock." The seer is said to have meant that the injustice of the father would be visited on the children.

The recollection of these stories was curiously revived in Cromarty in the spring of 1829 ; when a labourer employed in digging a pit on the eminence above the harbour, and within a few yards of the *Witch's Hole*, struck his mattock through a human skull, which immediately fell in pieces. A pair of shin-bones lay directly below it, and on digging a little further there were found the remains of two several skeletons and a second skull. From the manner in which the bones were blended together, it seemed evident that the bodies had been thrown into the same hole, with their heads turned in opposite directions, either out of carelessness or in studied contempt. And they had, apparently, lain undisturbed in this place for centuries. A child, by pressing its foot against the skull which had been raised entire, crushed it to pieces like the other ; and the whole of the bones had become so light and porous, that when first seen by the writer, some of the smaller fragments were tumbling over the sward before a light breeze, like withered leaves, or pieces of fungous wood.

CHAPTER VII.

"He was a veray parfit, gentil knight."—CHAUCER.

OF Sir Thomas Urquhart very little is known but what is re-lated by himself, and though as much an egotist as most men, he has related but little of a kind available to the biographer. But there are characters of so original a cast that their more prominent features may be hit off by a few strokes; and Sir Thomas's is decidedly of this class. It is impossible to mistake the small dark profile which he has left us, small and dark though it be, for the profile of any mind except his own. He was born in 1613, the eldest son of Sir Thomas Urquhart of Cromarty, and of Christian, daughter of Alexander Lord Elphinston. Of his earlier years there is not a single anecdote, nor is there anything known of either the manner or place in which he pursued his studies. Prior to the death of his father, and, as he himself expresses it, "before his brains were yet ripened for eminent undertakings," he made the tour of Europe. In travelling through France, Spain, and Italy, he was repeatedly complimented on the fluency with which he spoke the languages of these countries, and advised by some of the people to pass himself for a native. But he was too true a patriot to relish the proposal. He had not less honour, he said, by his own poor country than could be derived from any country whatever; for, however much it might be surpassed in riches and fertility—in honesty, valour, and learning, it had no superior. And this assertion he maintained at the sword's point, in single combat three several times, and at each time discomfited his antagonist. He boasts on another occasion, that not in all the fights in which he had ever been engaged, did he yield an inch-breadth to the enemy before the day of Worcester battle.

On the breaking out of the troubles in 1638, he took part with the King against the Covenanters, and was engaged in an obscure skirmish, in which he saw the first blood shed that flowed in the protracted quarrel, which it took half a century and two great revolutions to settle. In a subsequent skirmish, he succeeded, with eight hundred others, many of them "brave gentlemen," in surprising a body of about twelve hundred strong, encamped at Turriff, and broke up their array. And then marching with his friends upon Aberdeen, which was held by the Covenanters, he assisted in ejecting them, and in taking possession of the place. Less gifted with conduct than courage, however, the cavaliers suffered their troops to disperse, and were cooped up within the town by the "Earl Marischal of Scotland," who, hastily levying a few hundred men, came upon them, when, according to Spalding, they "were looking for nothing less;" and the "young laird of Cromartie," with a few others, were compelled to take refuge "aboard of Andrew Finlay's ship, then lying in the road," and "hastily hoisted sail for England." Urquhart had undertaken to be the bearer of despatches to Charles, containing the signatures of his associates and neighbours the leading anti-covenanters; and in the audience which he obtained of the monarch, he was very graciously received, and favoured with an answer, "which gave," he says, "great contentment to all the gentlemen of the north that stood for the king." In the spring of 1641 he was knighted by Charles at Whitehall, and his father dying soon after, he succeeded to the lands of Cromarty.

Never was there a proprietor less in danger of sinking into the easy apathetical indolence of the mere country gentleman; for, impressed with a belief that he was born to enlarge the limits of all science, he applied himself to the study of every branch of human learning, and, having *mastered what was already known*, and finding the *amount but little*, he seriously set himself to add to it. And first, as learning can be communicated only by the aid of language, "words being the signs of things," he deemed it evident that, if language be imperfect, learning must of necessity be so likewise; quite on the principle that a defect in the carved figure of a signet cannot fail of being transmitted to the image formed by it on the wax. The result of his inquiries on this subject differed only a very little from the conclusion which, when pursuing a similar course of study, the celebrated Lord Monboddo arrived at more than a hundred years after. His Lordship believed that all languages, except Greek, are a sort of vulgar dialects which have grown up rather through accident than design, and exhibit, in consequence, little else than a tissue of defects both in sound and sense. Greek, however, he deemed a perfect language; and he accounted for its superiority by supposing that, in some early age of the world, it had been constructed on philosophical principles, out of one of the old jargons, by a society of ingenious grammarians, who afterwards taught it to the common people. Sir Thomas went a little further; for, not excepting even the Greek, he condemned every language, ancient and modern, and set himself to achieve what, according to Monboddo, had been already achieved by the grammarians of Greece. And hence his ingenious but unfortunate work, "The Universal Language."

"A tree," he thus reasoned, "is known by its leaves, a stone by its grit, a flower by the smell, meats by the taste, music by the ear, colours by the eye," and, in short, all the several natures of things by the qualities or aspects with which they address themselves to the senses or the intellect. And it is from these obvious traits of similarity or difference that the several classes are portioned by the associative faculty into the corresponding cells of understanding and memory. But it is not thus with words in any of the existing languages. Things the most opposite in nature are often represented by signs so similar that they can hardly be distinguished, and things of the same class by signs entirely different. Language is thus formed so loosely and unskilfully, that the associative faculty cannot be brought to bear on it ;—one great cause why foreign languages are so difficult to learn, and when once learned, so readily forgotten. And there is a radical defect in the alphabets of all languages ; for in all, without exception, do the nominal number of letters fall far short of the real, a single character being arbitrarily made to represent a variety of sounds. Hence it happens that the people of one country cannot acquaint themselves by books alone with the pronunciation of another. The words, too, proper to express without circumvolution all the multiform ideas of the human mind, are not to be found in any one tongue ; and though the better languages have borrowed largely from each other to supply their several deficiencies, even the more perfect are still very incomplete. Hence the main difficulty of translation. Some languages are fluent without exactness. Hence an unprofitable wordiness, devoid of force and precision. Others, comparatively concise, are harsh and inharmonious. Hence, perhaps, the grand cause why some of the civilized nations (the Dutch for instance), though otherwise ingenious, make but few advances compared with others, in philology and the *belles-lettres.*

These, concluded Sir Thomas, are the great defects of language. In a perfect language, then, it is fundamentally necessary that there should .be classes of resembling words to represent the classes of resembling things—that every idea should have its sign, and every simple sound its alphabetical character. It is necessary, too, that there should be a complete union of sweetness, energy, and precision. Setting himself down in the old castle of Cromarty to labour on these principles for the benefit of all mankind, and the glory of his country, he constructed his Universal Tongue. There is little difficulty, when we remember where he wrote, in tracing the origin of his metaphor, when he says of the existing languages, that though they may be improved in structure " by the striking out of new light and doors, the outjetting of kernels, and the erecting of prickets and barbicans," they are yet restricted to a certain base, beyond which they must not be enlarged. In his own language the base was fitted to the superstructure. His alphabet consisted of ten vowels, and twenty-five consonants.　His radical classes of words amounted to two hundred and fifty, and, to use his own allegory, were the denizens of so many cities divided into streets, which were again subdivided into lanes, the lanes into houses, the houses into storeys, and the storeys into apartments.　It was impossible that the natives of one city should be confounded with those of another; and by prying into their component letters and syllables, the street, lane, house, storey, and apartment of every citizen, could be ascertained without a possibility of mistake.　Simple ideas were expressed by monosyllables, and every added syllable expressed an added idea.　So musical was this language, that for poetical composition it surpassed every other; so concise, that the weightiest thoughts could be expressed in it by a few syllables, in some instances by a single word; so precise, that even sounds and colours could be expressed by it in all their varieties of tone and shade; and so comprehensive, that there was no word in any language, either living or dead, that could not be translated into it without suffering the slightest change of meaning.　And, with all its rich variety of phrase, so completely was it adapted to the associative faculty, that it was possible for a boy of ten years thoroughly to master it in the short space of three months !　The entire work, consisting of a preface, grammar, and lexicon, was comprised in a manuscript of twelve hundred folio pages.

Laborious as this work must have proved, it was only one of a hundred great works completed by Sir Thomas before he had attained his thirty-eighth year, and all in a style so exquisitely original, that neither in subject nor manner had he been anticipated in so much as one of them. He had designed, and in part digested, four hundred more. A complete list of these, with such a description of each as I have here attempted of his Universal Language, would be, perhaps, one of the greatest literary curiosities ever exhibited to the world; but so unfortunate was he, as an author, that the very names of the greater number of the works he finished have died with himself, while the names of his projected ones were, probably, never known to any one else. He prepared for the press a treatise on Arithmetic, intended to remedy some defects in the existing system. The invention of what he terms the " Trissotetrail Trigonometry for the facilitating of calculations by representations of letters and syllables," was the subject of a second treatise; and the proving of the Equipollencie and Opposition both of Plain and Modal Enunciations by rules of Geometry (I use his own language, for I am not scholar enough to render it into common English), he achieved in a third. A fourth laid open the profounder recesses of the Metaphysics by a continued Geographical Allegory. He was the author also of ten books of Epigrams, in all about eleven hundred in number, which he " contryved, blocked, and digested," he says, " in a thirteen weeks tyme;" and of this work the manuscript still exists. It is said to contain much bad verse, and much exceptionable morality; but at least one of its stanzas, quoted by Dr. Irvine, in his elaborate and scholarlike Biographies of Scottish writers, possesses its portion of epigrammatic point.

> " A certain poetaster, not long since,
>   Said I might follow him in verse and prose;
> But, truly if I should, 'tis as a prince
>   Whose ushers walk before him as he goes."

In Blackwood's Magazine for 1820, in a short critique on the *Jewel*, it is stated that the writer had " good reasons to believe Sir Thomas to be the real author of that singular production,

A Century of Names, and Scantlings of Inventions, the credit
or discredit of which was dishonestly assumed by the Marquis
of Worcester." The " good" reasons are not given; nor am
I at all sure that they would be found particularly good; for
the Marquis is a well-known man; and yet, were intrinsic
evidence to be alone consulted, it might be held that either
this little tract was written by Sir Thomas, or, what might be
deemed less probable, that the world, nay, the same age and
island, had produced two Sir Thomases.[1] Some little weight,
too, might be attached to the facts, that many of his manuscripts were lost in the city of Worcester, with which place,
judging from his title, it is probable the Marquis may have
had some connexion, by residence or otherwise; and that the
" Century of Names" was not published until 1663, two years
after death had disarmed poor Sir Thomas of his sword and his
pen, and rendered him insensible to both his country's honour

[1] The resemblance between the inventions of Sir Thomas as described in the *Jewel*,
and of the Marquis as intimated in the *Century*, is singularly close. The following passages, selected chiefly for their brevity as specimens, may serve to show how very much
the minds that produced them must have been of a piece.

FROM THE JEWEL.

" In the denominations of the fixed stars,
the Universal Language affordeth the most
significant way imaginary; for by the single
word alone which represents the star you
shall know the magnitude, together with
the longitude and latitude, both in degrees
and minutes, of the star that is expressed
by it.

" Such as will hearken to my instruc-
tions, if some strange word be proposed to
them, whereof there are many thousands
of millions devisable by the wit of man,
which never hitherto by any breathing
have been uttered, shall be able, although
they know not the ultimate signification
thereof, to declare what part of speech it
is; or if a noun, to what predicament or
class it is to be reduced; whether it be the
sign of a real or natural thing, or somewhat
concerning mechanic trades in their tools
or terms; or if real, whether natural or
artificial, complete or incomplete."

FROM THE CENTURY.

" To write by a knotted silk string, so
that every knot shall signify any letter,
with a comma, full point, or interrogation,
and as legible as with pen and ink upon
white paper. The like by the smell, by
the taste, by the touch, by these three
senses, as perfectly, distinctly, and uncon-
fusedly, yea, as readily as by the sight.

" How to compose an universal character, methodical, and easy to be written, yet
intelligible in any language; so that if an
Englishman write it in English, a French-
man, Italian, Spaniard, Irish, Welch, being
scholars, yea, Grecian or Hebritan, shall as
perfectly understand it in their own tongue,
distinguishing the verbs from the nouns,
the numbers, tenses, and cases, as properly expressed in their own language as it
was in English."

and his own. If in reality the author of this piece, he must be regarded, it is said, as the original inventor of the steam-engine.

But the merit of the most curious of all his treatises no one has ventured to dispute with him—a work entitled " The True Pedigree and Lineal Descent of the Ancient and Honourable Family of Urquhart." It records the names of all the fathers of the family, from the days of Adam to those of Sir Thomas; and may be regarded as forming no bad specimen of the inverted climax—beginning with God, the creator of all things, and ending with the genealogist himself. One of his ancestors he has married (for he was a professed lover of the useful) to a daughter of what the Abbé Pluche deemed an Egyptian symbol of husbandry, and another to a descendant of what Bacon regarded as a personification of human fortitude. In his notice of the arms of the family he has surpassed all the heralds who have flourished before or since. The first whose bearings he describes is Esormon, sovereign prince of Achaia, the father of all such as bear the name of Urquhart, and the fifth from Japhet by lineal descent. His arms were three banners, three ships, and three ladies in a field; *or*, the crest, a young lady holding in her right hand a brandished sword, and in her left a branch of myrtle; the supporters, two Javanites attired after the soldier habit of Achaia; and the motto, Ταῦτα ἡ τρία ἀξιοθε-άτα—These three are worthy to behold. Heraldry and Greek were alike anticipated by the genius of this family. The device of Esormon was changed about six hundred years after, under the following very remarkable circumstances. Molin, a celebrated descendant of this prince, and a son-in-law of Deucalion and Pyrrha, accompanied Galethus, the Æneas of Scotland, to the scene of his first colony, a province of Africa, which in that age, as in the present, was infested with wild beasts. He excelled in hunting; and having in one morning killed three lions, he carried home their heads in a large basket, and presented it to his wife Panthea, then pregnant with her first child. Unconscious of what the basket contained, she raised the lid, and, filled with horror and astonishment by the apparition of the heads, she struck her hand against her left side, exclaiming, in the suddenness of her surprise, " O Hercules! what is this?" By a wonderful sympathy, the likeness of the three heads, grim and horrible as they appeared in the basket, was impressed on the left side of the infant, who afterwards became a famous warrior, and transferred to his shield the badge which nature had thus bestowed upon him. The external ornaments of the bearings remained unaltered until the days of Astorimon, who, after his victory over Ethus, changed the myrtle branch of the lady for one of palm, and the original motto for Εὐνοεῖτω, εὐλόγε, καὶ εὐπράττε—Mean, speak, and do well. Both the shield and the supporters underwent yet another change in the reign of Solvatious of Scotland, who, in admiration of an exploit achieved by the Urquhart and his two brothers in the great Caledonian forest, converted the lions' heads into the heads of bears, and the armed Javanites of Esormon into a brace of greyhounds. And such were the arms of the family in the days of Sir Thomas, as shown by the curious stone lintel now at Kinbeakie.

This singular relic, which has, perhaps, more of character impressed upon it than any other piece of sandstone in the kingdom, is about five feet in length, by three in breadth, and bears date A.M. 5612, A.C. 1651. On the lower and upper edges it is bordered by a plain moulding, and at the ends by belts of rich foliage, terminating in a chalice or vase. In the upper corner two knights in complete armour on horseback, and with their lances couched, front each other, as if in the tilt-yard. Two Sirens playing on harps occupy the lower. In the centre are the arms—the charge on the shield three bears' heads, the supporters two greyhounds leashed and collared, the crest a naked woman holding a dagger and palm, the helmet that of a knight, with the beaver partially raised, and so profusely mantled that the drapery occupies more space than the shield and supporters, and the motto MEANE WEIL, SPEAK WEIL, AND DO WEIL. Sir Thomas's initials, S. T. V. C., are placed separately, one letter at the outer side of each supporter, one in the centre of the crest, and one beneath the label; while the names of the more celebrated heroes of his genealogy, and the eras in which they flourished, occupy, in the following inscription, the space between the figures:—ANNO ASTORIMONIS, 2226. ANNO VOCOMPOTIS, 3892. ANNO MOLINI, 3199. ANNO RODRICI, 2958. ANNO CHARI, 2219. ANNO LUTORCI, 2000. ANNO ESORMONIS, 3804. It is melancholy enough that this singular exhibition of family pride should have been made in the same year in which the family received its death-blow—the year of Worcester battle.

During the eventful period which intervened between the death of Sir Thomas's father and this unfortunate year, he was too busily engaged with science and composition to take an active part in the affairs of the kingdom. " In the usual sports of country gentlemen, he does not seem," says Dr. Irvine, " to have taken any great share ;" and a characteristic anecdote which he relates in his " Logopandacteision," shows that he rated these simply by what they produced, estimated at their money value, and accordingly beneath the care of a man born to extend the limits of all human knowledge. " There happened," he says, " a gentleman of very good worth to stay awhile at my house, who one day, amongst many others, was pleased in the deadest time of all the winter, with a gun upon his shoulder, to search for a shot of some wild-fowl ; and after he had waded through many waters, taken excessive pains in quest of his game, and by means thereof had killed some five or six moor-fowls and partridges, which he brought along with him to my house, he was, by some other gentlemen who chanced to alight at my gate as he entered in, very much commended for his love of sport ; and as the fashion of most of our countrymen is not to praise one without dispraising another, I was highly blamed for not giving myself in that kind to the same exercise, having before my eyes so commendable a pattern to imitate. I answered, though the gentleman deserved praise for the evident proof he had given that day of his inclination to thrift and laboriousness, that nevertheless I was not to blame, seeing, whilst he was busied about that sport, I was employed in a diversion of another nature, such as optical secrets, mysteries of natural philosophie, reasons for the variety of colours, the finding out of the longitude, the squaring of a circle, and wayes to accomplish all trigonometrical calculations by signes, without tangents, with the same comprehensiveness of computation ; which, in the estimation of learned men, would be accounted worth six hundred thousand partridges and as many moorfowls. That night past—the next morning I gave sixpence to a footman of mine to try his fortune with the gun during the time I should disport myself in the breaking of a young horse ; and it so fell out, that by I had given myself a good heat by riding, the boy returned with a dozen of wildfowls, half moorfowl half partridge ; whereat, being exceedingly well pleased, I alighted, gave him my horse to care for, and forthwith entered in to see my gentlemen, the most especiall whereof was unable to rise out of his bed by reason of the gout and siatick, wherewith he was seized through his former day's toil."

Sir Thomas, though he had taken part with the king, was by no means a cavalier of the extreme class. His grandfather, with all his ancestors for centuries before, had been Papists ; and he himself was certainly no Presbyterian, and indeed not a man to contend earnestly about religion of any kind. He hints somewhat broadly in one of his treatises, that Tamerlane might possibly be in the right in supposing God to be best pleased with a diversity of worship. But though lax in his religious opinions, he was a friend to civil liberty ; and loved his country too well to be in the least desirous of seeing it sacrificed to the ambition of even a native prince. And so we find him classing in one sentence, the doctrine "*de jure divino*" with "*piæ fraudes*" and "political whimsies," and expressing as his earnest wish in another, that a free school and standing library should be established in every parish of Scotland. But if he liked ill the tyranny and intolerance of Kings and Episcopalians, he liked the tyranny and intolerance of Presbyterian churchmen still worse. And there was a circumstance which rendered the Consistorial government much less tolerable to him than the Monarchical. The Monarchical recognised him as a petty feudal prince, vested with a prerogative not a whit less kingly in his own little sphere than that which it challenged for itself; while the Consistorial pulled him down to nearly the level of his vassals, and legislated after the same fashion for both.

He found, too, that unfortunately for his peace, the churchmen were much nearer neighbours than the King. He was patron, and almost sole heritor of the churches of Cromarty, Kirkmichael, and Cullicuden, and in desperate warfare did he involve himself with all the three ministers at once. Two of them were born vassals of the house; an ancestor of one of these "had shelter on the land, by reason of slaughter committed by him, when there was no refuge for him anywhere else in Scotland;" and the other owed his admission to his charge solely to the zeal of Sir Thomas, by whom he was inducted in opposition to the wishes of both the people and the clergy. And both ministers, prior to their appointment, had faithfully promised, as became good vassals, to remain satisfied with the salaries of their immediate predecessors. Their party triumphed, however, and the promise was forgotten. In virtue of a decree of Synod, they sued for an augmentation of stipend; Sir Thomas resisted; and to such extremities did they urge matters against him, as to "outlaw and declare him rebel, by open proclamation, at the market-cross of the head town of his

G

own shire." He joined issue with Mr. Gilbert Anderson, the minister of Cromarty, on a different question. The church he regarded as exclusively his own property; and the minister, who thought otherwise, having sanctioned one of his friends to erect a desk in it, Sir Thomas, who disliked the man, pulled it down. There was no attempt made at replacing it; but for several Sabbaths together, all the worst parts of Mr. Anderson's sermons were devoted entirely to the benefit of the knight; who was by much too fond of panegyric not to be affected by censure. Even when a prisoner in the Tower, and virtually stripped of all his possessions, he continued to speak of the " aconital bitterness" of the preacher in a style that shows how keenly he must have felt it.

On the coronation of Charles II. at Scone, he quitted the old castle, to which he was never again to return, and joined the Scottish army: carrying with him, among his other luggage, three huge trunks filled with his hundred manuscripts. He states that on this occasion he " was his own paymaster, and took orders from himself." The army was heterogeneously composed of Presbyterians and Cavaliers; men who had nothing in common but the cause which brought them together, and who, according to Sir Thomas, differed even in that. He has produced no fewer than four comparisons, all good, and all very original, to prove that the obnoxious Presbyterians were rebels at heart. They make use of kings, says he, as we do of card kings in playing at the hundred, discard them without ceremony, if there be any chance of having a better game without them; —they deal by them as the French do by their *Roi de la fève*, or king of the bean—first honour them by drinking their health, and then make them pay the reckoning; or as players at nine-pins do by the king Kyle, set them up to have the pleasure of knocking them down again; or, finally, as the wassailers at Christmas serve their king of Misrule, invest him with the title for no other end than that he may countenance all the riots and disorders of the family. He accuses, too, some of the Presbyterian gentlemen, who had been commissioned to levy troops for the army, of the practices resorted to by the redoubtable Falstaff, when intrusted with a similar commission; and of returning homewards when matters came to the push, out of an unwillingness to "hazard their precious persons, lest they should seem to trust to the arm of flesh." Poor Sir Thomas himself was not one of the people who, in such circumstances, are readiest at returning home. At any rate he stayed long enough on the disastrous field of Worcester to be taken prisoner. Indifferent, however, to personal risk or suffering, he has detailed only the utter woe which befell his hundred manuscripts.

He had lodged, prior to the battle, in the house of a Mr. Spilsbury, " a very honest sort of man, who had an exceeding good woman to his wife;" and his effects, consisting of " scarlet cloaks, buff suits, arms of all sorts, and seven large portmantles full of precious commodity," were stored in an upper chamber. Three of the " portmantles," as has been said already, were filled with manuscripts in folio, " to the quantity of six score and eight quires and a half, divided into six hundred forty and two quinternions, the quinternion consisting of five sheets, and the quire of five-and-twenty." There were, besides, law-papers and bonds to the value of about three thousand pounds sterling. After the total rout of the king's forces, the soldiers of Cromwell went about ransacking the houses; and two of them having broken into Mr. Spilsbury's house, and finding their way to the upper chamber, the scarlet cloaks, the buff suits, the seven " portmantles," and the hundred manuscripts fell a prey to their rapacity. The latter had well-nigh escaped, for at first the soldiers merely scattered them over the floor; but reflecting, after they had left the chamber, on the many uses to which they might be applied, they returned and bore them out to the street. Some they carried away with them, some they distributed among their comrades, and the people of the town gathered up the rest. One solitary quinternion, containing part of the preface to the Universal Language, found its way into the kennel, and was picked out two days after by a Mr. Broughton, "a man of some learning," who restored it to Sir Thomas. His Genealogy was rescued from the tobacco-pipes of a file of musketeers, by an officer of Colonel Pride's regiment, and also restored. But the rest he never saw. He was committed to the Tower, with some of the other Scottish gentlemen taken at Worcester; and a body of English troops were garrisoned in the old castle, " upon no other pretence but that the stance thereof was stately, and the house itself of a notable good fabric and contrivance." So oppressive were their exactions, that though he had previously derived from his lands an income of nearly a thousand pounds per annum (no inconsiderable sum in the days of the Commonwealth), not a single shilling found its way to the Tower.

The ingenuity which had hitherto been taxed for the good of mankind and the glory of his country, had now to be exerted for himself. First he published his Genealogy, to convince Cromwell and the Parliament that a family "which Saturn's scythe had not been able to mow in the progress of all former ages, ought not to be prematurely cut off;" but neither Cromwell nor the Parliament took any notice of his Genealogy. Next he published, in a larger work entitled the Jewel, a prospectus of his Universal Language: Cromwell thought there were languages enough already. He described his own stupendous powers of mind; Cromwell was not in the least astonished at their magnitude. He hinted at the vast discoveries with which he was yet to enrich the country; Cromwell left him to employ them in enriching himself. In short, notwithstanding the much he offered in exchange for liberty and his forfeited possessions, Cromwell disliked the bargain; and so he remained a close prisoner in the Tower. It must be confessed that, with all his ingenuity he was little skilled to conciliate the favour of the men in power. They had beheaded Charles I., and he yet tells them how much he hated the Presbyterians for the manner in which they had treated that unfortunate monarch; and though they would fain have dealt with Charles II. after the same fashion, he assures them, that in no virtue, moral or intellectual, was that prince inferior to any of his hundred and ten predecessors. Besides the Genealogy and the Jewel, he published, when in the Tower, a translation of the three first books of Rabelais, which has been described by a periodical critic as the " finest monument of his genius, and one of the most perfect transfusions of an author, from one language into another, that ever man accomplished." And it is remarked, with reference to this work, by Mr. Motteux, that Sir Thomas " possessed learning and fancy equal to the task which he had undertaken, and that his version preserves the very style and air of the original." What is known of the rest of his history may be summed up in a few words. Having found means to escape out of prison, he fled to the Continent, and there died on the eve of the Restoration (indeed, as is said, out of joy at the event), in his forty-eighth year.

" The character of Sir Thomas Urquhart," says a modern critic, "was singular in the extreme. To all the bravery of the soldier and learning of the scholar, he added much of the knight-errant, and more of the *visionnaire* and projector. Zealous for the honour of his country, and fully determined to wage war, both with his pen and his sword, against all the defaulters who disgraced it—credulous yet sagacious—enterprising but rash, he appears to have chosen the Admirable Crichton as his pattern and model for imitation. For his learning he may be denominated the Sir Walter Raleigh of Scotland, and his pedantry was the natural fruit of erudition deeply engrained in his mind. To this I may add, he possessed a disposition prone to strike out new paths in knowledge, and a confidence in himself that nothing could weaken or disturb. In short, the characters of the humorist, the braggadocio, the schemer, the wit, the pedant, the patriot, the soldier, and the courtier, were all intermingled in his, and, together, formed a character which can hardly ever be equalled for excess of singularity, or excess of humour—for ingenious wisdom, or entertaining folly." He is described by another writer as " not only one of the most curious and whimsical, but one of the most powerful also, of all the geniuses our part of the island has produced."

He was unquestionably an extraordinary man. There occur in some-characters anomalies so striking, that, on their first appearance, they surprise even the most practised in the study of human nature. By a careful process of analysis, however, we may arrive, in most instances, at what may be regarded as the simple elements which compose them, and see the mystery explained. But it is not thus with the character of Sir Thomas. Anomaly seems to have formed its very basis, and the more we analyse the more inexplicable it appears. It exhibits traits so opposite, and apparently so discordant, that the circumstance of their amazing contrariety renders him as decidedly an original as the Caliban of Shakspere.

His inventive powers seem to have been of a high order. The new chemical vocabulary, with all its philosophical ingenuity, is constructed on principles exactly similar to those which he divulged more than a hundred years prior to its invention, in the preface to his Universal Language. By what process could it be anticipated that the judgment which had enabled him to fix upon these principles, should have suffered him to urge in favour of that language the facility it afforded in the making of anagrams ! As a scholar, he is perhaps not much overrated by the critic whose character of him I have just transcribed. It is remarked of the Greek language by Monboddo, that, " were there nothing else to convince him of its being a work of philosophers and grammarians, its dual number would of itself be sufficient ; for, as certainly as the principles of body are the point, the line, and the surface, the principles of number are the monad and the duad, though philosophers only are aware of the fact." His Lordship, in even this—one of the most refined of his speculations—was anticipated by Sir Thomas. He, too, regarded the duad, "not as number, but as a step towards number—as a medium between multitude and unity;" and he has therefore assigned the dual its proper place in his Universal Language. And is it not strikingly anomalous, that, with all this learning, he should not only have failed to detect the silly fictions of the old chroniclers, but that he himself should have attempted to impose on the world with fictions equally extravagant! We find him, at one time, seriously pleading with the English Parliament that he had a claim, as the undoubted head and representative of the family of Japhet, to be released from the Tower. We see him at another producing solid and powerful arguments to prove that a union of the two kingdoms would be productive of beneficial effects to both. When we look at his literary character in one of its phases, and see how unconsciously he lays himself open to ridicule, we wonder how a writer of such general ingenuity should be so totally devoid of that sense of the incongruous which constitutes the perception of wit. But, viewing him in another, we find that he is a person of exquisite humour, and the most successful of all the translators of Rabelais. We are struck in some of his narratives (his narrative of the death of Crichton, for instance) by a style of description so gorgeously imaginative, that it seems to partake in no slight degree of the grandeur and elevation of epic poetry. We turn over a few of the pages in which these occur, and find some of the meanest things in the language. And his moral character seems to have been equally anomalous. He would sooner have died in prison than have concealed, by a single falsehood, the respect which he entertained for the exiled Prince, at the very time when he was fabricating a thousand for the honour of his family. Must we not regard him as a kind of intellectual monster—a sort of moral centaur! His character is wonderful, not in any of its single parts, but in its incongruity as a whole. The horse is formed like other animals of the same species, and the man much like other men; but it is truly marvellous to find them united.

CHAPTER VIII.

" ———Times
Whose echo rings through Scotland to this hour."—WORDSWORTH.

PRIOR to the Reformation there were no fewer than six chapels in the parish of Cromarty. The site of one of these, though it still retains the name of the Old Kirk, is now a sand-bank, the haunt of the crab and the sea-urchin, which is covered every larger tide by about ten feet of water; the plough has passed over the foundations of two of the others; of two more the only vestiges are a heap of loose stones, and a low grassy mound; and a few broken fragments of wall form the sole remains of the sixth and most entire. The very names of the first three have shared the fate of the buildings themselves; two of the others were dedicated to St. Duthac and St. Bennet; and two fine springs, on which time himself has been unable to effect any change, come bubbling out in the vicinity of the ruins, and bear the names of their respective saints. It is not yet twenty years since a thorn-bush, which formed a little canopy over the spring of St. Bennet, used to be covered anew every season with little pieces of rag, left on it as offerings to the saint, by sick people who came to drink of the water; and near the chapel itself, which was perched like an eyry on a steep solitary ridge that overlooks the Moray Firth, there was a stone trough, famous, about eighty years before, for virtues derived also from the saint, like those of the well. For if a child was carried away by the fairies, and some mischievous unthriving imp left in its place, the parents had only to lay the changeling in this trough, and, by some invisible process, their child would be immediately restored to them.   It was termed the fairies' cradle ; and was destroyed shortly before the rebellion of 1745, by Mr. Gordon, the minister of the parish, and two of his elders. The last, and least dilapidated of the chapels, was dedicated to St. Regulus; and there is a tradition, that at the Reformation a valuable historical record, which had belonged to it—the work probably of some literary monk or hermit—was carried away to France by the priest.   I remember a very old woman who used to relate, that when a little girl, she chanced, when playing one day among the ruins with a boy a few years older than herself, to discover a small square recess in the wall, in which there was a book ; but that she had only time to remark that the volume was a very tattered one, and apparently very old, and that there were beautiful red letters in it, when the boy, laying claim to it, forced it from her.   What became of it afterwards she did not know, and, unconscious of the interest which might have attached to it, never thought of making any inquiry.

There does not survive a single tradition of the circumstances which, in this part of the country, accompanied the great event that consigned the six chapels to solitude and decay.   One may amuse one's-self, however, in conceiving of the more interesting of these, and, with history and a little knowledge of human nature for one's guide, run no great risk of conceiving amiss. The port of Cromarty was one of considerable trade for the age and country, and the people of the town were Lowland Scots. A more inquisitive race live nowhere.   First there would come to them wild vague reports, by means of the seamen and merchants, of the strange doctrines which had begun to disturb the Continent and the sister kingdom.   Shreds of heretic sermons would be whispered over their ale ; and stories brought from abroad, of the impositions of the priests, would be eked out, in some instances with little corroborative anecdotes, the fruit of an experience acquired at home. For there were Liberals even then, though under another name;—a certain proportion of the people of Scotland being born such in every age of its independence. Then would come the story of the burning of good Patrick Hamilton, pensionary of the neighbouring abbey of Fearn; and everybody would be exceedingly anxious to learn the particular nature of his crime. Statements of new doctrines, and objections urged against some of the old, would in consequence be eagerly listened to, and as eagerly repeated. Then there would come among them two or three serious, grave people, natives of the place, who would have acquired, when pursuing their occupations in the south, as merchants or mechanics, a knowledge, not merely speculative, of the new religion. A traveller of a different cast would describe with much glee to groups of the younger inhabitants, the rare shows he had seen acted on the Castle-hill of Cupar; and producing a black-letter copy of "The Thrie Estaites" of Davy Lindsay, he would set all his auditors a-laughing at the expense of the Church. One of the graver individuals, though less openly, and to a more staid audience, would also produce a book, done into plain English, out of a very old tongue, by one Tyndale, and still more severe on the poor priests than even "The Thrie Estaites." They would learn from this book that what they were beginning to deem a rational, but at the same time new religion, was in reality the old one; and that Popery, with all its boasted antiquity, was by far the more modern of the two. In the meantime, the priests of the chapels would be the angriest men in the parish; —denouncing against all and sundry the fire and fagots of this world, and the fire without fagots of the next; but one of them, a good honest man, neither the son of a churchman himself, nor yet burdened with a family of his own, would set himself, before excommunicating any one, to study the old, newly-translated book, that he might be better able to cope with the maligners of his Church. Before half completing his studies, however, his discourses would begin to assume a very question-able aspect.　　Little would they contain regarding the Pope, and little concerning the saints; and more and more would he press upon his hearers the doctrines taught by the Apostles.　　Anon, however, he would assume a bolder style of language; and sometimes conclude, after saying a great deal about the spiritual Babylon, and the Man of sin, by praying for godly John Knox, and all the other ministers of the Evangel.　　In short, the honest priest would prove the rankest heretic in the whole parish.　　And thus would matters go on from bad to worse.　　A few grey heads would be shaken at the general defection, but these would be gradually dropping away; and the young themselves would be growing old without changing their newly-acquired opinions. They would not all be good Christians;—for every one should know it is quite a possible thing to be a Protestant, sound enough for all the purposes of party, without being a Christian at all;—but they would almost all be reformers; and when the state should at length set itself to annihilate root and branch of the old establishment, and to build up a new one on the broad basis of the kingdom, not a parish in the whole of it would enter more cordially into the scheme than the parish of Cromarty.

But however readily the people might have closed with the doctrines of the Reformation, they continued to retain a good deal of the spirit of the old religion.　　Having made choice of a piece of land on the edge of the ridge which rises behind the houses as a proper site for their church, they began to collect the materials.　　It so chanced, however, that the first few stones gathered for the purpose, being thrown down too near the edge of the declivity, rolled to the bottom; the circumstance was deemed supernaturally admonitory; and the church, after due deliberation, was built at the base instead of the top of the ridge, on exactly the spot where the stones had rested.　　The first Protestant minister of the parish was a Mr. Robert Wil-liamson.　　His name occurs oftener than once in Calderwood's

Church History; and his initials, with those of his wife, are still to be seen on a flat triangular stone in the eastern part of the town, which bears date 1593. It is stated by Calderwood, that "Jesuits having libertie to passe thorough the countrey in 1583, during the time of the Earle of Huntlie's lieutenantrie, great coldness of religion entered in Ross;" and by an act of council passed five years after, this Robert Williamson, and "John Urquhart, tutor of Cromartie," were among the number empowered to urge matters to an extremity against them.

There awaited Scotland a series of no light evils in the short-sighted policy which attempted to force upon her a religion which she abhorred. The surplice and the service-book were introduced into her churches; and the people, who would scarcely have bestirred themselves had merely their civil rights been invaded, began to dread that they could not, without being unhappy in more than the present world, conform to the religion of the state. And so they set themselves seriously to inquire whether the power of kings be not restricted to the present world only. They learned, in consequence, that not merely is such the case, but that it has yet other limitations; and the more they sought to determine these, the more questionable did its grounds become. The spirit manifested on this occasion by the people of this part of the country, is happily exemplified by Spalding's narrative of a riot which took place at the neighbouring Chanonry of Ross, in the spring of 1638. The service-book had been quietly established by the bishop two years before; but the more thoroughly the people grew acquainted with it, the more unpopular it became. At length, on the second Sunday of March, just as the first bell had rung for sermon, but before the ringing of the second, a numerous party of schoolboys broke into the cathedral, and stripped it in a twinkling of all the service books. Out they rushed in triumph, and, procuring a lighted coal and some brushwood, they marched off in a body to the low sandy promontory beneath the town, to make a bonfire of the whole set. But a sudden shower extinguishing the coal, instead of burning they tore the books into shreds, and flung the fragments into the sea. The bishop went on with his sermon; but it was more than usually brief; and such were the feelings exhibited at its close by the people, that, taking hastily to his horse, he quitted the kingdom. "A very busy man was he esteemed," says the annalist, " in the bringing in of the service-book, and therefore durst he not, for fear of his life, return again to Scotland." In short, the country was fully awakened; and before the close of the following month, the National Covenant was subscribed in the shires of Ross, Cromarty, and Nairn.

Some of the minor events which took place in the sheriffdom of Cromarty, on the triumph of Presbyterianism, have been detailed, as recorded by Sir Thomas, in the foregoing chapter. Even on his own testimony, most men of the present day will not feel disposed to censure very severely the churchmen of his district. It must be confessed, however, that the principles of liberty, either civil or ecclesiastical, were but little understood in Scotland in the middle of the seventeenth century; the parties which divided it deeming themselves too exclusively in the right to learn from the persecutions to which they were in turn subjected, that the good old rule of doing as we would be done by, should influence the conduct of politicians as certainly as that of private men. And there is a simple fact which ought to convince us, however zealous for the honour of our church, that the Presbyterian synod of Ross, which Sir Thomas has termed " a promiscuous knot of unjust men," was by no means a very exemplary body. Five-sixths of its members conformed at the Restoration, and became curates ; and as they were notoriously intolerant as Episcopalians, it is not at all probable that they should have been strongly characterized by liberality during the previous period, when they had found it their interest to be Presbyterians.

The restoration of Charles, and the appointment of Middleton as his commissioner for Scotland, were followed by the fatal act which overturned Presbyterianism, and set up Episcopacy in its place. It is stated by Wodrow, that Middleton, previous to the bringing in of this act, had been strengthened in the resolution which led to it, by Mackenzie of Tarbat, and Urquhart of Cromarty ; and that the latter, who had lately " counterfeited the Protestor," ended miserably some time after. In what manner he ended, however, is not stated by the historian, but tradition is more explicit. On the death of Sir Thomas, he was succeeded by his brother Alexander, who survived him only a year, and dying without male issue, the estate passed to Sir John Urquhart of Craigfintrie, the head of a branch of the family which had sprung from the main stock about a century before. This Sir John was the friend and counsellor of Middleton. About eleven years after the passing of the act, he fell into a deep melancholy, and destroyed himself with his own sword in one of the apartments of the old castle. The sword, it is said, was flung into a neighbouring draw-well by one of the domestics, and the stain left by his blood on the walls and floor of the apartment, was distinctly visible at the time the building was pulled down.

So well was the deprecated act received by the time-serving Synod of Ross, that they urged it into effect against one of their own body, more than a year before the ejection of the other nonconforming clergymen. In a meeting of the Synod which took place in 1661, the person chosen as moderator was one Murdoch Mackenzie ;—a man so strong in his attachments that he had previously sworn to the National Covenant no fewer than fourteen times, and he had now fallen desperately in love with the Bishopric of Moray. One of his brethren, however, an unmanageable, dangerous person, for he was uncompromisingly honest, and possessed of very considerable talent, stood directly in the way of his preferment. This member, the celebrated Mr. Hogg of Kiltearn, had not sworn to the Covenant half so often as his superior, the Moderator, but then so wrong-headed was he as to regard his few oaths as binding ; and he could not bring himself to like Prelacy any the better for its being espoused by the king. And so his expulsion was evidently a matter of necessity. The Moderator had nothing to urge against his practice,—for no one could excel him in the art of living well ; but his opinions lay more within his reach ; and no sooner had the Synod met, than, singling him out, he demanded what his thoughts were of the Protestors—the party of Presbyterians who, about ten years before, had not taken part with the king against the Republicans. Mr. Hogg declined to answer ; and on being removed, that the Synod might deliberate, the Moderator rose and addressed them. Their brother of Kiltearn, he said, was certainly a great man—a very great man—but as certainly were the Protestors opposed to the king ; and if any member of Synod took part with them, whatever his character, it was evidently the duty of the other members to have him expelled. Mr. Hogg was then called in, and having refused, as was anticipated, judicially to disown the Protestors, sentence of deposition was passed against him. But the consciences of the men who thus dealt with him, betrayed in a very remarkable manner their real estimate of his conduct. It is stated by Wodrow, on the authority of an eye-witness, that sentence was passed with a peculiar air of veneration, as if they were ordaining him to some higher office ; and that the Moderator was so deprived of his self-possession as to remind him, in a consolatory speech, that " our Lord Jesus Christ had suffered great wrong from the Scribes and Pharisees."

Mackenzie received the reward of his zeal shortly after in an appointment to the Bishopric of Moray ; and one Paterson, a man of similar character, was ordained Bishop of Ross. On the order of council, issued in the autumn of 1662, for all

ministers of parishes to attend the diocesan meetings, and take
the newly-framed oaths, while in some of the southern districts of the kingdom only a few ministers attended, in the
diocese of Ross there were but four absent, exclusive of Mr.
Hogg.   These four were, Mr. Hugh Anderson of Cromarty,
Mr. John Mackilligen of Alness, Mr. Andrew Ross of Tain, and
a Mr. Thomas Ross, whose parish is not named in the list.
And they were all in consequence ejected from their charges.
Mr. Anderson, a nephew of Sir Thomas's opponent, Mr. Gilbert,
who was now dead, retired to Moray, accompanied by his
bedral, who had resolved on sharing the fortunes of his pastor;
and they returned together a few years after to a small estate,
the property of Mr. Anderson, situated in the western extremity
of the parish.   Mr. Mackilligen remained at Alness, despite of
the council and the bishops, who had enacted that no nonconforming minister should take up his abode within twenty miles
of his former church.   Mr. Ross of Tain resided within the
bounds of the same Presbytery; and Mr. Fraser of Brea, a
young gentleman of Cromartyshire, who was ordained to the
ministry about ten years after the expulsion of the others, had
his seat in the parish of Resolis.   In short, as remarked by
Wodrow, there was more genuine Presbyterianism to be found
on the shores of the Bay of Cromarty, notwithstanding the
general defection, than in any other part of the kingdom north
of the Tay.

And the current of popular feeling seems to have set in
strongly in its favour about the year 1666.   Towards the close
of this year, Paterson the bishop, in a letter to his son, describes
the temper of the country about him as very cloudy; and
complains of a change in the sentiments of many who had previously professed an attachment to Prelacy.   Mr. Mackilligen,
a faithful and active preacher of the forbidden doctrines, seems
to have given him so much trouble, that he even threatened to
excommunicate him, but the minister regarding his threat in

H

the proper light, replied to it by comparing him to Balaam the wicked prophet, who went forth to curse Israel, and to Shimei the son of Gera, who cursed David.   The joke spread, for as such was it regarded, and Paterson, who had only the sanctity of his office to oppose to the personal sanctity of his opponent, deemed it prudent to urge the threat no further : he had the mortification of being laughed at for having urged it so far. There is a little hollow among the hills, about three miles from the house of Fowlis, and not much farther from Alness, in the gorge of which the eye commands a wide prospect of the lower lands, and the whole Firth of Cromarty.   It lies, too, on the extreme edge of the cultivated part of the country, for beyond there stretches only a brown uninhabited desert ; and in this hollow the neighbouring Presbyterians used to meet for the purpose of religious worship.   On some occasions they were even bold enough to assemble in the villages.   In the summer of 1675, Mr. Mackilligen, assisted by his brethren of Tain and Cromarty, and the Laird of Brea, celebrated the Communion at Obsdale, in the house of the Lady Dowager of Fowlis.   There was an immense concourse of people ; and "so plentiful was the effusion of the Spirit," says the historian whom I have so often had occasion to quote, "that the oldest Christians present never witnessed the like."   Indisputably, even from natural causes, the time must have been one of much excitement ; and who that believes the Bible, will dare affirm that God cannot comfort his people by extraordinary manifestations, when deprived of the common comforts of earth for their adherence to him ?   One poor man, who had gone to Obsdale merely out of curiosity, was so affected by what he heard, that when some of his neighbours blamed him for his temerity, and told him that the bishop would punish him for it by taking away his horse and cow, he assured them that in such a cause he was content to lose not merely all his worldly goods, but his head also.   A party had been despatched, at the instance of the bishop, to take

Mackilligen prisoner; but, misinformed regarding the place where the meeting was held, they proceeded to his house at Alness, and spent so much time in pillaging his garden, that before they reached Obsdale he had got out of their way. But he fell into the hands of his enemy, the bishop, in the following year, and during his long imprisonment on the Bass Rock—for to such punishment was he subjected—he contracted a disease of which he died. Mr. Ross of Tain, and Mr. Fraser of Brea, were apprehended shortly after, and disposed of in the same manner.

Nor was it only a few clergymen that suffered in this part of the country for their adherence to the church. Among the names of the individuals who, in the shires of Ross and Cromarty, were subjected to the iniquitous fine imposed by Middleton on the more rigid Presbyterians, I find the name of Sir Robert Munro of Fowlis, the head of a family which ranks among the most ancient and honourable in the kingdom. Sir John Munro, son of Sir Robert, succeeded to the barony in 1668. His virtues, and the persecutions to which he was subjected, are recorded by the pen of Doddridge :—"The eminent piety of this excellent person exposed him," says this writer, "to great sufferings in the cause of religion in those unhappy and infamous days, when the best friends to their country were treated as the worst friends to the government. His person was doomed to long imprisonment for no pretended cause but what was found against him in the matters of his God ; and his estate, which was before considerable, was harassed by severe fines and confiscations, which reduced it to a diminution much more honourable, indeed, than any augmentation could have been, but from which it has not recovered to this day."

But, perhaps, a brief narrative of the sufferings of a single individual may make a stronger impression on the reader than any general detail of those of the party. Mr. James Fraser of Brea was born in the western part of the shire of Cromarty, in

the year 1639. On the death of his father, whom he lost
while in his infancy, he succeeded to the little property of about
£100 per annum, of which the name, according to the fashion
of Scotland, is attached to his own. His childhood was passed
much like the childhood of most other people; but with this
difference, that those little attempts at crime which serve to
identify the moral nature of children with that of men, and
which, in our riper years, are commonly either forgotten altogether, or regarded with an interest which owes nought of its
intensity to remorse, were considered by him as the acts of a
creature accountable to the Great Judge for even its earliest
derelictions from virtue. But this trait belongs properly to his
subsequent character. In his seventeenth year, after a youth
spent unhappily, in a series of conflicts with himself, for he was
imbued with a love of forbidden pleasures, and possessed of a
conscience exquisitely tender, a change came over him, and he
became one of the excellent few who live less for the present
world than for the future. As he was not wedded by the prejudices of education to any set of religious opinions, he had,
with only the Scriptures for his guide, to frame a creed for himself; and having come in contact, in Edinburgh, with some
Quakers, he was well-nigh induced to join with them. But on
more serious consideration, he deemed some of their tenets not
quite in unison with those of the Bible. He attended, for some
time after the Restoration, the preaching of the curates; but,
profiting little by their doctrines, he deliberated whether he did
right in hearing them, and concluded in the negative, in the
very year in which all such conclusions were declared treason by
act of Parliament. In short, by dint of reasoning and reading,
he landed full in Presbyterianism, at a time when there was
nothing to be gained by it, and a great deal to be lost. And
not merely did he embrace it for himself, but deeming it the
cause of God, he came forward in this season of wrong and
suffering, when the bad opposed it, and the timid shrunk from

it, to preach it to the people. He believed himself called to the ministerial office in a peculiar manner, by the Great Being who had fitted him for it ; and the simple fact that he did not, in Scotland at least, gain a single sixpence by all his preaching until after the Revolution, ought surely to convince the most sceptical that he did not mistake on this occasion the suggestions of interest for those of duty. He began to preach the forbidden doctrines in the year 1672 ; and he was married shortly after to a lady to whom he had been long attached.

The sufferings to which he had been subjected prior to his marriage affected only himself. He had been fined and exposed to ridicule ; and he had had to submit to loss and imposition, out of a despair of finding redress from corrupt judges, whose decisions would have been prompted rather by the feelings with which they regarded his principles than by any consideration of the merits of his cause. No sooner, however, had he married, and become a preacher, than he was visited by evils greater in themselves, and which he felt all the more deeply from the circumstance that their effects were no longer confined to himself. He was summoned before councils for preaching without authority, and in the fields, and denounced and outlawed for not daring to appear. But he persevered, notwithstanding, wandering under hiding from place to place, and preaching twice or thrice every week to all such as had courage enough to hear him. He was among the number intercommuned by public writ ; all the people of Scotland, even his own friends and relatives, being charged, under the severest penalties, not to speak to him, or receive him into their houses, or minister even the slightest comfort to his person. And yet still did he persevere on the strength of the argument urged by St. Peter before the Jewish Sanhedrim. The lady he married was a person every way worthy of such a husband. "In her," I use his own simple and expressive language, " did I behold as in a glass the Lord's love to me ; and so effectually did she sweeten the sorrows of my pilgrimage, that I have often been too nearly led to exclaim, It is good for me to be here!" But she was lent him only for a short season. Four years after his marriage, when under hiding, word was brought him that she lay sick of a fever; and hurrying home in "great horror and darkness of mind," he reached her bedside only to find that she had departed, and that he was left alone.

His sorrow at the bereavement oppressed, but it could not overwhelm him; for, with an energy rendered more intense by a sense of desolateness, and a feeling that the world had become as nothing to him, he applied afresh to what he deemed his bounden duty, the preaching of the Word. He was diligent in ministering to the comfort of many who were less afflicted than himself; and enveloped in the very flames of persecution, he confirmed, by his exhortations, such as were shrinking from their approach. So well was his character understood by the prelates, that he was one of three expressly named in an act of council as peculiarly obnoxious, and a large sum of money was offered to any who would apprehend him. Great rewards, too, were promised on the same account by the Archbishop of St. Andrews, out of his private purse; and after a series of hair-breadth escapes, he at length fell into his hands through the treachery of a servant. The questions put to him on his trial, with his replies to them, are given at full length by Wodrow. Without in the least compromising his principles, he yet availed himself of every legal argument which the circumstances of his case admitted; and such was the ingenuity of his defence, that he was repeatedly complimented on the score of ability by the noblemen on the bench. He was charged, however, with a breach of good manners; for, while he addressed his other judges with due respect, he replied to the accusations of the archbishop as if they had been urged against him by merely a private individual. In answer to the charge, he confessed that he was but a rude man, and hinted, with some humour, that he had surely been brought before their lordships for some other purpose than simply to make proof of his breeding. And, after all, there was little courtesy lost between himself and the archbishop. He had been apprehended near midnight, and before sunrise next morning, the servant of the latter was seen standing at the prison gate, instructing the jailer that the prisoner should be confined apart, and none suffered to have access to him. When the court met, the archbishop strove to entrap him, with an eagerness which only served to defeat its object, into an avowal of the sentiments with which he regarded the king and his ministers; and failing to elicit these, for the preacher was shrewd and sagacious, he represented him to the other members of council as a person singularly odious and criminal, and an enemy to every principle of civil government. He was a schismatic too, he affirmed—a render asunder of the Church of Christ! To the charge that he was a preacher of sedition, Mr. Fraser replied with apostolic fervour, that in "none of his discourses had he urged aught disloyal or traitorous; but that as the Spirit enabled him, had he preached repentance towards God, and faith towards Jesus Christ, and no other thing but what was contained in the Prophets and the New Testament. And so far," he added, "was he from being terrified or ashamed to own himself a minister of Christ, that although of no despicable extraction, yet did he glory most to serve God in the gospel of His Son, and deem it the greatest honour to which he had ever attained." After trial he was remanded to prison, and awakened next morning by the jailer, for he had slept soundly, that he might prepare for a journey to the Bass. He was escorted by the way by a party of twelve horsemen and thirty foot, and delivered up on landing in the island to the custody of the governor.

Here a new series of sufferings awaited him, not perhaps so harassing in themselves as those to which he had recently been subjected—for punishment in such cases is often less severe than the train of persecution which leads to it; but he felt them all the more deeply, because he could no longer, from his situation, exert that energy of mind which had enabled him to divest, on former occasions, an evil of more than half its strength, by meeting it, as it were, more than half-way. He had now to wait in passive expectation until the evil came. There were a number of other prisoners confined to the Bass for their attachment to Presbyterianism; and the governor, a little-minded, capricious man, who loved to display the extent of his authority, by showing how many he could render unhappy, would sometimes deny them all intercourse with each other, by closely confining them to their separate cells. At times, too, when permitted to associate together, some of the profaner officers would break in upon them, and annoy them with the fashionable wit and blasphemy of the period. A dissolute woman was appointed to wait upon them, and scandalous stories circulated at their expense; all the letters brought them from the land were broken open and made sport of by the garrison; they were neither allowed to eat nor to worship together; and though their provisions and water were generally of the worst kind, they had sometimes to purchase them—even the latter—at an exorbitant price. But there were times at which the preacher could escape from all his petty vexations. In the higher part of the island there are solitary walks, which skirt the edge of the precipices, and command an extensive view of the neighbouring headlands and the ocean. On these, when his jailers were in their more tolerant moods, would he be permitted to saunter for whole hours; indulging, as the waves were breaking more than a hundred yards beneath him, and the sea-fowl screaming over him, in a not unpleasing melancholy—musing much on the future, with all its doubtful probabilities, or " looking back on the days of old, when he joyed with the wife of his youth." And there was a considerable part of his time profitably spent in the study of Greek and Hebrew. He besides read divinity, and wrote a treatise on faith, with several other miscellanies : and at length, after an imprisonment of two years and a half, during which period his old enemy the archbishop had suffered the punishment which there was no law to inflict upon him, he was set at liberty; and he quitted his prison with not less zeal, and with more learning than he had brought into it.

He still deemed preaching as much his duty as before, and the state regarded it as decidedly a crime; and so he had to resume his wandering, unsettled life of peril and hardship; "labouring to be of some use to every family he visited." Falling sick of an ague, contracted through this mode of living, he was cited before the council, at the instance of some of his old friends the bishops; who, reckoning on his inability to appear on the day named, took this way of having him outlawed a second time. But they had miscalculated; for no sooner had he received the citation, than dragging himself from his bed, he set out on his journey to Edinburgh. Legal oppression he respected as little as he had done six years before; but he was now differently circumstanced—one of his friends, on his liberation from the Bass, having bound himself as his surety; and sooner would he have died by the way than have subjected him to any loss. When the day arrived, he presented himself at the bar of the council; and defended himself with such ability and spirit, that his lay judges were on the eve of acquitting him. Not so the bishops; and the matter, after some debate, being wholly referred to their judgment, he was sentenced to be imprisoned at Blackness until he had paid a fine of five thousand merks, and given security that he should not again preach in Scotland. To Blackness he was accordingly sent; and there he remained in close confinement, and subjected, as he had been at the Bass, to the caprice of a tyrannical governor, for about seven weeks; when he was set at liberty on condition that he should immediately quit the kingdom. He

passed therefore into England; and he soon found—for the
Christian is a genuine cosmopolite—" that a good Englishman
was more truly his countryman than a wicked Scot." He was
much esteemed by English people of his own persuasion; and
though he had at first resolved to forbear preaching out of the
dread of being reckoned a " barbarian," for he could not divest
himself of his Scotticisms, he yielded to the solicitations of his
newly-acquired friends; and soon attained among them, as he
had done at home, the character of being a powerful and useful
preacher. But bonds and imprisonment awaited him even there.
On the execution of Russell and Sydney, he was arrested on the
suspicion of being one of their confederates; and on refusing to
take what was termed the Oxford Oath, he was committed to
Newgate, where he was kept for six months. But from his
previous experience of the prisons of Scotland, he seems, with
Goldsmith's sailor, to have deemed Newgate a much better sort
of place than it is usually esteemed;—his apartment was large
and lightsome, and the jailers were all very kind. Resuming,
on his release, his old mode of living, he continued to preach
and study by turns, until the Revolution; when, returning to
Scotland, he was invited by the people of Culross to preside
over them as their pastor;—a fit pastor for a parish which,
during the reign of Prelacy, had suffered and resisted more than
almost any other in the kingdom. In this place he continued
until his death; grateful for all the mercies bestowed upon him,
and few men could reckon them better; but peculiarly grateful
that, in a season of hot persecution, he had been enabled to take
part with God.

Nor were strong-minded men, like Fraser of Brea, the only
persons who espoused this cause in the day of trouble, and
dared to suffer for it. There is a quiet passive fortitude in the
better kind of women, which lies concealed, as it ought, under
a cover of real gentleness and seeming timidity, until called forth
by some occasion which renders it a duty to resist; and this ex-

cellent spirit was exhibited during this period by at least one lady of Cromarty. She was a Mrs. Gordon, the wife of the parish minister ;—a lady who, at an extreme old age, retained much of the beauty of youth—a smooth unwrinkled forehead, shaded by a profusion of black glossy hair without the slightest tinge of grey: and it was said of her, so exquisite was her complexion, that, when drinking a glass of wine, her neck and throat would assume the ruddy hue of the liquid—an imaginary circumstance, deemed characteristic at one time, by the common people of Scotland, of the higher order of beauties, and which is happily introduced by Allan Cunningham into one of the most pleasing of his ballads :—

> " Fu' white white was her bonny neck,
>   Twist wi' the satin twine ;
>   But ruddie ruddie grew her hawse,
>   While she sipp'd the bluid-red wine."

Mrs. Gordon could scarcely have attained to her eighteenth year at the Revolution ; and yet she had been exposed to suffering on the score of religion, in the previous troubles. There was a story among the people, that her ears had been cut off ; it was even observed, that her tresses were always so arranged as to conceal the supposed mutilation ; and some of the wilder spirits of the place used to call her *Luggie,* in allusion to the story ; but she was too highly respected for the name to take. When a very old woman, she was one day combing her hair in the presence of a little girl, who was employed in dressing up the apartment in which she sat, and who threw at her from time to time a very inquisitive glance. " Come here, Maggie," said the lady, who guessed the cause of her solicitude ; " you are a curious little girl, and have heard that I have lost my ears —have you not ? Here they are, however," she continued, shading back her hair as she spoke, and displaying two very pretty ones ; " wicked men once threatened to cut them off, and a knife was sharpened for the purpose, but God permitted them not."

CHAPTER IX.

"The scart bears weel wi' the winter's cauld,
  The aik wi' the gurly win';
But the bonny wee burds, and the sweet wee flowers,
  Were made for the calm an' the sun."—OLD BALLAD.

THE southern Sutor terminates, where it overhangs the junction
of the Cromarty and Moray Firths, in a noble precipice, which,
planting its iron feet in the sea, rears its ample forehead a hundred yards over it. On the top there is a moss-covered, partially
wooded knoll, which, commanding from its abrupt height and
semi-insular situation a wide and diversified prospect, has been
known from time immemorial to the town's-people as " the Look-
out."   It is an exquisite little spot, sweet in itself, and sublime
in what it commands ;—a fine range of forest scenery stretches
along the background, while in front the eye may wander over
the hills of seven different counties, and so vast an extent of
sea, that, on the soberest calculation, we cannot estimate it under
a thousand square miles. Nor need there be any lack of pleasing
association to heighten the effect of a landscape which, among
its other scenes of the wild and the wonderful, includes the
bleak moor of Culloden, and the " heath near Forres."   It is,
however, to the immense tract of sea which it overlooks that
the little knoll owes its deepest interest, and when, after a storm
from the west has scattered the shipping bound for port, and
day after day has gone by without witnessing their expected return—there are wistful eyes that turn from it to the wide waste
below, and anxious hearts that beat quicker and higher, as sail
after sail starts up, spark-like, on the dim horizon, and grows

into size and distinctness as it nears the shore. Nor is the rock beneath devoid of an interest exclusively its own.

It is one of those magnificent objects which fill the mind with emotions of the sublime and awful; and the effect is most imposing when we view it from below. The strata, strangely broken and contorted, rise almost vertically from the beach. Immense masses of a primary trap crop out along their bases, or wander over the face of the precipice in broad irregular veins, which contrast their deep olive-green with the ferruginous brown of the mass. A whitened projection, which overhangs the sea, has been for untold ages the haunt of the cormorant and the sea-mew; the eagle builds higher up, and higher still there is a broad inaccessible ledge in a deep angle of the rock, on which a thicket of hip and sloe-thorn bushes and a few wild apple-trees have taken root, and which, from the latter circumstance, bears among the town's-boys the name of the *apple-yardie*. The young imagination delights to dwell amid the bosky recesses of this little spot, where human foot has never yet trodden, and where the crabs and the wild berries ripen and decay unplucked and untasted. There was a time, however, when the interest which attached to it owed almost all its intensity to the horrible. The eastern turret of the old castle of Cromarty has, with all its other turrets, long since disappeared; but the deep foliage of the ledge mantles as thickly as ever, and the precipice of the Look-out rears its dark front as proudly over the beach. They were frightfully connected—the shelf and the turret—in the associations of the town's-people for more than a hundred years; the one was known as the Chaplain's Turret—the other as the Chaplain's Lair; but the demolition of the castle has dissolved the union, and there are now scarcely a dozen in the country who know that it ever existed.

I have said that the proprietor of the lands of Cromarty, in the early part of the reign of Charles II., was a Sir John Urquhart of Craigfintrie, celebrated by Wodrow, though the celebrity be of no enviable character, as the person whose advice, strengthened by that of Sir George Mackenzie of Tarbat, led Commissioner Middleton to introduce the unhappy Act which overturned Presbyterianism in Scotland. He was a shrewd, strong-minded man, thoroughly acquainted with all the worse and some of the better springs of human action—cool, and cautious, and skilful, in steering himself through the difficulties of so unsettled a period by the shifts and evasions of a well-balanced expediency. The current of the age had set in towards religion, and Sir John was by much too prudent to oppose the current. There were few men who excelled him in that most difficult art of computation, the art of estimating the strength of parties ; perhaps he was all the more successful in his calculations from his never suffering himself to be disturbed in them by an over-active zeal ; and believing, with Tamerlane and Sir Thomas, that the Deity usually declares for the stronger party, Sir John was always religious enough to be of the stronger party too.

About twelve years before his death, which took place in 1673, there resided in his family a young licentiate of the Scottish Church, a nephew of his own, who officiated as chaplain. Dallas Urquhart was naturally a soft-tempered, amiable man, of considerable attainments, and of no inferior powers of mind, but his character lacked the severer virtues ; and it was his fate to live in an age in which a good-natured facility of disposition was of all qualities the worst fitted to supply their place. He was deemed a person of more than ordinary promise in an age when the qualifications of the Presbyterian minister were fixed at least as high as in any after period ;— there was a charm, too, in his pliant and docile disposition, which peculiarly recommended him to his older friends and advisers; for even the wise are apt to overvalue whatever flatters themselves, and to decide regarding that modest facility which so often proves in after life the curse of its possessor, according to an estimate very different from that by which they rate the wilful, because partially developed strength which thwarts and opposes them. Dallas therefore had many friends; but the person to whom he was chiefly attached was the young minister of Cromarty—the "good Maister Hugh Anderson," whose tombstone, a dark-coloured slab, roughened with uncouth sculpture, and a neat Latin inscription, may still be seen in the eastern gable of the parish church, and to whom I have already referred as one of the few faithful in this part of the country in a time of fiery trial. The two friends had passed through college together, associates in study and recreation, and, what is better still, for they were both devout young men, companions in all those acts of religious fellowship which renders Christianity so true a nurse of the nobler affections. And yet no two persons could be less similar in the original structure of their minds. Dallas was gentle-tempered and imaginative, and imbued, through a nature decidedly intellectual, with a love of study for its own sake. His friend, on the contrary, was of a bold energetic temperament, and a plain, practical understanding, which he had cultivated rather from a sense of duty than under the influence of any direct pleasure derived in the process. But it is probable that had they resembled one another more closely, they would have loved one another less. Friendship, if I may venture the metaphor, is a sort of ball-and-socket connexion. It seems to be a first principle in its economy that its agreements should be founded in dissimilarity—the stronger with the weaker—the softer with the more rugged. And perhaps, in looking round to convince ourselves of the fact, we have but to note how the sexes—formed for each other by God himself—have been created, not after a similar, but after a diverse pattern—and that their natures piece together, not because they were made to resemble, but to correspond with each other.

The friends were often together; the huge old castle, grey

with the lichens of a thousand years, towered on its wooded eminence immediately over the town ; and the little antique manse, with its narrow serrated gables, and with the triangular tablets of its upper windows, rising high in the roof, occupied, nest-like, an umbrageous recess so directly below, that the chaplain in his turret was scarcely a hundred yards distant from the minister in his study.   There hardly passed a day in which Dallas did not spend an hour or two in the manse ; at times speculating on some abstruse scholastic question with his friend the clergyman, whom he generally found somewhat less than his match on such occasions ; at times still more pleasingly engaged in conversing, though on somewhat humbler terms, with his friend's only sister.   Mary Anderson was a sylph-looking young creature, rather below the middle size, and slightly though finely formed.   Her complexion, which was pale and singularly transparent, indicated no great strength of constitution ; but there was an easy grace in all her motions, that no one could associate with the weakness of positive indisposition, and an expression in her bright speaking eyes and beautiful forehead, that impressed all who knew her rather with the idea of an active and powerful spirit than of a delicate or feeble frame. There was an unpretending quietness in her manners, and a simple good sense in all she said and did on ordinary occasions, that seemed to be as much the result of correct feeling as of a discriminating intellect ; but there were depths in the character beyond the reach of the ordinary observer—powers of abstract thought which only a superior mind could fully appreciate, and a vigorous but well-regulated imagination that could bedeck the perfections of the moral world with all that is exquisite among the forms and colours of the natural.   No one ever seemed less influenced by the tender feelings than Mary Anderson ; she loved her brother's friend—loved to study, to read, to converse with him, but in no respect did her regard for him seem to differ from her regard for her brother himself.   She was his friend—a tender and attached one, it is true—but his friend only. But the young chaplain, whose nature it was to cleave to everything nobler and more powerful than himself, was of a different temper, and he had formed a deep though silent attachment to the highly-gifted maiden of the manse.

Troublesome times came on ; the politic and strong-minded proprietor was no longer known as the friend of the Presbyterian Church, and the comparatively weak and facile chaplain wavered in all the agonies of irresolution, under the fascinating influences of the massier and wilier character. All the persuasive powers of Sir John were concentrated on the conversion of his nephew. Acts of kindness, expressions of endearment and good-will, and a well-counterfeited zeal for the interests of true religion formed with him merely a sort of groundwork for arguments of real cogency so far as they went, and facts which, though of partial selection, could not be well disputed. He had passed, he said, over the ground which Dallas now occupied, and was thus enabled to anticipate some of his opinions on the subject ; he, too, had once regarded Christianity and the Presbyterian form of it as identical, and associated the excellencies of the one with the peculiarities of the other. He now saw clearly, however ; and his nephew, he was assured, would soon see it too, that they were things as essentially different as soul and body, and that the Presbyterian form—the Presbyterian *body*, he might say—was by no means the best which the true religion could inhabit ! He pointed out what he deemed the peculiar defects of Presbyterianism ; and summed up with consummate skill the various indications of the subdued and unresisting mood which at this period formed that of the entire country.

"Men of all classes," he said, " have been wearied in the long struggle of twenty years, from which they have but just escaped, and stand in need of rest. They see, too, that they have been contending, not for themselves, but for others—

I

striving to render kings less powerful, that Churchmen might become more so. They see that they have thus injured the character of a body of men, valuable in their own sphere, but dangerous when invested with the power of the magistrate; and that they have so weakened the hands of Government, that to escape from anarchy they have •to fling themselves into the arms of comparative despotism. But there are better times coming; and the wiser sort of men are beginning to perceive that religion must work more effectually under the peaceable protection of a paternal Government, than when united to a form which cannot exist without provoking political heats and animosities. Not a few of our best men are more than prepared for the movement. You already know something of Leighton; I need not say what sort of a man *he* is;—and young Burnet and Scougall are persons of resembling character. But there are many such among the belied and persecuted Episcopalians; and does it not augur well, that since the one Church must fall, we should have materials of such value for building up the other? I cannot anticipate much opposition to the change. A few good men of narrow capacity, such as your friend Anderson, will not acquiesce in it till they are made to distinguish between form and spirit, which may take some time; and leaders in the Church, who have become influential at the expense of their country's welfare, must necessarily be hostile to it for another çause; for no man willingly parts with power. But certain I am it can meet with no effectual opposition."

Dallas had little to urge in reply. Sir John had ever been kind to him, nor was his disposition a cold or ungrateful one; and, naturally facile and diffident besides, he had been invariably in the habit of yielding up his own judgment, in matters of a practical tendency, to the more mature and powerful judgment of his uncle. He could feel, however, that on this occasion there was something criminal in his acquiescence; but his weakness overcame him. He passed sleepless nights, and days of restless inquietude; at times half resolved to seek out Sir John and say that, having cast in his lot with the sinking Church, he could not quit her in the day of trouble—at times groaning under a despairing sense of his thorough inability to oppose him—yielding to what seemed to be the force of destiny, and summing up the various arguments so often pressed upon him, less with the view of ascertaining their real value, than of employing them in his own defence. Meanwhile whole weeks elapsed ere he could muster up resolution enough to visit his friend the clergyman; and the report had gone far and wide that he had declared with his uncle for the court religion. He at length stole down one evening to the manse with a sinking heart, and limbs that trembled under him.

Mary was absent on a visit; her brother he found sitting moodily in his study. The minister had but just returned from a Presbytery, at which he had to contend single-handed against the arguments of Sir John, and the votes of all the others; and, under the influence of the angry feelings awakened in a conflict so hopeless and unequal, and irritated by the gibes and taunts of his renegade brethren, he at once denounced Dallas as a time-server and an apostate. The temper of the chaplain gave way, and he retorted with a degree of spirit which might have given a different colour to his after life, had it been exerted during his earlier interviews with Sir John. The anger of the friends, heightened by mutual reproaches, triumphed over the affection of years; and, after a scene of bitter contention, they parted with the determination of never meeting again. Dallas felt not the ground, as, with throbbing pulses and a flushed brow, he hastily scaled the ascent which led to the castle; and when, turning round from beside the wall to look at the manse, he thought of Mary, and thought, too, that he could now no longer visit her as before, his heart swelled almost to bursting. " But I am like a straw on the current," he said,

" and must drift wherever the force of events carries me. Coward that I am! Why do I live in a world for which I am so wretchedly unfitted?"

He had now passed the Rubicon. His first impressions he had resisted; and the feebler suggestions which afterwards arose in his mind, only led him to entrench himself the more strongly in the arguments of Sir John. Besides, he had declared himself a friend to Episcopacy, and thus barred up his retreat by that shrinking dread of being deemed wavering and inconsistent, which has so often merely the effect of rendering such as lie openest to the imputation firm in the wrong place. There was a secret bitterness in his spirit, that vented itself in caustic remarks on all whom he had once admired and respected —all save Mary and her brother, and of them he never spoke. His former habits of application were broken up; yet, though no longer engrossed by the studies which it had once seemed the bent of his nature to pursue, he remained as indifferent as ever to the various pursuits of interest and ambition which engaged his uncle. After passing a day of restless inactivity among his books, he walked out a little before sunset in the direction of the old chapel of St. Regulus, and, ere he took note of where his wanderings led him, found himself among the graves.

It was a lovely evening of October. The ancient elms and wild cherry-trees which surrounded the burying-ground still retained their foliage entire, and the elms were hung in gold, and the wild cherry-trees in crimson, and the pale yellow tint of the straggling and irregular fields on the hill-side contrasted strongly with the deepening russet of the surrounding moor. The tombs and the ruins were bathed in the yellow light of the setting sun; but to the melancholy and aimless wanderer the quiet and gorgeous beauty of the scene was associated with the coming night and the coming winter, with the sadness of inevitable decay and the gloom of the insatiable grave. He passed moodily onward, and, on turning an angle of the chapel, found Mary Anderson seated among the ruins, on the tombstone of her mother, whom she had lost when a child. There was a slight flush on her countenance as she rose to meet him, but she held out her hand as usual, though the young man thought, but it might not be so, that the grasp was less kindly.

"You have been a great stranger of late, Dallas," she said; "how have we been so unhappy as to offend you?"

The young chaplain looked as if he could have sunk into the ground, and was silent.

"Can it be true," resumed the maiden, "that you have left us in our distress, and gone over to the prelates?"

Dallas stammered out an apology, and reminded her that Christianity, not Presbyterianism, formed the basis of their friendship. "The Church," he said, "had too long paid an overweening regard to mere forms; it was now full time to look to essentials. What mattered it whether men went to heaven under the jurisdiction of a Presbyterian or of an Episcopalian Church?" He passed rapidly over the arguments of Sir John.

"You are deceived, my dear friend," said Mary. "Look at these cottages that glitter to the setting sun on the hillside. Eighty years ago their inmates were the slaves of gross superstition—creatures who feared and worshipped they knew not what; and, with no discipline of purity connected with their uncertain beliefs, they could, like mere machines, be set in motion, either for good or ill, at the will of their capricious masters. These cottages, Dallas, are now inhabited by thinking men; there are Bibles in even the humblest of them—in even yonder hovel where the old widow lives—and these Bibles are read and understood. We may hear even now the notes of the evening psalm! Wist ye how the change was wrought? or what it was that converted mere animal men into rational creatures? Was it not that very Church which you have, alas! so rashly forsaken, and now denounce as intolerant? A strange intolerance, surely, that delivers men from the influence of their grosser nature, and delights to arm its vassals with a power before which all tyranny must eventually be overthrown. Be not deceived, Dallas! Men sometimes suffer themselves to be misled by theories of a perfect but impossible freedom—impossible, because unsuited to the low natures and darkened minds of those on whom they would bestow it.—and then submit in despair to the quiet of a paralysing despotism, because they cannot realize what they have so fondly imagined. Know ye not that none but the wise and the good can be truly free—that the vile and the ignorant are necessarily slaves under whatever form of government they may chance to live? See you not that the deprecated sway of the Scottish Church has been in truth but a paternal tutelage—that her children were feeble in mind, and rude and untoward, when she first laid the hand of her discipline upon them—and that she has now well-nigh trained them up to be men? And think you that these, our poor countrymen, already occupy that place which He who died for them has willed they should attain to? or that the *many*, no longer a brute herd, but moral and thoughtful, and with the Bible in their hands, are to remain the willing, unresisting slaves of the *few*? No, Dallas! when men increase in goodness and knowledge, there must be also an onward progress towards civil liberty; and the political bias which you denounce as unfavourable to religion, is merely the onward groping of this principle. A strange intolerance, surely, that has already broken the fetters of the bondsman; they still clank about him, but being what he now is—intelligent and conscientious—they must inevitably drop off, let his master fret as he may, and leave him a freeman."

There was a pause, during which Dallas doggedly fixed his eyes on the ground. "I have viewed the subject," he at length said, " with different eyes. And of this I am sure, there are in the Episcopal Church truly excellent men who cannot exist without doing good."

" But look round you," said Mary, " and say whether the great bulk of those who are now watching as on tiptoe to swell its ranks, are of the class you describe? Can you shut your eyes to the fact, that there is a winnowing process going on in the one Church, and that the chaff and dust are falling into the other? But, Dallas," she said, laying her hand on her breast, " I can no longer dispute with you; and 'twould be unavailing if I could, for it is not argument but strength that you want—strength to resist the influence of a more powerful but less honest mind than your own. There is assuredly a time of trouble coming; but I feel, Dallas, that your escape from it cannot be more certain than mine."

Dallas, who had hitherto avoided her glance, now regarded her with an expression of solicitude and alarm. She was thin—much thinner than usual—and her cheeks were crimsoned by a flush of deadly beauty. The anger which she had excited—for she had convinced him of his error, despite of his determination not to be convinced, and he was necessarily angry—vanished in a moment. He grasped her hand, and bursting into tears, " O Mary!" he exclaimed, " I am a weak, worthless thing—pity me—pray for me; but no, it were vain, it were vain; I am lost, and for ever!"

The maiden was deeply affected, and strove to console him. " Retrace your steps," she said, " in the might of Him whose strength is perfected in weakness, and all shall yet be well. My poor brother has mourned for your defection as he would have done for your death; but he loves you still, and deeply regrets that an unfortunate quarrel should have estranged you from him. Come and see him as usual; he has a keen temper, but need I tell you that he has an affectionate heart? And I did not think, Dallas, that you could have so soon forgotten myself; but come, that I may have my revenge."

The friends parted, and at this time neither of them thought they had parted for ever. But so it was. Facile and wavering as nature had formed the young chaplain, he yet indulged in a pride that, conscious of weakness, would fain solace itself with at least an outside show of strength and consistency; and he could not forget that he had now chosen his side. Weeks and months passed, and the day arrived on which, at the instance of the Bishop of Ross, the nonconforming minister of Cromarty was to be ejected from his parish.

It was early in December. There had been a severe and still increasing snow-storm for the two previous days; the earth was deeply covered; and a strong biting gale from the north-east was now drifting the snow half-way up the side-walls of the manse. The distant hills rose like so many shrouded spectres over the dark and melancholy sea—their heads enveloped in broken wreaths of livid cloud; nature lay dead; and the very firmament, blackened with tempest, seemed a huge burial vault. The wind shrieked and wailed like an unhappy spirit among the turrets and chimneys of the old castle. Dallas undid the covering of the shot-hole that looked down on the manse, and then hastily shutting it, flung himself on his bed, where he lay with his face folded in the bedclothes. Ere he had risen, the shades of evening, deepened by a furious snow-shower, had set in. He again unbolted the shot-hole and looked out. The flakes flew so thickly that they obscured every nearer object, and wholly concealed the more remote: even the manse had disappeared; but there was a faint gleam of light flickering in that direction through the shower; and as the air cleared, the chaplain could see that it proceeded from two lighted candles placed in one of the windows. Dreading he knew not what, he descended the turret stair, and on entering the hall found one of the domestics, an elderly woman, preparing to quit it. "This," he said, addressing her, "is surely no night for going abroad, Martha?" "Ah, no!" replied the woman, "but I am going to the manse; there is distress there. Mary Anderson died this morning, and it will be a thin lyke-wake." "Mary

Anderson! thin lyke-wake!" said Dallas, repeating her words as if unconscious of their meaning: "I'll go with you." And, as if moved by some impulse merely mechanical, he followed the woman.

The dead chamber was profusely lighted, according to the custom of the country, and the bed, with every other piece of furniture which it contained, was hung with white. The bereaved brother had shut himself up in his room, where he might be heard at times as if struggling with inexpressible anguish in the agony of prayer; and only two elderly women, one of them the nurse of Mary, watched beside the corpse. With an unsteady step and pallid countenance, on the lineaments of which a quiescent but settled despair was frightfully impressed, Dallas entered the apartment and stood fronting the bed. The nurse greeted him in a few brief words, expressive of their mutual loss; but he saw her not—heard her not. He saw only the long white shroud with its fearfully significant outline—heard only the beatings of his own heart. The eyes of the good old woman filled with tears as she gazed on him, and, slowly rising, she uncovered the face of the dead. He bent forward; *there* was the open and beautiful forehead, and there the exquisite features, thin and wasted 'tis true, but lovely as ever. A faint smile still lingered on the lip; it was a smile that called upon thoughts and feelings the most solemn and holy, and whispered of the joys of immortality from amid the calm and awful sublimity of death. "Ah, my bairn!" said the woman, "weel and lang did she loe you, and meikle did she grieve for you and pray for you, when ye went o'er to the prelates. But her griefs are a' ended now. Ken ye, Dallas, that for years an' years she loed ye wi' mair than a sister's luve, an' that if she didna just meet wi' your hopes, it was only because she kent o'er weel she was to die young?" Dallas struck his open palm against his forehead; a convulsive emotion shook his frame; and, bursting into tears, he flung himself out of the room.

The funeral passed over; and the brother of Mary quitted the parish a homeless and solitary man, with a grieving heart but an unbroken spirit. He had to mourn both for the dead and the estranged; and found that the low insults and cruel suspicions of the persecutor dogged him wherever he went. But his state was one of comfort compared with that of his hapless friend. Grief, terror, and remorse, lorded it over the unfortunate chaplain by turns, and what was at first but mere inquietude had become anguish. He was sitting a few weeks after the interment in the eastern turret, his eyes fixed vacantly on the fire, which was dying on the hearth at his feet, when Sir John abruptly entered, and drew in his seat beside him.

"Your fire, nephew," he said, as he trimmed it, "very much resembles yourself. There is no lack of the right material, but it wants just a little stirring, and is useless for want of it. It is no time, Dallas, to be loitering away life when a bright prospect of honourable ambition and extensive usefulness is opening full before you. Wot you not that our neighbour the bishop, now a worn-out old man, has been confined to his room for the last week? And should my cousin of Tarbat and myself agree in recommending a successor, as we unquestionably shall, where think you lies the influence powerful enough to thwart us? But a diocese so important, nephew, can be the prize of no indolent dreamer."

"Uncle," said Dallas, in a tone of deep melancholy, "do you believe that those who have been once awakened to the truth may yet fall away and perish?"

"Why perplex yourself with such questions?" replied his uncle. "We are creatures intended for both this world and the next; each demands that certain duties be performed; and of all men, woe to him of a musing and speculative turn, who, though not devoid of a sense of duty, fails in the requirements of the present state. His thoughts become fiends to torment him. But up, nephew, and act, and you will find that all will be well."

"Act! How?—to what purpose?—how read you the text —'It is impossible for those who were once enlightened, if they fall away, to renew them to repentance?' I have fallen—fallen for ever."

"Dallas," said Sir John, "what wild thought has now possessed you?—You are but one of many thousands;—I know not a more hopeful clergyman of your standing connected with the Church."

"Wretched, wretched Church, if it be so! But what are her ministers? Trees twice dead, plucked up by the roots—wandering stars, for whom is reserved the blackness of darkness for ever. Yes, I am as hopeful as most of her ministers. Mary Anderson told me what was coming; and, hypocrite that I am! I believed her, and yet denied that I did. I saw her last night;—she was beautiful as ever—but ah! there was no love, no pity in her eye—and the wide, wide gulf was between us."

"Dearest nephew, why talk so wildly?" exclaimed Sir John.

"Uncle, you have ever been kind to me," replied Dallas; "but you have ruined—no, wretched creature! 'twas I, myself, who have ruined my soul; and there is neither love nor gratitude in the place to which I am going. O leave to myself! my thoughts become more fearful when I embody them in words;—leave me to myself! and, uncle, while there is yet space, seek after that repentance which is denied to me. Avoid the unpardonable sin."

The strong mind of Sir John was prostrated before the fearfully excited feelings of his nephew, as a massy barrier of iron may be beaten down by a cannon-ball; and he descended the turret stair rebuked and humbled by an energy more potent than his own—as if, for the time, he and the young man had exchanged characters. Next morning Dallas was nowhere to be found.

He was seen about sunrise, by a farmer of the parish, passing hurriedly along the ridge of the hill. The man, a staid

Presbyterian, with whom he had once loved to converse, had saluted him as he passed, and then paused for half a second in the expectation that, as usual, he would address him in turn; but he seemed wholly unconscious of his presence. His face, he said, was of a deadly paleness, and his pace, though hurried, seemed strangely unequal, as if he were exhausted by indisposition or fatigue. The day wore on; and towards evening, Sir John, who could no longer conceal the anxiety which he felt, ordered out all his domestics in quest of him. But the night soon fell dark and rainy, and the party was on the eve of relinquishing the search, when, in passing along the edge of the *Look-out,* one of the servants observed something white lying on the little grassy bank which surmounts the precipice. It was an open Bible—the gift, as the title-page intimated, of Mary Anderson to Dallas Urquhart. Sir John struck his clenched fist against his forehead.

"Gracious heaven!" he exclaimed, "he has destroyed himself!—to the foot of the rock—to the foot of the rock;—and haste! for the tide is fast rising. But stay—let me forward— I will lead the way myself."

And, passing through his terrified attendants, he began to descend by a path nearly invisible in the darkness, and which, winding along the narrow shelves of the precipice, seemed barely accessible even by day to the light foot of the schoolboy. There was only one of the many who now thronged the rock edge who had courage enough to follow him—a tall spare man, wrapped up in a dark-coloured cloak. As the path became narrower and more broken, and overhung still more and more fearfully the dizzy descent, the stranger, who passed lightly and steadily along, repeatedly extended his arm to the assistance of the knight, who, through agitation and the stiffness incident to a period of life considerably advanced, stumbled frightfully as he hurried down. They reached the shore in safety together. All was dismally solitary. They could see only the dark rock towering over them, and the line of white waves which were tumbling over the beach, and had now begun to lash the base of the precipice.

"Alas! my poor lost friend!" ejaculated the stranger—"lost, alas! for ever, when I had hoped most for thy return. Wretched, unfortunate creature! with little care of thine own for the things of this world, and yet ever led away by those who worshipped them as their only god—alas! alas! how hast thou perished!"

"Spare me, Hugh Anderson!" said Sir John, "spare me!— do not, I implore you, add to the anguish of this miserable night!"

They walked together in silence to where the waves barred their further progress, and then returned to the top of the precipice. The search was renewed in the morning, but as ineffectually as on the preceding night—there was no trace of the body. Seasons passed away; Sir John, as has been already related, perished by his own act; Episcopacy fell; and Hugh Anderson, now a greyheaded elderly man, was reappointed, after the lapse of more than thirty years, to his old charge, the parish of Cromarty.

He had quitted it amid the snow-wreaths of a severe and boisterous winter; he returned to it after a storm of wind and snow from the sea had heaped the beach with wrack and tangle, and torn their mantles of ivy from some of the higher precipices. He revisited with anxious solicitude the well-remembered haunts endeared to him by so many fond, yet mournful recollections— Mary's favourite walks—the cliffs which he had so often scaled with Dallas—and the path which he had descended in the darkness with the hapless Sir John. He paused at the foot of the precipice—the storm had swept fiercely over its iron forehead, and an immense bush of ivy, that had fallen from the ledge of the *apple-yardie*, lay withering at its base. His eye caught something of unusual appearance amid the torn and broken foliage—it was a human skull, bleached white by the rains and the sunshine of many seasons, and a few disjointed and fractured bones lay scattered near it.   Painfully did he gather them up, and painfully scooping out with his pointed stick a hollow in the neighbouring bank, he shed, as he covered them up from the sight, the last tears that have fallen to the memory of the lost Dallas Urquhart.

CHAPTER X.

" A mighty good sort of man."—BONNEL THORNTON.

THE Episcopalian minister of Cromarty was a Mr. Bernard Mackenzie, a quiet, timid sort of man, with little force of character, but with what served his turn equally well, a good deal of cunning. He came to the parish in the full expectation of being torn to pieces, and with an aspect so wo-begone and miserable—for his very countenance told how unambitious he was of being a martyr—that the people pitied instead of insulting him; and, in the course of a few weeks, he had not an ill-wisher among them, however disaffected some of them were to his Church. No one could be more conversant than the curate with the policy of submission, or could become all things to all men with happier effect. The people, who, like the great bulk of the people everywhere, were better acquainted with the duties of ministers than with their own, were liberal in giving advices, and no person could be more submissive in listening to these than the curate. Some of them, too, had found out the knack of being religious without being moral, and the curate was by much too polite to hint to them that the knack was a bad one. And thus he went on, suiting himself to every event, and borrowing the tone of his character from those whom it was his duty rather to lead than to follow, until the great event of the Revolution, which he also surmounted by taking the oath of allegiance that recognised William as king both in fact and in law. With all his policy, however, he could not help dying a few years after, when he was succeeded by the old ejected minister, Hugh Anderson.

The curate of the neighbouring parish of Nigg, a Mr. James Mackenzie, was in some respects a different sort of person. He was nearly as quiet and submissive as his namesake of Cromarty, and he was not much more religious; for when, one Sunday morning, he chanced to meet the girls of a fishing village returning home laden with shellfish, he only told them that they should strive to divide the day so as to avail themselves both of the church and the ebb. He was, however, a simple, benevolent sort of man, who had no harm in him, and never suspected it in others; and so little was he given to notice what was passing around him, as to be ignorant, it was said, of the exact number of his children, though it was known to every one else in the parish that they amounted to twenty. They were all sent out to nurse, as was customary at the period; and when the usual term had expired, and they were returned to the manse, it proved a sad puzzle to the poor curate to recollect their names. On one occasion, when the whole twenty had gathered round his table, there was a little red-cheeked girl among them, who having succeeded in climbing to his knee, delighted him so much with her prattle, that he told her, after half smothering her with kisses, that "gin she were a bairn o' his, he would gie her a tocher o' three hunder merk mair nor ony o' the lave." "Then haud ye, gudeman," said his wife; "for as sure as ye're sitting there, it's yer ain Jenny." The descendants of the curate, as might be anticipated from the number of his children, are widely spread over the country, and exhibit almost every variety of fortune and cast of mind. One of them, a poor pauper, died a few years ago in the last extreme of destitution and wretchedness; another, an eminent Scottish lawyer, now sits on the Bench as one of the Lords of Session. One of his elder sons was grandfather to the celebrated Henry Mackenzie of Edinburgh, and the great-great-grandchild of the little prattling Jenny is the writer of these chapters.

The bulk of the people of Nigg had just as little religion as

their pastor.   Every Sunday forenoon they attended church,
but the evening of the day was devoted to the common athletic
games of the country.   A robust active young fellow, named
Donald Roy, was deemed their best club-player; and, as the
game was a popular one, his Sabbath evenings were usually
spent at the club.   He was a farmer, and the owner of a small
herd of black cattle.   On returning home one Sabbath evening,
after vanquishing the most skilful of his competitors, he found
the carcase of one of his best cattle lying across the threshold,
where she had dropt down a few minutes before.   Next Sabbath
he headed the club-players as usual, and on returning at the
same hour, he found the dead body of a second cow lying in
exactly the same place.   " Can it be possible," thought he,
" that the Whigs are in the right after all?"   A challenge, however, had been given to the club-players of a neighbouring parish,
and, as the game was to be played out on the following Sabbath,
he could not bring himself to resolve the question.   When the
day came, Donald played beyond all praise, and, elated by the
victory which his exertions had at length secured to his parish,
he was striding homeward through a green lane, when a fine cow,
which he had purchased only a few days before came pressing
through the fence, and flinging herself down before him, expired
at his feet with a deep horrible bellow.   " This is God's judgment!" exclaimed Donald, " the Whigamores are in the right;
—I have taken *His* day, and he takes *my* cattle."   He never
after played at the club; and, such was the change effected on
his character, that shortly after the Revolution he was ordained
an elder of the church, and he became afterwards one of the
most notable worthies of the North.   There are several stories
still extant regarding him, which show that he must have latterly
belonged to that extraordinary class of men (now extinct) who,
living, as it were, on the extreme verge of the natural world, and
seeing far into the world of spirits, had in their times of darkness to do battle with the worst inmates of the latter, and saw

K

in their seasons of light the extreme bounds of the distant and the future. This class comprised at one time some of the stanchest champions of the Covenant, and we find at its head the celebrated Donald Cargill and Alexander Peden.

Some of the stories told of Donald Roy, which serve to identify him with this class, are worthy of being preserved. On one occasion, it is said, that when walking after nightfall on a solitary road, he was distressed by a series of blasphemous thoughts, which came pouring into his mind despite of all his exertions to exclude them. Still, however, he struggled manfully, and was gradually working himself into a better frame, when looking downwards he saw what seemed to be a black dog trotting by his side. "Ah!" he exclaimed, "and so I have got company; I might have guessed so sooner." The thing growled as he spoke, and bounding a few yards before him, emitted an intensely bright jet of flame, which came streaming along the road until it seemed to hiss and crackle beneath his feet. On he went, however, without turning to the right hand or to the left, and the thing bounding away as before, stood and emitted a second jet. "Na, na, it winna do," said the imperturbable Donald : "ye first tried to loose my haud o' my Master, and ye would now fain gie me a fleg; but I ken baith him and you ower weel for that." The appearance, however, went on bounding and emitting flame by turns, until he had reached the outer limits of his farm, when it vanished.

About seventy years after the Revolution, he was engaged in what is termed *proofing* the stacks of a corn-yard, on the hilly farm of Castle-Craig. There were other two men with him employed in handing down and threshing the sheaves. The day was exceedingly boisterous, and towards evening there came on a heavy snow-storm. "Our elder," said one of the men to his companion, "will hae deep stepping home through the snaw thraves; he would better stay at the Craig—will you no ask him?" "Man, look," said the other, "what is he about?—

look—look!" At a little distance, in a waste corner of the barn, sat the elder, his broad blue bonnet drawn over his brow, his eyes fixed on the wall; and ever and anon would he raise his hands and then clasp them together, as if witnessing some scene of intense and terrible interest. At times, too, he would mutter to himself like a deranged person; and the men, who had dropped their flails and stood looking at him, could hear him exclaiming in a rapid but subdued tone of voice, "Let her drive—let her drive!—Dinna haud her side to the sea." Then striking his palms together, he shouted out, "She's o'er—she's o'er!—O the puir widows o' Dunskaith!—but God's will be done." "Elder," said one of the men, "are ye no weel?—ye wald better gang in till the house." "No," said Donald, "let's awa to the burn o' Nigg;—there has been ill enough come o' this sad night already—let's awa to the burn, or there'll be more." And rising from his seat with the alacrity of his club-playing days, though he was now turned of ninety, he strode out into the storm, followed by the two men. "What's that?" asked one of the men, pointing, as he reached the burn, to a piece of red tartan which projected from the edge of an immense wreath, "Od! but it's our Jenny's *brottie* sticking out thro' the snaw:—An' oh! but here's Jenny hersel'." The poor woman, who had been visiting a friend at the other end of the parish, had set out for Castle-Craig at the beginning of the storm, and exhausted with cold and fatigue, had sunk at the side of the stream a few minutes before. She was carried to the nearest cottage, and soon recovered. And the following morning afforded a sad explanation of the darker vision—the wreck of a Dunskaith boat, and the dead bodies of some of the crew being found on the beach below the Craig.

A grand-daughter of the elder who was married to a respectable Cromarty tradesman, was seized in her thirtieth year by a dangerous fever, and her life despaired of. At the very crisis of the disease her husband was called by urgent business

to the parish of Tarbat.   On passing through Nigg he waited
on Donald, and, informing him of her illness, expressed his fears
that he would not again see her in life.   " Step in on your
coming back," said the elder, " and dinna tine heart—for she's
in gude hands."   The husband's journey was a hurried one,
and in less than three hours after he had returned to the cottage
of Donald, who came out to meet him.   " Come in, Robert,"
he said, " and cool yoursel' ; ye hae travelled ower hard ;—
come in, and dinna be sae distressed, for there's nae cause.
Kettie will get o'er this, and live to see the youngest o' her
bairns settled in the world, and doing for themselves."   And his
prediction was accomplished to the very letter.   The husband,
on his return, found that the fever had abated in a very remarkable manner a few hours before ; and in less than a week
after, his wife had perfectly recovered.   More than forty years
from this time, when the writer was a little kilted urchin of five
summers, he has stood by her knee listening to her stories of
Donald Roy.   " And now," has she said, after narrating the
one in which she herself was so specially concerned, " all my
bairns are doing for themselves, as the good man prophesied ;
and I have lived to tell of him to you, my little curious boy,
the bairn of my youngest bairn."   I have little of the pride of
family in my disposition ; and, indeed, cannot plume myself
much on the score of descent, for, for the last two hundred
years, my ancestors have been merely shrewd honest people, who
loved their country too well to do it any discredit ; but I am
unable to resist the temptation of showing that I can claim
kindred with the good old seer of Nigg, and the Addison of
Scotland.

There is a still more wonderful story told of Donald Roy
than any of these.   On one of the days of preparation set apart
by the Scottish Church previous to the dispensation of the
sacrament, it is still customary in the north of Scotland for the
elders to address the people on what may be termed the internal

evidences of religion, tested by their own experience. The day dedicated to this purpose is termed *the day of the men;* and so popular are its duties, that there are none of the other days which the clergyman might not more safely set aside. When there is a lack of necessary talent among the elders of a parish, they are called dumb elders, and their places are supplied on the day of the men by the more gifted worthies of the parishes adjoining. Such a lack occurred about a century ago in the eldership of Urray, a semi-Highland parish, near Dingwall;—and at the request of the minister to the Session of Nigg, that some of the Nigg elders, who at that time were the most famous in the country, should come and officiate in the room of his own, Donald Roy and three other men were despatched to Urray. They reached the confines of the parish towards evening, and when passing the house of a gentleman, one of the heritors, they were greeted by the housekeeper, a woman of Nigg, who insisted on their turning aside and spending the evening with her. Her mistress, she said, was a stanch Roman Catholic, but one of the best creatures that ever lived, and, if the thing was possible, a Christian;—her master was a kind, good-natured man, of no religion at all; she was a great favourite with both, and was very sure that any of her friends would be made heartily welcome to the best their hall afforded. Donald's companions would have declined the invitation, as beneath the dignity of men of independence and elders of the Church; but he himself, though quite as much a Whig as any of them, joined with the woman in urging them to accept of it. "I am sure," he said, "we have been sent here for some special end, and let us not suffer a silly pride to turn us back without our errand."

There was one of the closets of the house converted by the lady into a kind of chapel or oratory. A small altar was placed in the centre; and the walls were hollowed into twelve niches, occupied by little brass images of the apostles. The lady was on the eve of retiring to this place to her evening devotions, when the housekeeper came to inform her of her guests, and to request that they should be permitted to worship together, after the manner of their Church, in one of the out-houses. Leave was granted, and the lady retired to her room. Instead, however, of kneeling before the altar as usual, she seated herself at a window. And first there rose from the out-house a low mellow strain of music, swelling and sinking alternately, like the murmurs of the night wind echoing through the apartments of an old castle. When it had ceased she could hear the fainter and more monotonous sounds of reading. Anon there was a short pause, and then a scarcely audible whisper, which heightened, however, as the speaker proceeded. Donald Roy was engaged in prayer. There were two wax tapers burning on the altar, and as the prayer waxed louder the flames began to stream from the wicks, as if exposed to a strong current of air, and the saints to tremble in their niches. The lady turned hastily from the window, and as she turned, one of the images toppling over, fell upon the floor ; another and another succeeded, until the whole twelve were overthrown. When the prayer had ceased, the elders were summoned to attend the lady. " Let us take our Bibles with us," said Donald ; " Dagon has fallen, and the ark o' the Bible is to be set up in his place." And so it was ;—they found the lady prepared to become a willing convert to its doctrines ; and on the following morning the twelve images were flung into the Conan. Rather more than twenty years ago a fisherman, when dragging for salmon in a pool of the river in the immediate neighbourhood of Urray, drew ashore a little brass figure, so richly gilt, that for some time it was supposed to be of gold ; and the incident was deemed by the country people an indubitable proof of the truth of the story.

Donald Roy, after he had been for full sixty years an elder of the Church, was compelled by one of those high-handed acts of ecclesiastical intrusion, which were unfortunately so common in Scotland about the middle of the last century, to quit it for ever ; and all the people of the parish following him as their leader, they built for themselves a meeting-house, and joined the ranks of the Secession.   Such, however, was their attachment to the National Church, that for nearly ten years after the outrage had been perpetrated, they continued to worship in its communion, encouraged by the occasional ministrations of the most distinguished divine of the north of Scotland in that age, Mr. Fraser of Alness, the author of a volume on Sanctification, still regarded as a standard work by our Scottish theologians. The presbytery, however, refusing to tolerate the irregularity, the people were at length lost to the Established Church ; and the dissenting congregation which they formed still exists as one of the most numerous and respectable in that part of the kingdom.   We find it recorded by Dr. Hetherington in his admirable Church History, that "great opposition was made by the pious parishioners to the settlement of the obnoxious presentee, and equal reluctance manifested by the majority of the presbytery to perpetrate the outrage commanded by the superior courts. But the fate of Gillespie was before their eyes ; and, under a strong feeling of sorrow and regret, four of the presbytery repaired to the church at Nigg to discharge the painful duty.   The church was empty ; not a single member of the congregation was to be seen.   While in a state of perplexity what to do in such a strange condition, one man appeared who had it in charge to tell them. ‘ That the blood of the people of Nigg would be required of them if they should *settle a man to the walls of the kirk.*’   Having delivered solemnly this appalling message, he departed, leaving the presbytery astonished and paralysed. And proceeding no further at the time, they reported the case to the General Assembly of the following year ; by whom, however, the intrusion of the obnoxious presentee was ultimately compelled."   I need scarce say, that the one man who on this occasion paralysed the presbytery and arrested the work of intrusion for the day, was the venerable patriarch of Nigg, at this time considerably turned of eighty. He died in the month of January 1774, in the 109th year of his age and the 84th of his eldership, and his death and character were recorded in the newspapers of the time.

In bearing Donald company into an age so recent, I have wandered far from the era of the curates, and must now return. Their time-serving dogmas seem to have had no very heightening effect on the morals of the burghers of Cromarty. Prior to the year 1670, the town was a royal burgh, and sent its commissioner to the Convention, and its representative to Parliament. For the ten years previous, however, its provost and bailies had set themselves with the most perfect unanimity to convert its revenues into gin and brandy, the favourite liquors of the period ; and then to contract heavy debts on its various properties, that they might carry on the process on a more extensive scale. And in this year, when the whole was absorbed, they made over their lands to Sir John Urquhart, the proprietor, " in consideration," says the document in which the transaction is recorded, " of his having instantly advanced, paid, and delivered to them 5000 merks Scots, for outredding them of their necessary and most urgent affairs." The burgh was disfranchised shortly after by an act of the Privy Council, in answer to a petition from Sir John and the burghers. There is a tra· dition, that in the previous ten years of license, in which the leading men of Cromarty were so successful in imitating the leading men of the kingdom, the council met regularly once a day in the little vaulted cell beneath the cross, to discuss the affairs of the burgh ; and so sorely would they be exhausted, it is said, by a press of business and the brandy, that it was generally found necessary to carry them home at night. But it was all for the good of the place ; and so perseveringly were they devoted to its welfare, that their last meeting was prolonged for three days together.

Sir John did not long enjoy this accession to his property, destroying himself in a fit of melancholy, as has been already related, three years after.   He was succeeded by his son Jonathan, the last of the Urquharts of Cromarty ; for, finding the revenues of his house much dilapidated by the misfortunes of Sir Thomas, and perceiving that all his father's exertions had failed to improve them, he brought the estate to sale, when it was purchased by Sir George Mackenzie of Tarbat.   This accomplished courtier and able man was the scion of a family that, in little more than a century, had buoyed itself up, by mere dint of talent, from a state of comparative obscurity into affluence and eminence.   The founder, Roderick Mackenzie, was second son of Colin of Kintail, a Highland chief of the sixteenth century, whose eldest son, Kenneth, carried on the line of Seaforth.   Roderick, who, says Douglas in his Scottish Peerage, was a man of singular prudence and courage, and highly instrumental in civilizing the northern parts of the kingdom, was knighted by James VI., and left two sons, John and Kenneth.   John, the elder, was created a baronet in the succeeding reign, and bequeathed at his death his lands to his son George of Tarbat, the purchaser of the lands of Cromarty.   Sir George was born in the castle of Loch-Slin, near Tain, in the year 1630, and devoted a long life to the study of human affairs, and the laws and antiquities of the kingdom.   He was one of those wary politicians who, according to Dryden, neither love nor hate, but are honest as far as honesty is expedient, and never glaringly vicious, because it is impolitic to be wicked over-much.   And never was there a man more thoroughly conversant with the intrigues of a court, or more skilful in availing himself of every chance combination of circumstances.   Despite of the various changes which took place in the government of the country, he rose gradually into eminence and power during the reigns of Charles and James, and reached, in the reign of Anne, when he was made secretary of state and Earl of Cromartie, the apex of his ambition.   He found leisure, in the course of a very busy life, to write two historical dissertations of great research—the one a vindication of Robert III. of Scotland from the charge of bastardy, the other an account of the Gowrie conspiracy. He wrote, besides, a *Synopsis Apocalyptica*, and recorded several interesting facts regarding the formation of peat-moss, which we find quoted by Mr. Rennie of Kilsyth in his elaborate essay. He is the writer, too, of a curious letter on the second sight, addressed to the Honourable Robert Boyle, which may be found in an appendix attached to the fifth volume of Pepys' Memoirs. On his death, which took place in 1714, his eldest son John succeeded to his titles and the lands of Tarbat, and his second son Sir Kenneth to the estate of Cromarty.

Some time ago, when on a journey in Easter Ross, I had to take shelter from a sudden shower in an old ruinous building, which had once been the dwelling-house of Lord Cromartie's chamberlain. The roof was not yet gone, but the floors had fallen, and the windows were divested of the frames. Miscellaneous heaps of rubbish were spread over the pavement; and in one of the corners there was a pile of tattered papers, partially glued to the floor and to each other by the rain, which pattered upon them through the crevices of the roof. The first I examined was written in a cramp old hand, and bore date 1682. At the bottom was the name of the writer, George Mackenzie. The next, which was dated nineteen years later, was in the same hand, but still more cramp. It was signed Tarbat. The third was scarcely legible, but I could decipher the word Cromartie, appended to it as a signature. Alas! I exclaimed, for the sagacious statesman. He was, I perceive, becoming old as he was growing great; and I doubt much whether the honours of his age, when united to its infirmities, were half so productive of happiness as the hopes and high spirits of his youth. And what *now* is the result of all his busy hours, if they were not completely satisfactory *then?* Here are a few sybilline-like leaves, the sole records, perhaps, of his common everyday affairs; his literary labours fill a few inches of the shelves of our older libraries ; and a few unnoticed pages in the more prolix histories of our country tell all the rest. Life would not be worth one's acceptance if it led to nothing better ; and yet of all the mere men of the world who ever designed sagaciously, and laboured indefatigably, how very few have been so fortunate as the Earl of Cromartie !

There was no very immediate effect produced by the Revolution in the parish of Cromarty, and indeed but little in the north of Scotland. The Episcopalian clergymen in this quarter were quite as unwilling to relinquish their livings, as the Presbyterians had been twenty-eight years before ; and setting themselves to reconcile, as they best could, their interest with what they deemed their duty, they professed their willingness to recognise William as their king in *fact*, though not in *law*. To meet this sophism, William demanded, in what was termed the Assurance Oath, a recognition of his authority as not only actual but legitimate ; and a hundred Episcopal ministers, who complied by taking the oath, were allowed to retain their livings without being restricted to the jurisdiction of courts of Presbytery. So large a proportion of these fell to the share of the northern counties, that in that part of the kingdom which extends from the Firth of Beauly to John o' Groat's, and from sea to sea, there was only one presbytery, consisting, for several years after the Revolution, of only eight clergymen.

The next political event of importance which agitated the kingdom was the Union. And there was at least one of the people of Cromarty who regarded it with no very complacent feeling. He was a Mr. William Morrison, the parish schoolmaster. I have seen a manuscript of 230 pages, written by this person between the years 1710 and 1713, containing a full detail of his religious experience ; and as a good deal of his religion consisted in finding fault, and a good deal more in the vagaries of a wild imagination, though the residue seems to have been sincere, he has introduced into his pages much foreign matter, of a kind interesting to the local antiquary. He was one of that class who read the Bible in a way it can be made to prove anything ; and he deemed it directly opposed not only to the Union and the Abjuration Oath of the succeeding reign, but to the very Act of Toleration, which secured to the poor curates the privilege of being, like himself, the open opponents of both. "May we not truly account," says he, "for the deadness and carnality of the Church at this present time (1712), by the great hand many of its members had in carrying on the late Union, of sorrowful memory, whereby our country's power to act for herself, both as to religion and libertie, is hung under the belt of idolatrous England ? Woe unto thee, Scotland, for thou hast sold thy birth-right ! Woe unto thee for the too too much Erastian-like obedience of the most part of thy Church to the laws of the men of this generation—men who, having established a tolleration for all sorts of wickedness, have set up Baal's altars beside the altars of the Lord ! Woe unto thee for that Shibboleth, the oath of abjuration, which the Lord hath permitted to try thy pulse to see how it did beat towards him ! Alreadie hath thy Church, through its unvaliant, faint, cowardly, and, I am bold to say, ungodly spirit, suffered woful encroachments to be made on Christ's truths in this kingdom, and yet all under a biassing pretence of witt and policy—leaving not only hoofs in Egypt, but also many of the best of the flock of God's revealed injunctions. Art thou not discouraged and beaten back, O Church ! from thy duty, by the sounding of the shaking leaf of a parliament of the worms of the earth, that creep, peep, and cry, appearing out of their holes and dens in this time in Scotland's dark night, when only such creatures come abroad in their native shapes and colours. For if the sun did now as clearly shine on the land as at former times, they would not so appear. It is in the night-time that evil spirits and wild beasts seize on folk, and cry in the streets to fleg and flichter them ; and such as they find most feared and apprehensive they haunt most. And so, O Scotland ! is thy Church affeared and flichtered with the scriekings and worryings of an evil Parliament."

CHAPTER XI.

" Give us, for our abstractions, solid facts—
    For our disputes, plain pictures."—WORDSWORTH.

RELIGION never operates on the human mind without stamping upon it the more prominent traits of its own character, and very rarely without being impressed in turn (if I may so speak) with some of the peculiar traits of every individual mind on which it acts.   Like a chemical test applied to a heterogeneous mixture, it meets much that it must necessarily repel, and much also with which it combines.   And we find it not only accommodating itself in this way to the peculiarities of character, but, in many instances, even adding a new force to these.   In the mind of the deep thinker it is moulded into a sublime and living philosophy, and he cannot subsist under its influence without thinking more deeply, and becoming more truly a philosopher than before.   And what does it prove to the ignorant and credulous man ?—no superstition certainly, and yet so exceedingly akin in some of its effects to superstition, that we find it lending, as if it were really such, an indirect sanction to at least the less heterodox of his superstitious beliefs—the wonders of Revelation moulding themselves into a kind of corroborate evidences of whatever else of the supernatural he had previously credited.   It imparts a higher tone of ecstasy to the raptures of the enthusiast—furnishes the visionary with his brightest dreams —gives a more intense gaiety to the joyous—a deeper gravity to the serious—and not unfrequently a darker gloom to the melancholy.   Like the most fervent of its apostles, and in much the same degree, it becomes all things to all men. And if this hold true in individual character, its truth is not less strikingly apparent in the character of an age or country. The school-master of Cromarty and the elder of Nigg belonged each to a numerous class ; and the brief notices of these men which I have introduced into the foregoing chapter, may properly enough be regarded rather as national than individual in their character. No one who has perused the more popular writings of the Cove-nanters—Naphtali, the Hind let Loose, the Tracts of Peter Walker, and the older editions of the Scots Worthies—will fail of recognising, in my quotation from Morrison, the self-same spirit which animated the writers of these volumes, or be dis-posed to question the propriety of classing Donald Roy with our Cargills, Pedens, and Rutherfords.

The aspect of religion, when thus amalgamated with the en-thusiasm or the superstitions of a country, is always in accordance with the direction which that enthusiasm has taken, or with the peculiar cast of these superstitions, or with the nature of the circumstances and events by which they were modified or produced. These last (circumstances and events) must be re-garded as primary agents in this process of amalgamation ; and they may be divided into three distinct classes. In the first are great political convulsions, which agitate and unsettle the minds of whole communities. In every period of the history of every country, there exists a certain quantum of superstition and en-thusiasm—a certain proportion of the men who see visions and dream dreams ; but in times of quiet, when every visionary has his own distinct province assigned to him by some little chance peculiar to himself, the quantum is variously directed ; and thus, flowing in a thousand obscure channels, it can have no marked effect on the body of a people. But it is not thus in times of convulsion, when all men look one way, are interested in the same events, and direct their energies on the same objects. The quantum, swollen in bulk by the workings of these storms of the people, flows also in one channel; and thus, to a force increased in all its details, there is added a collective impetus. Hence its overmastering power. No one acquainted with English history need be reminded of those times of the Commonwealth, in which, through an atmosphere of lightning and tempest, whole hordes of visionaries gazed on what they deemed a still brighter, but more placid future, and called each one on his own little sect to rejoice in the prospect. And the first French Revolution was productive of similar effects. I need not refer to the singular interest elicited in our own country among the humbler people by the wild predictions of Brothers, or to the many soberer dreamers who were led, by the general excitement and portentous events of the period, to interpret amiss a surer word of prophecy. No one intimate enough with human nature to recognise its impulses and passions in their various disguises of belief and opinion, can be ignorant that there is a superstition of scepticism as surely as of credulity, or fail of identifying the wild infidelity of the French Commonwealth with the almost equally wild fanaticism of the English one.

The second class of circumstances includes famine, pestilence, and persecution; and, in particular, the effects of the last are strikingly singular. In the others, the mind, unsettled by suffering and terror, ceases to deduce the evils which are overwhelming it from the old fixed causes which govern the universe, and sends out imagination in quest of the new. Demons are abroad—death itself becomes a living spirit, voices of lamentation are heard in the air—spectres seen on the earth. In such circumstances, however, the very prostration of the mind sets limits to its delusions; the inventive powers are rather passive than active; but it is not so in seasons of persecution, when our fellow-creature—man—is the visible cause of the evils to which we are subjected, and the combative principle, maddened by oppression, is roused into an almost preternatural activity. Hence, and from the energy of excitement and the melancholy of suffering, the persecuted enthusiast becomes more enthusiastical, and the superstitions of the credulous assume a darker aspect. Even the true religion seems impressed with a new character. As Solomon has well expressed it— "wise men become mad;" and, seen through the medium of their disturbed imaginations, the common traits of character and of circumstance are exaggerated into the supernatural. The oppression which is grinding them to the earth, assumes for their destruction a visible form, and a miraculous control over the laws of nature. The evil spirit is no longer formidable merely from his power of biassing the will, and obliterating the better feelings of the heart; for, assuming a still more terrible character, and a real and tangible shape, he assails them in their hiding-places—the cavern and the desert. Even their human enemies, charmed against the stroke of sword and bullet, are rendered invulnerable by the same power. And there are miracles wrought also in their behalf. Their places of hiding are discovered by the persecutor; but a sudden blindness falls on him, and he cannot avail himself of the discovery. They are pursued on the hill-side by a troop of horse; but, when exhausted in the flight, a thick cloud is dropped over them, and they escape. The enemy is removed by judgments sudden and fearful. Their curse becomes terribly potent. There is a power given them of reading the inmost thoughts of the heart; they have visions of the distant—revelations of the future. These, however, are but the traits of a comparatively sober enthusiasm, which persecution cannot altogether goad into madness. In some of the wilder instances we see even the moral principle unsettled. The Huguenots of Languedoc, when driven to their mountains by the intolerance of Louis XIV., were headed by two leaders, a young man whom they named David, and a prophetess whom they termed the Great Mary. These leaders exercised over them a despotic authority; and, when any of them proved refractory, they were condemned by the prophetess without form of trial, and put to death by their infatuated companions. A few years after the battle of Bothwell Bridge, a small party of Covenanters, of whom the greater part, says Walker, " were serious and very gracious souls, though they then stumbled and fell," assembled in a moor near Stirling, and burned their Bibles. Is it not probable that the terrible feuds which convulsed Jerusalem during the siege of Titus, aggravating in a tenfold degree the horrors of war and famine, were in part the effects of a similar frenzy?

The third class of circumstances is of a quieter, but not less influential character. When a false religion gives place in any country to the true, there is commonly a mass of what may be termed neutral superstitions which survive the change. Thor and Woden are dethroned and forgotten, but the witch, the fairy, and the seer, the ghost of the departed, and the wraith of the dying, the spirits of the moor and the forest, of the sea and the river, remain as potent as before. The great national colossuses of heathenism are prostrated before the genius of Christianity; but the little idols of the household can be vanquished by only philosophy and the arts. For religion, as has been already remarked, instead of militating against the minor superstitions, lends them, in at least the darker ages, the support of what seems a corroborative evidence. And as, from natural causes, they must still be receiving fresh accessions of strength in every country in which they have taken root, and which remains unvisited by the arts, the testimony of the heathen fathers regarding them is confirmed by what is deemed the experience of the Christian children. The visions of the seer are as distinct as ever, the witch as malignant, the spectre as terrible. Enthusiasm and superstition go hand in hand together as before, and under the supposed sanction of a surer creed. The one works miracles, the other inspires a belief in them; the one predicts, the other traces the prediction to its

L

fulfilment; the one calls up the spirits of the dead, the other sees them appear, even when uncalled.

From a peculiar circumstance in the past state of this country, its traditional history presents us both with the appearance assumed by superstition when thus connected with religion, and the very similar aspect which it bears when left to itself. The country had its two distinct tribes of people, believers in nearly the same superstitions, but as unlike as can well be imagined in their degree of religious feeling. No pagan of the past ages differed more in this respect from the Christians of the present, than the clansmen of the Highland host did from the poor Covenanters, on whom they were turned loose by the Archbishop of St. Andrews. And yet neither Peden nor Cargill, nor any of the other prophets of the Covenant, were favoured with clearer revelations of the future than some of the Highland seers. What was deemed prophecy in the one class, was reckoned indeed merely the second-sight in the other; but there seems to be little danger of error in referring what are so evidently the same effects to the same causes. Donald Roy's vision of the foundering boat, and of the woman perishing in the snow, is quite in character with the visions of the seers. Peden was forty miles from Bothwell Bridge on the day of the battle; but he saw his friends " fleeing and falling before the enemy, with the hanging and hashing, and the blood running like water." " Oh the monzies! the monzies!" he exclaimed on another occasion, when foretelling a bloody invasion of the French which was to depopulate the country, " See how they run! see how they run! they are at our firesides, slaying men, women, and children." " Be not afraid," said Bruce of Anwoth, in a sermon preached on the day the battle of Killiecrankie was stricken; " be not afraid, I see the enemy scattered, and Claverhouse no longer a terror to God's people. This day I see him killed—lying a corpse!" But there is no lack of such instances, nor of the stories of second-sight with which they may be so clearly identified. The Tracts of Peter Walker, and the Lives of the Scots Worthies, abound with the former; some very striking specimens of the latter may be found in Pepys' Correspondence with Lord Rea.

Kenneth Ore, a Highlander of Ross-shire, who lived some time in the seventeenth century, may be regarded as the Peden of the class whom I have described as superstitious without religion. It is said, that when serving as a field labourer with a wealthy clansman, who resided somewhere near Brahan Castle, he made himself so formidable to the clansman's wife by his shrewd sarcastic humour, that she resolved on destroying him by poison. With this design, she mixed a preparation of noxious herbs with his food, when he was one day employed in digging turf in a solitary morass, and brought it to him in a pitcher. She found him lying asleep on one of those conical fairy hillocks which abound in some parts of the Highlands; and her courage failing her, instead of awakening him, she set down the pitcher by his side, and returned home. He awoke shortly after, and, seeing the food, would have begun his repast, but feeling something press coldly against his heart, he opened his waistcoat, and found a beautiful smooth stone, resembling a pearl, but much larger, which had apparently been dropped into his breast while he slept. He gazed at it in admiration, and became conscious as he gazed that a strange faculty of seeing the future as distinctly as the present, and men's real designs and motives as clearly as their actions, was miraculously imparted to him. And it was well for him that he should have become so knowing at such a crisis; for the first secret he became acquainted with was that of the treachery practised against him by his mistress. But he derived little advantage from the faculty ever after, for he led, it is said till extreme old age, an unsettled, unhappy kind of life—wandering from place to place, a prophet only of evil, or of little trifling events, fitted to attract notice when they occurred, merely from the circumstance of their having been foretold.

There was a time of evil, he said, coming over the Highlands, when all things would appear fair and promising, and yet be both bad in themselves, and the beginnings of what would prove worse. A road would be opened among the hills from sea to sea, and a bridge built over every stream; but the people would be degenerating as their country was growing better; there would be ministers among them without grace, and maidens without shame; and the clans would have become so heartless, that they would flee out of their country before an army of sheep. Moss and muir would be converted into corn-land, and yet hunger press as sorely on the poor as ever. Darker days would follow, for there would arise a terrible persecution, during which a ford in the river Oickel, at the head of the Dornoch Firth, would render a passage over the dead bodies of men, attired in the plaid and bonnet; and on the hill of *Finnbheim*, in Sutherlandshire, a raven would drink her full of human blood three times a day for three successive days. The greater part of this prophecy belongs to the future; but almost all his minor ones are said to have met their fulfilment. He predicted, it is affirmed, that there would be dram-shops at the end of almost every furrow; that a cow would calve on the top of the old tower of Fairburn; that a fox would rear a litter of cubs on the hearth-stone of Castle Downie; that another animal of the same species, but white as snow, would be killed on the western coast of Sutherlandshire; that a wild deer would be taken alive at Fortrose Point; that a rivulet in Western Ross would be dried up in winter; and that there would be a deaf Seaforth. But it would be much easier to prove that these events have really taken place than that they have been foretold. Some of his other prophecies are nearly as equivocal, it has been remarked, as the responses of the old oracles, and true merely in the letter, or in some hidden meaning capable of being elicited by only the events which they anticipated. He predicted, it is said, that the ancient Chanonry of Ross, which is still standing, would fall "full of Mackenzies;" and as the floor of the building has been used, for time immemorial, as a burying-place by several powerful families of this name, it is supposed that the prophecy cannot fail, in this way, of meeting its accomplishment. He predicted, too, that a huge natural arch near the Storhead of Assynt would be thrown down, and with so terrible a crash that the cattle of Led-more, a proprietor who lived twenty miles inland, would break from their fastenings at the noise. It so happened, however, says the story, that some of Led-more's cattle, which were grazing on the lands of another proprietor, were housed within a few hundred yards of the arch when it fell. The prophet, shortly before his death, is said to have flung the white stone into a lake near Brahan, uttering as his last prediction, that it would be found many years after, when all his prophecies would be fulfilled, by a lame humpbacked mendicant.

There is a superstitious belief which, in the extent to which it has been received, ranks next in place to that enthusiasm which inspired the visionary and the prophet; and it was alike common in the past age to the Highlander and the Presbyterian. I allude to the belief that evil spirits have a power of assuming visible forms, in which to tempt and affright the good, and sometimes destroy the bad—a belief as old, at least, as the days of St. Dunstan, perhaps much older. For it seems probable that Satan is merely a successor in the class of stories which illustrates this belief to the infernal deities; indeed, in some of our more ancient Scottish traditions, nearly the old designation of one of these is retained. The victims of Flowden were summoned at the Cross of Edinburgh in the name of Platcock, *i.e.*, Pluto. There is but one story of this class which I at present remember in the writings of Walker—that of Peden in the cave of Galloway; the author of Waverley,

however, in referring to the story, attests the prevalence of the
belief.   The autobiographies of Methodists of the last century
abound with such; they form, too, in this part of the country
(for the story of Donald Roy and the dog is but one of a
thousand) the most numerous class of our traditions.   Out of
this multitudinous class I shall select, by way of specimen, two
stories which belong to the low country party, and two others
peculiar to the Highlands.

Not much more than thirty years ago, a Cromarty fisherman
of staid, serious character, who had been visiting a friend in the
upper part of the parish, was returning home after nightfall by
the Inverness road.   The night was still and calm, and a thick
mantle of dull yellowish clouds, which descended on every side
from the centre to the horizon, so obscured the light of the moon,.
though at full, that beyond the hedges which bounded the road
all objects seemed blended together without colour or outline.
The fisherman was pacing along in one of his happiest moods;
his mind occupied by serious thoughts, tempered by the feelings
of a genial devotion, when the stillness was suddenly broken by
a combination of the most discordant sounds he had ever heard.
At first he supposed that a pack of hounds had opened in full
cry in the field beside him; and then, for the sounds sunk as
suddenly as they had risen, that they were ranging the moors
on the opposite side of the hill.   Anon there was a fresh burst,
as if the whole pack were baying at him through the hedge.
He thrust his hand into his pocket, and drew out a handful of
crumbs, the residue of his last sea stock; but as he held them
out to the supposed dogs, instead of open throats and glaring
eyes, he saw only the appearance of a man, and the sounds
ceased.   "Ah!" thought he, "here is the keeper of the pack;
—I am safe."   He resumed his walk homewards, the figure
keeping pace with him as he went, until, reaching a gap in the
hedge, he saw it turning towards the road.   He paused to await
its coming up; but what was his astonishment and horror to

see it growing taller and taller as it neared the gap, and then, dropping on all fours, assume the form of a horse. He hurried onwards; the horse hurried too. He stood still; the horse likewise stood. He walked at his ordinary pace; the horse walked also, taking step for step with him, without either outstripping him or falling behind. It seemed an ugly misshapen animal, bristling all over with black shaggy hair, and lame of a foot. It accompanied him until he reached the gate of a burying-ground, which lies about two hundred yards outside the town; where he was blinded for a moment by what seemed an intensely bright flash of lightning; and, on recovering his sight, he found that he was alone. There is a much older but very similar story told of a man of Ferindonald, who, when on a night journey, is said to have encountered the evil one in five different shapes, and to have lost his senses through fright a few hours *after;* but this story, unlike the one related, could be rationally enough accounted for, by supposing the man to have lost his senses a few hours *before*.

The parishes of Cullicuden and Kiltearn are situated on opposite sides of the bay of Cromarty; and their respective manses, at the beginning of the last century, nearly fronted each other; the waters of the bay flowing between. Their clergymen, at that period, were much famed for the sanctity of their lives, and their diligence in the duties of their profession; and, from the similarity of their characters, they became strongly attached. They were both hard students; and, for at least two hours after midnight, the lights in their closet windows would be seen as if twinkling at each other across the Firth. When the light of the one was extinguished, the other regarded it as a signal to retire to rest. "But, how now," thought the minister of Kiltearn, as one night, in answer to the accustomed sign, he dropped the extinguisher on his candle, "how now are the sleeping watchmen to fulfil their duties? Would it not be better that, like sentinels, we should relieve each other by turns? There would then be at

all times within the bounds appointed us, open eyes and a praying heart." He imparted the thought to his friend ; and ever
after, as long as they lived, the one minister never retired to
bed until the casement of the other had given evidence that he
had risen to relieve him. A few years after this arrangement had
taken place, a parishioner of Cullicuden, who had been detained
by business till a late hour in some of the neighbouring parishes,
was walking homeward over the solitary Maolbuoy, when he
was joined by a stranger gentleman, who seemed journeying in
the same direction, and entered with him into conversation. He
found him to be one of the most intelligent, amusing men he had
ever met with. He seemed to know everything ; and though
he was evidently no friend to the Church, he did nothing worse
than laugh at it. The man of Cullicuden felt more than half
inclined to laugh at it too, and more than half convinced by the
ludicrous stories of the stranger, that its observances were merely
good jokes. In this mood they reached the extreme edge of the
Maolbuoy, where it borders on Cullicuden, when the stranger
made a full stop. "Our road runs this way," said the man.
"Ah !" replied the stranger, "but I cannot accompany you:
see you that ?" pointing, as he spoke, to a faint twinkling light on
the opposite side of the bay—"The watchman is stationed there,
and I dare not come a step further." It was only from this
confession that the Cullicuden man learned the true character of
his companion.

The merely superstitious stories of this class are generally of a
wilder and more imaginative cast than those which have sprung
up within the pale of the Church ; and the chief actor in them
is presented to us in a more imposing attitude, and in some instances bears rather a better character. Somewhat less than a
century ago (I am wretchedly uncertain in my dates), the ancient
castle of Ardvrock in Assynt was tenanted by a dowager lady
—a wicked old woman, who had a singular knack of setting the
people in her neighbourhood together by the ears. A gentle-

man who lived with his wife at a little distance from the castle, was lucky enough to escape for the first few years; but on the birth of a child his jealousy was awakened by some insinuations dropped by the old lady, and he taxed his wife with infidelity, and even threatened to destroy the infant. The poor woman in her distress wrote to two of her brothers, who resided in a distant part of the country; and in a few days after they both alighted at her gate. They remonstrated with her husband, but to no effect. "We have but one resource," said the younger brother, who had been a traveller, and had spent some years in Italy; "let us pass this evening in the manner we have passed so many happy ones before, and visit to-morrow the old lady of Ardvrock. I will confront her with perhaps as clever a person as herself; and whatever else may come of our visit, we shall at least arrive at the truth." On the morrow they accordingly set out for the castle—a grey, whinstone building, standing partly on a low moory promontory, and partly out of a narrow strip of lake which occupies a deep hollow between two hills. The lady received them with much seeming kindness, and replied to their inquiries on the point which mainly interested them with much apparent candour. "You can have no objection," said the younger brother to her, "that we put the matter to the proof, by calling in a mutual acquaintance?" She replied in the negative. The party were seated in the low-browed hall of the castle, a large, rude chamber, roofed and floored with stone, and furnished with a row of narrow, unglazed windows, which opened to the lake. The day was calm, and the sun riding overhead in a deep blue sky, unspecked by a cloud. The younger brother rose from his seat on the reply of the lady, and bending towards the floor, began to write upon it with his finger, and to mutter in a strange language; and as he wrote and muttered, the waters of the lake began to heave and swell, and a deep fleece of vapour, that rose from the surface like an exhalation, to spread over the face of the heavens.

At length a tall black figure, as indistinct as the shadow of a man by moonlight, was seen standing beside the wall. "Now," said the brother to the husband, "put your questions to *that*, but make haste;" and the latter, as bidden, inquired of the spectre, in a brief tremulous whisper, whether his wife had been faithful to him. The figure replied in the affirmative: as it spoke, a huge wave from the lake came dashing against the wall of the castle, breaking in at the hall windows; a tremendous storm of wind and hail burst upon the roof and the turrets, and the floor seemed to sink and rise beneath their feet like the deck of a ship in a tempest. "He will not away from us without his *bountith*," said the brother to the lady, "whom can you best spare?" She tottered to the door, and as she opened it, a little orphan girl, one of the household, came rushing into the hall, as if scared by the tempest. The lady pointed to the girl: "No, not the orphan!" exclaimed the appearance; "I dare not take her." Another immense wave from the lake came rushing in at the windows, half filling the apartment, and the whole building seemed toppling over. "Then take the old witch herself!" shouted out the elder brother, pointing to the lady—"take her."—"She is mine already," said the shadow, "but her term is hardly out yet; I take with me, however, one whom your sister will miss more." It disappeared as it spoke, without, as it seemed, accomplishing its threat; but the party, on their return home, found that the infant, whose birth had been rendered the occasion of so much disquiet, had died at the very time the spectre vanished. It is said, too, that for five years after the grain produced in Assynt was black and shrivelled, and that the herrings forsook the lochs. At the end of that period the castle of Ardvrock was consumed by fire, kindled no one knew how; and luckily, as it would seem, for the country, the wicked lady perished in the flames; for after her death things went on in their natural course—the corn ripened as before, and the herrings returned to the lochs.

The other Highland story of this class is, if possible, of a still wilder character.

The river Auldgrande, after pursuing a winding course through the mountainous parish of Kiltearn for about six miles, falls into the upper part of the Firth of Cromarty. For a considerable distance it runs through a precipitous gulf of great depth, and so near do the sides approach each other, that herd-boys have been known to climb across on the trees, which, jutting out on either edge, interweave their branches over the centre. In many places the river is wholly invisible : its voice, however, is ever lifted up in a wild, sepulchral wailing, that seems the lament of an imprisoned spirit. In one part there is a bridge of undressed logs thrown over the chasm. "And here," says the late Dr. Robertson in his statistical account of the parish, " the observer, if he can look down on the gulf below without any uneasy sensation, will be gratified by a view equally awful and astonishing. The wildness of the steep and rugged rocks—the gloomy horror of the cliffs and caverns, inaccessible to mortal tread, and where the genial rays of the sun never yet penetrated—the waterfalls, which are heard pouring down in different places of the precipice, with sounds various in proportion to their distances—the hoarse and hollow murmuring of the river, which runs at the depth of one hundred and thirty feet below the surface of the earth—the fine groves of pines, which majestically climb the sides of a beautiful eminence that rises immediately from the brink of the chasm—all these objects cannot be contemplated without exciting emotions of wonder and admiration in the mind of every beholder."

The house and lands of Balconie, a beautiful Highland property, lie within a few miles of the chasm. There is a tradition that, about two centuries ago, the proprietor was married to a lady of very retired habits; who, though little known beyond her narrow circle of acquaintance, was regarded within that circle with a feeling of mingled fear and respect. She was singularly reserved, and it was said spent more of her waking hours in solitary rambles on the banks of the Auldgrande, in places where no one else would choose to be alone, than in the house of Balconie. Of a sudden, however, she became more social, and seemed desirous to attach to herself, by acts of kindness and confidence, one of her own maids, a simple Highland girl; but there hung a mysterious wildness about her—a sort of atmosphere of dread and suspicion—which the change had not removed; and her new companion always felt oppressed, when left alone with her, by a strange sinking of the vital powers—a shrinking apparently of the very heart—as if she were in the presence of a creature of another world. And after spending with her, on one occasion, a whole day, in which she had been more than usually agitated by this feeling, and her ill-mated companion more than ordinarily silent and melancholy, she accompanied her at her bidding, as the evening was coming on, to the banks of the Auldgrande.

They reached the chasm just as the sun was sinking beneath the hill, and flinging his last gleam on the topmost boughs of the birches and hazels which then, as now, formed a screen over the opening. All beneath was dark as midnight. "Let us approach nearer the edge," said the lady, speaking for the first time since she had quitted the house. "Not nearer, ma'am," said the terrified girl; "the sun is almost set, and strange sights have been seen in the *gully* after nightfall." "Pshaw," said the lady, "how can you believe such stories? come, I will show you a path which leads to the water: it is one of the finest places in the world; I have seen it a thousand times, and must see it again to-night. Come," she continued, grasping her by the arm, "I desire it much, and so down we must go." "No, lady!" exclaimed the terrified girl, struggling to extricate herself, and not more startled by the proposal than by the almost fiendish expression of mingled anger and fear which now shaded the features of her mistress, "I shall swoon with terror

and fall over." "Nay, wretch, there is no escape!" replied the
lady, in a voice heightened almost to a scream, as, with a
strength that contrasted portentously with her delicate form,
she dragged her, despite of her exertions, towards the chasm.
"Suffer me, ma'am, to accompany you," said a strong masculine
voice from behind; "your surety, you may remember, must be a
willing one." A dark-looking man, in green, stood beside them;
and the lady, quitting her grasp with an expression of passive
despair, suffered the stranger to lead her towards the chasm.
She turned round on reaching the precipice, and, untying from
her belt a bunch of household keys, flung them up the bank
towards the girl; and then, taking what seemed to be a farewell look of the setting sun, for the whole had happened in so
brief a space that the sun's upper disk still peeped over the hill,
she disappeared with her companion behind the nearer edge of
the gulf. The keys struck, in falling, against a huge granitic
boulder, and sinking into it as if it were a mass of melted wax,
left an impression which is still pointed out to the curious visitor.
The girl stood rooted to the spot in utter amazement.

On returning home, and communicating her strange story,
the husband of the lady, accompanied by all the males of his
household, rushed out towards the chasm; and its perilous
edge became a scene of shouts, and cries, and the gleaming of
torches. But, though the search was prolonged for whole days
by an eager and still increasing party, it proved fruitless. *There*
lay the ponderous boulder impressed by the keys; immediately
beside it yawned the sheer descent of the chasm; a shrub, half
uprooted, hung dangling from the brink; there was a faint line
drawn along the green mould of the precipice a few yards lower
down; and that was just all. The river at this point is hidden by
a projecting crag, but the Highlanders could hear it fretting and
growling over the pointed rocks, like a wild beast in its den;
and as they listened and thought of the lady, the blood curdled
at their hearts. At length the search was relinquished, and

they returned to their homes to wonder, and surmise, and tax
their memories, though in vain, for a parallel instance. Months
and years glided away, and the mystery was at length assigned
its own little niche among the multitudinous events of the past.

About ten years after, a middle-aged Highlander, the servant
of a maiden lady who resided near the Auldgrande, was engaged
one day in fishing in the river, a little below where it issues
from the chasm. He was a shrewd fellow, brave as a lion and
kindly-natured withal, but not more than sufficiently honest ;
and his mistress, a stingy old woman, trusted him only when
she could not help it. He was more than usually successful
this day in his fishing ; and picking out some of the best of the
fish for his aged mother, who lived in the neighbourhood, he
hid them under a bush, and then set out for his mistress with
the rest. " Are you quite sure, Donald," inquired the old lady
as she turned over the contents of his basket, " that this is the
whole of your fishing ?—where have you hid the rest ?" " Not
one more, lady, could I find in the burn." " O Donald !" said
the lady. " No, lady," reiterated Donald, " devil a one !"
And then, when the lady's back was turned, off he went to the
bush to bring away the fish appropriated to his mother. But the
whole had disappeared ; and a faintly marked track, spangled
with scales, remained to show that they had been dragged
apparently by some animal along the grass in the direction of
the chasm.

The track went winding over grass and stone along the edge of
the stream, and struck off, as the banks contracted and became
more steep and precipitous, by a beaten path which ran along
the edge of the crags at nearly the level of the water, and which,
strangely enough, Donald had never seen before. He pursued
it, however, with the resolution of tracing the animal to its den.
The channel narrowed as he proceeded ; the stream which, as
he entered the chasm, was eddying beneath him in rings of
a mossy brown, became one milky strip of white, and, in the

language of the poet, " boiled, and wheeled, and foamed, and thundered through ;" the precipices on either hand, beetled in some places so high over his head as to shut out the sky, while in others, where they receded, he could barely catch a glimpse of it through a thick screen of leaves and bushes, whose boughs meeting midway, seemed twisted together like pieces of basket work. From the more than twilight gloom of the place, the track he pursued seemed almost lost, and he was quite on the eve of giving up the pursuit, when, turning an abrupt angle of the rock, he found the path terminate in an immense cavern. As he entered, two gigantic dogs, which had been sleeping one on each side of the opening, rose lazily from their beds, and yawning as they turned up their slow heavy eyes to his face, they laid themselves down again. A little further on there was a chair and table of iron apparently much corroded by the damps of the cavern. Donald's fish, and a large mass of leaven prepared for baking, lay on the table ; in the chair sat the lady of Balconie.

Their astonishment was mutual. " O Donald !" exclaimed the lady, "what brings you here?" "I come in quest of my tish," said Donald, " but, O lady ! what *keeps* you here ? Com away with me, and I will bring you home ; and you will be lady of Balconie yet." " No, no !" she replied, " that day is past ; I am fixed to this seat, and all the Highlands could not raise me from it." Donald looked hard at the iron chair ; its ponderous legs rose direct out of the solid rock as if growing out of it, and a thick iron chain red with rust, that lay under it, communicated at the one end to a strong ring, and was fastened round the other to one of the lady's ankles. " Besides," continued the lady, " look at these dogs.——Oh ! why have you come here ? The fish you have denied to your mistress in the name of my jailer, and his they have become ; but how are you yourself to escape !" Donald looked at the dogs. They had again risen from their beds, and were now eyeing him with a keen

vigilant expression, very unlike that with which they had regarded him on his entrance. He scratched his head. "'Deed,
mem," he said, "I dinna weel ken ;—I maun first durk the
twa tykes, I'm thinking." "No," said the lady, "there is but
one way ; be on the alert." She laid hold of the mass of
leaven which lay on the table, flung a piece to each of the dogs,
and waved her hand for Donald to quit the cave. Away he
sprang ; stood for a moment as he reached the path to bid farewell to the lady ; and after a long and dangerous scramble
among the precipices, for the way seemed narrower, and steeper,
and more slippery than when he had passed by it to the cave, he
emerged from the chasm just as the evening was beginning to
darken into night. And no one, since the adventure of Donald,
has seen aught of the lady of Balconie.

CHAPTER XII.

Fu' mony a schriek that waefu' nicht
  Raise fra the boisterous main,
An' vow'd was mony a bootless vow,
  An' praied war' praiers vaine.

An' sair-pyned widows moned forlorn
  For mony a wearie daye,
An' maidens, ance o' blithsome mood,
  Tined heart and dwyned away.—OLD BALLAD.

THE headland which skirts the northern entrance of the Firth
is of a bolder character than even the southern one—abrupt,
stern, and precipitous as that is.   It presents a loftier and more
unbroken wall of rock; and where it bounds on the Moray
Firth there is a savage magnificence in its cliffs and caves, and
in the wild solitude of its beach, which we find nowhere equalled
on the shores of the other.   It is more exposed, too, in the
time of tempest.   The waves often rise, during the storms of
winter, more than a hundred feet against its precipices, festoon-
ing them, even at that height, with wreaths of kelp and tangle;
and for miles within the bay, we may hear, at such seasons, the
savage uproar that maddens amid its cliffs and caverns coming
booming over the lashings of the nearer waves like the roar of
artillery.   There is a sublimity of desolation on its shores, the
effects of a conflict maintained for ages, and on a scale so
gigantic.   The isolated, spire-like crags that rise along its base,
are so drilled and bored by the incessant lashings of the surf,
and are ground down to shapes so fantastic, that they seem but
the wasted skeletons of their former selves; and we find almost
every natural fissure in the solid rock hollowed into an immense

M

cavern, whose very ceiling, though the head turns as we look up to it, owes evidently its comparative smoothness to the action of the waves.

One of the most remarkable of these recesses occupies what we may term the apex of a lofty promontory. The entrance, unlike that of most of the others, is narrow and rugged, though of great height, but it widens within into a shadowy chamber, perplexed, like the nave of a cathedral, by uncertain cross lights, that come glimmering into it through two lesser openings which perforate the opposite sides of the promontory. It is a strange ghostly-looking place; there is a sort of moonlight greenness in the twilight which forms its noon, and the denser shadows which rest along its sides; a blackness so profound that it mocks the eye, hangs over a lofty passage which leads from it, like a corridor, still deeper into the bowels of the hill; the light falls on a sprinkling of half-buried bones, the remains of animals that, in the depth of winter, have creeped into it for shelter and to die; and when the winds are up, and the hoarse roar of the waves comes reverberated from its inner recesses, it needs no over-active fancy to people its avenues with the shapes of beings long since departed from every gayer or softer scene, but which still rise uncalled to the imagination in those by-corners of nature which seem dedicated, like this cavern, to the wild, the desolate, and the solitary.

A few hundred yards from where the headland terminates towards the south, there is a little rocky bay, which has been known for ages to the seafaring men of the town as the *Cova-Green*. It is such a place as we are sometimes made acquainted with in the narratives of disastrous shipwrecks. First, there is a broad semicircular strip of beach, with a wilderness of insulated piles of rock in front; and so steep and continuous is the wall of precipices which rises behind, that, though we may see directly overhead the grassy slopes of the hill, with here and there a few straggling firs, no human foot ever gained the nearer edge. The bay of Cova-Green is a prison to which the sea presents the only outlet ; and the numerous caves which open along its sides, like the arches of an amphitheatre, seem but its darker cells. It is in truth a wild impressive place, full of beauty and terror, and with none of the squalidness of the mere dungeon about it. There is a puny littleness in our brick and lime receptacles of misery and languor, which speaks as audibly of the feebleness of man as of his crimes or his inhumanity ; but here all is great and magnificent—and there is much, too, that is pleasing. Many of the higher cliffs which rise beyond the influence of the spray, are tapestried with ivy ; we may see the heron watching on the ledges beside her bundle of withered twigs, or the blue hawk darting from her cell ; there is life on every side of us—life in even the wild tumbling of the waves, and in the stream of pure water which, rushing from the higher edge of the precipice in a long white cord, gradually untwists itself by the way, and spatters ceaselessly among the stones over the entrance of one of the caves. Nor does the scene want its old story to strengthen its hold on the imagination.

Some time early in the reign of Queen Anne, a fishing yawl, after vainly labouring for hours to enter the Bay of Cromarty during a strong gale from the west, was forced at nightfall to relinquish the attempt, and take shelter in the Cova-Green. The crew consisted of but two persons—an old fisherman and his son. Both had been thoroughly drenched by the spray, and chilled by the piercing wind, which, accompanied by thick snow-showers, had blown all day through the opening from off the snowy top of Ben-Wevis ; and it was with no ordinary satisfaction that, as they opened the bay on their last tack, they saw the red gleam of a fire flickering from one of the caves, and a boat drawn up on the beach.

" It must be some of the Tarbat fishermen," said the old man, " wind-bound like ourselves ; but wiser than us, in having made provision for it. I'll feel willing enough to share their fare with them for the night."

"But see," remarked the younger, "that there be no unwillingness on the other side. I am much mistaken if that be not the boat of my cousins the Macinlas! Hap what may, however, the night is getting worse, and we have no choice of quarters. Hard up your helm, father, or we shall barely clear the Skerries;—there now—every nail an anchor!"

He leaped ashore, carrying with him the small hawser attached to the stem, known technically as the *swing*, which he wound securely round a jutting crag, and then stood for a few seconds until the old man, who moved but heavily along the thwarts, had come up to him. All was comparatively calm under the lee of the precipices; but the wind was roaring fearfully in the woods above, and whistling amid the furze and ivy of the higher cliffs; and the two boatmen as they entered the cave could see the flakes of a thick snow-shower, that had just begun to descend, circling round and round in the eddy.

The place was occupied by three men—two of them young, and rather ordinary-looking persons; the third, a greyheaded old man, apparently of great muscular strength, though long past his prime, and of a peculiarly sinister cast of countenance. A keg of spirits, which was placed before them, served as a table. There were little drinking-measures of tin on it; and the mask-like, stolid expressions of the two younger men, showed that they had been indulging freely. The elder was comparatively sober. A fire, composed mostly of fragments of wreck and drift wood, threw up its broad cheerful flame towards the roof; but so spacious was the cavern, that, except where here and there a whiter mass of stalactites, or bolder projection of cliff, stood out from the darkness, the light seemed lost in it. A dense body of smoke, which stretched its blue level surface from side to side, and concealed the roof, went rolling onwards like an inverted river. On the entrance of the fishermen, the three boatmen within started to their feet, and one of the younger, laying hold of the little cask, pitched it hurriedly into a dark corner of the cave.

"Ay, ye do well to hide it, Gibbie!" exclaimed the savage-looking old man in a bitter ironical tone, as he recognised the intruders; "here are your good friends, William and Ernest Beth come to see if they cannot rob and maltreat us a second time. Well! they had better try."

There could not be a more luckless meeting. For years before had the crew of the little fishing-yawl been regarded with the bitterest hatred by the temporary inmates of the cave; nor was old Eachen of Tarbat one of the class whose resentments may be safely slighted. He had passed the first thirty years of his life among the buccaneers of South America; he had been engaged in its latter seasons among the smugglers, who even at this early period infested the eastern coasts of Scotland; and Eachen, of all his associates, whether smugglers or buccaneers, had ever been deemed one of the fiercest and most unscrupulous. On his return from America the country was engaged in one of its long wars with Holland, and William Beth, the elder fisherman, who had served in the English fleet, was lying in a Dutch prison at the time, and a report had gone abroad that he was dead. He had inherited some little property from his father in the neighbouring town—a house and a little field, which in his absence was held by an only sister; who, on the report of his death, was of course regarded as a village heiress, and whose affections, in that character, Eachen of Tarbat had succeeded in engaging. They were married, but the marriage had turned out singularly ill; Eachen was dissipated and selfish, and of a harsh, cruel temper; and it was the fate of his poor wife, after giving birth to two boys, the younger inmates of the cave, to perish in the middle of her days, a care-worn, heart-broken creature. Her brother William had returned from Hol-

land shortly before, and on her death claimed and recovered his
property from her husband; and from that hour Eachen of
Tarbat had regarded him with the bitterest malice. A second
cause of dislike, too, had but lately occurred. Ernest Beth,
William's only son, and one of his cousins, the younger son of
Eachen, had both fixed their affections on a lovely young girl,
the toast of a neighbouring parish; and Ernest, a handsome
and high-spirited young man, had proved the successful lover.
On returning with his mistress from a fair, only a few weeks
previous to this evening, he had been waylaid and grossly insulted by his two cousins; and the insult he might perhaps
have borne for the sake of what they seemed to forget—his
relationship to their mother; but there was another whom they
also insulted, and that he could not bear; and as they were
mean enough to take odds against him on the occasion, he had
beaten the two spiritless fellows that did so.

The old fisherman had heard the ominous remark of the
savage as if he heard it not. "We have not met for many
years, Eachen," he said—"not since the death of my poor sister,
when we parted such ill friends; but we are shortlived creatures
ourselves, surely our anger should be shortlived too. I have
come to crave from you a seat by your fire."

"It was no wish of mine, William Beth," said Eachen, "that
we should ever meet; but there is room enough for us all beside
the fire."

He resumed his seat; the two fishermen took their places
fronting him, and for some time neither party exchanged a
word.

"This is but a gousty lodging-place," at length remarked the
old fisherman, looking round him; "but I have seen worse,
and I wish the folk at hame kent we were half sae snug."

The remark seemed directed to no one in particular, and there
was no reply. In a second attempt he addressed himself to the
old man.

"It has vexed me, Eachen," he said, "that our young folk, were it but for my sister's sake, should not be on mair friendly terms; an' we ourselves too—why suld we be enemies?" The old man, without deigning a reply, knit his grey shaggy brows, and looked doggedly at the fire.

"Nay, now," continued the fisherman, "we are getting auld men now, Eachen, an' wald better bury our hard thoughts o' ane anither afore we come to be buried oursels."

Eachen fixed his keen scrutinizing glance on the speaker, there was a tremulous motion of the upper lip as he withdrew it, and a setting of the teeth; but the tone of his reply savoured more of sullen indifference than of passion.

"William Beth," he said, "ye have tricked my boys out o' the bit property that suld have come to them by their mither; it's no so long since they barely escaped being murdered by your son. What more want you? But, mayhap, ye think it better that the time should be passed in making boss professions of good-will than employed in clearing off an old score."

"Ay," hiccuped out the elder of the two sons, "the houses might come my way then; an', if Helen Henry were to lose her ae joe, the tither might hae the better chance."

"Look ye, uncle," exclaimed the younger fisherman, "your threat might be spared. Our little property was my grandfather's and of right descended to his only son. As for the affair at the tryst, I dare either of my cousins to say the quarrel was of my seeking. I have no wish to raise my hand against the sons or the husband of my aunt; but if forced to it, you will find that neither my father nor myself are wholly at your mercy." He rose to his feet as he spoke.

"Whisht, Ernest," said the old fisherman calmly, "sit down; your uncle maun hae ither thoughts. It is now twenty years, Eachen," he continued, "since I was called to my sister's death-bed. You cannot forget what passed there. There had been grief and hunger beside that bed. I'll no say you were willingly unkind. Few folk are that but when they have some purpose to serve by it, and you could have none; but you laid no restraint on a harsh temper, and none on a craving habit, that forgets everything but itself, and sae my poor sister perished in the middle of her days, a wasted heart-broken thing. I have nae wish to hurt you. We baith passed our youth in a bad school, and I owre aften feel I havena unlearned a' my own lessons to wonder that you suldna have unlearned a' yours. But we're getting old men, Eachen, why suld we die fools? and fools we maun die, if we die enemies."

"You are likely in the right," said the stern old man. "But ye were aye a luckier man than me, William—luckier for this warld, I'm sure—maybe luckier for the next. I had aye to seek, and that without finding, the good that came in your gate o' itsel'. Now that age is coming upon us, ye get a snug rental frae the little house and croft, and I have naething; and ye have character and credit, but wha wald trust me, or wha cares for me? Ye hae been made *an elder* o' the kirk, too, I hear, and I am still a reprobate; but we were a' born to be just what we are, an' sae we maun submit. And your son, too, shares in your luck: he has heart and hand, and my whelps have neither; and the girl Henry, that scouts that sot there, likes him; but what wonder of that!—William Beth, we needna quarrel; but for peace' sake let me alone—we have naething in common, and friends we canna and winna be."

"We had better," whispered Ernest to his father, "not sleep in the cave to-night."

But why record the quarrels of this unfortunate evening? An hour or two passed away in disagreeable bickerings, during which the patience of even the old fisherman was well-nigh worn out, and that of Ernest had failed him altogether. And at length they both quitted the cave, boisterous as the night was, and it was now stormier than ever; and heaving off their boat till she rode at the full length of her swing from the shore, they sheltered themselves under the sail. The Macinlas returned next evening to Tarbat; but though the wind moderated during the day, the yawl of William Beth did not enter the Bay of Cromarty. Weeks passed away during which the clergyman of the place corresponded regarding the missing fishermen with all the lower ports of the Firth, but they had disappeared as it seemed for ever; and Eachen Macinla, in the name of his sons, laid claim to their property, and entered a second time into possession of the house and the little field.

Where the northern headland of the Firth sinks into the low sandy tract that nearly fronts the town of Cromarty, there is a narrow grassy terrace raised but a few yards over the level of the beach. It is sheltered behind by a steep undulating bank —for though the rock here and there juts out, it is too rich in vegetation to be termed a precipice. To the east, the coast retires into a semicircular rocky recess, terminating seawards in a lofty, dark-browed precipice, and bristling throughout all its extent with a countless multitude of crags that at every heave of the wave break the surface into a thousand eddies. Towards the west, there is a broken and somewhat dreary waste of sand. The terrace itself, however, is a sweet little spot, with its grassy slopes that recline towards the sun, partially covered with thickets of wild-rose and honeysuckle, and studded in their season with violets and daisies, and the delicate rock geranium. Towards its eastern extremity, with the bank rising immediately behind, and an open space in front which seemed to have been cultivated at one time as a garden, there stood a picturesque little cottage. It was that of the widow of William Beth. Five years had now elapsed since the disappearance of her son and husband, and the cottage bore the marks of neglect and decay. The door and window, bleached white by the sea winds, shook loosely to every breeze; clusters of chickweed luxuriated in the hollows of the thatch, or mantled over the eaves; and a honeysuckle, that had twisted itself round the chimney, lay withering in a tangled mass at the foot of the wall. But the progress of decay was more marked in the widow than in her dwelling. She had had to contend with grief and penury;—a grief not the less undermining in its effects from the circumstance of its being sometimes suspended by hope—a penury so extreme, that every succeeding day seemed as if won by some providential interference from absolute want. And she was now, to all appearance, fast sinking in the struggle. The autumn was well-nigh over; she had been weak and ailing for months before; and she had now become so feeble as to be confined for days together to her bed. But happily, the poor solitary woman had at least one attached friend in the daughter of a farmer of the parish, a young and beautiful girl, who, though naturally of no melancholy temperament, seemed to derive almost all she enjoyed of pleasure from the society of the widow.

Autumn we have said was near its close. The weather had given indications of an early and severe winter; and the widow, whose worn-out and delicate frame was affected by every change of atmosphere, had for a few days been more than usually indisposed. It was now long past noon, and she had but just risen. The apartment, however, bore witness that her young friend had paid her the accustomed morning visit; the fire was blazing on a clean, comfortable-looking hearth, and every little piece of furniture was arranged with the most scrupulous care. Her devotions were hardly over when the well-known tap and light foot of her friend Helen Henry were again heard at the door.

" To-morrow, mother," said Helen, as she took her seat beside her, " is Ernest's birthday. Is it not strange that, when our minds make pictures of the dead, it is always as they looked best, and kindliest, and most lifelike? I have been seeing Ernest all day long, as when I saw him on his *last* birthday."

"Ah, my bairn!" said the widow, grasping her young friend by the hand, "I see that, sae lang as we continue to meet, our thoughts will be aye running the ae way. I had a strange dream last night, an' must tell it you. You see yon rock to the east, in the middle o' the little bay, that now rises through the back draught o' the sea, like the hulk o' a ship, an' is now buried in a mountain o' foam. I dreamed I was sitting on that rock, in what seemed a bonny simmer's morning. The sun was glancin' on the water, an' I could see the white sand far down at the bottom, wi' the reflection o' the little waves aboon running over it in long curls o' gowd. But there was no way of leaving the rock, for the deep waters were round an' round me; an' I saw the tide covering ae wee bittie after anither, till at last the whole was covered. An' yet I had but little fear, for I remembered that baith Ernest an' William were in the sea afore me; an' I had the feeling that I could hae rest nowhere but wi' them. The water at last closed o'er me, an' I sank frae aff the rock to the sand at the bottom. But death seemed to have no power given him to hurt me, an' I walked as light as ever I had done on a gowany brae, through the green depths o' the sea. I saw the silvery glitter o' the trout an' the salmon shining to the sun, far, far aboon me, like white pigeons i' the lift; and around me there were crimson star-fish, an' sea-flowers, and long trailing plants that waved in the tide like streamers; an' at length I came to a steep rock wi' a little cave like a tomb in it. Here, I said, is the end o' my journey—William is here, an' Ernest. An' as I looked into the cave, I saw there were bones in it, an' I prepared to take my place beside them. But, as I stooped to enter, some one called on me, an', on looking up, there was William. 'Lillias,' he said, 'it is not night yet, nor is that your bed; you are to sleep, not with me, but, lang after this, with Ernest; haste you home, for he is waiting for you.' 'Oh, take me to him!' I said; an' then all at once I found mysel' on the shore dizzied and blinded wi' the bright sunshine; for at the cave there was a darkness like that o' a simmer's gloamin; an' when I looked up for William, it was Ernest that stood before me, lifelike and handsome as ever; an' you were beside him."

The day had been gloomy and lowering, and though there was little wind, a tremendous sea, that as the evening advanced rose higher and higher against the neighbouring precipice, had been rolling ashore since morning. The wind now began to blow in long hollow gusts among the cliffs, and the rain to patter against the widow's casement.

" It will be a storm from the sea," she said; " the scarts an' gulls hae been flying landward sin' daybreak, an' I hae never seen the ground-swell come home heavier against the rocks. Waes me for the puir sailors that maun bide under it a' !"

" In the long stormy nights," said her companion, " I cannot sleep for thinking of them; though I have no one to bind me to them now. Only look how the sea rages among the rocks as if it were a thing of life—that last wave rose to the crane's nest. And look, yonder is a boat rounding the rock with only one man in it. It dances on the surf as if it were a cork, and the little bit sail, so black and wet, seems scarcely bigger than a napkin. Is it not bearing in for the boat-haven below ?"

" My poor old eyes," replied the widow, " are growing dim, an' surely no wonder; but yet I think I should ken that boat-man. Is it no Eachen Macinla o' Tarbat ?"

" Hard-hearted old man !" exclaimed the maiden, " what can be taking him here ? Look how his skiff shoots in like an arrow on the long roll o' the surf !—and now she is high on the beach. How cruel it was of him to rob you of your little property in the very first of your grief ! But see, he is so worn out that he can hardly walk over the rough stones. Ah me, he is down !—wretched old man, I must run to his assistance ; but no, he has risen again. See, he is coming straight to the house ; and now he is at the door." In a moment after, Eachen entered the cottage.

" I am perishing, Lillias," he said, " with cold and hunger, an' can gang nae farther—surely ye'll no shut your door on me in a night like this ?"

The poor widow had been taught in a far different school. She relinquished to the worn-out boatman her seat by the fire, now hurriedly heaped with fresh fuel, and hastened to set before him the simple viands which her cottage afforded.

As the night darkened, the storm increased. The wind roared among the rocks like the rattling of a thousand carriages over a paved street ; and there were times when, after a sudden pause, the blast struck the cottage as if it were a huge missile flung against it, and pressed on its roof and walls till the very floor rocked, and the rafters strained and quivered like the beams of a stranded vessel. There was a ceaseless patter of mingled rain and snow—now lower, now louder ; and the fearful thunderings of the waves as they raged among the pointed crags, were mingled with the hoarse roll of the stones along the beach. The old man sat beside the fire fronting the widow and her companion, with his head reclined nearly as low as his knee, and his hands covering his face. There was no attempt at conversation. He seemed to shudder every time the blast yelled along the roof, and as a fiercer gust burst open the door, there was a half-muttered ejaculation.

" Heaven itsel' hae mercy on them ! for what can man do in a night like this ?"

" It is black as pitch !" exclaimed the maiden, who had risen to draw the bolt, " and the drift flees so thick that it feels to the hand like a solid snow-wreath. And, oh, how it lightens !"

" Heaven itsel' hae mercy on them !" again ejaculated the old man. " My two boys," said he, addressing the widow, " are at the far Firth ; an' how can an open boat live in a night like this !"

There seemed something magical in the communication—something that awakened all the sympathies of the poor bereaved woman ; and she felt she could forgive him every unkindness.

"Waes me !" she exclaimed, " it was in such a night as this, an' scarcely sae wild, that my Ernest perished."

The old man groaned and wrung his hands.

In one of the pauses of the hurricane there was a gun heard from the sea, and shortly after a second. "Some puir vessel in distress," said the widow, " but, alas ! where can succour come frae in sae terrible a night ? There is help only in Ane ! Waes me ! would we no better light up a blaze on the floor, an', dearest Helen, draw off the cover frae the window ? My puir Ernest has told me that my light has aften showed him his bearings frae the deadly bed o' Dunskaith. That last gun," for a third was now heard booming over the mingled roar of the sea and the wind, "cam frae the very rock edge. Waes me ! maun they perish, an' sae near ?" Helen hastily lighted a bundle of mire-fir, that threw up its red sputtering blaze half-way to the roof, and dropping the covering, continued to wave it opposite the window. Guns were still heard at measured intervals, but apparently from a safer offing ; and the last, as it sounded faintly against the wind, came evidently from the interior of the bay.

"She has escaped," said the old man ; " it's a feeble hand that canna do good when the heart is willing ;—but what has mine been doing a' life lang ?" He looked at the widow and shuddered.

Towards morning the wind fell, and the moon in her last quarter rose red and glaring out of the Firth, lighting the melancholy roll of the waves, and the broad white belt of surf that skirted the shore. The old fisherman left the cottage, and sauntered along the beach. It was heaped with huge wreaths of kelp and tangle, uprooted by the storm, and in the hollow of the rocky bay lay the scattered fragments of a boat. Eachen stooped to pick up a piece of the wreck, in the fearful expectation of finding some known mark by which to recognise it; when the light fell full on the swollen face of a corpse, that seemed staring at him from out a wreath of sea-weed. It was that of his eldest son; and the body of the younger, fearfully gashed and mangled by the rocks, lay a few yards further to the east.

The morning was as pleasant as the night had been boisterous; and, except that the distant hills were covered with snow, and that a heavy swell continued to roll in from the sea, there remained scarce any trace of the recent tempest. Every hollow of the neighbouring hill had its little runnel, formed by the rains of the previous night, that now splashed and glistened to the sun. The bushes round the cottage were well-nigh divested of their leaves; but their red berries—hips and haws, and the juicy fruit of the honeysuckle—gleamed cheerfully to the light, and a warm steam of vapour, like that of a May morning, rose from the roof and the little mossy platform in front. But the scene seemed to have something more than merely its beauty to recommend it to a young man, drawn apparently to the spot, with many others, by the fate of the two unfortunate fishermen, and who now stood gazing on the rocks, and the hill, and the cottage, as a lover on the features of his mistress. The bodies had been carried to an old storehouse, which may still be seen, a short mile to the west; and the crowds that, during the early part of the morning, had been perambulating the beach, gazing at the wreck, and discussing the various probabilities of the accident, had gradually dispersed. But this solitary individual, whom no one knew, remained behind. He was a tall and somewhat swarthy, though very handsome man, of about seven-and-twenty, with a slight scar on the left cheek; and his dress, which was plain and neat, was distinguished from that of the common seaman by three narrow strips of gold lace on the upper part of one of the sleeves. He had twice stepped towards the cottage door, and twice drawn back, as if influenced by some unaccountable feeling—timidity, perhaps, or bashfulness ; and yet the bearing of the man gave little indication of either. But at length, as if he had gathered heart, he raised the latch and went in.

The widow, who had had many visitors that morning, seemed to be scarcely aware of his entrance ; she was sitting on a low seat beside the fire, her face covered with her hands, while the tremulous rocking motion of her body showed that she was still brooding over the distresses of the previous night. Her companion, who, without undressing, had thrown herself across the bed, was fast asleep. The stranger seated himself beside the fire, which seemed dying amid its ashes, and, turning sedulously from the light of the window, laid his hand gently on the widow's shoulder. She started and looked up.

"I have strange news for you," he said. "You have long mourned for your husband and your son; but though the old man has been dead for years, your son Ernest is still alive, and is now in the harbour of Cromarty. He is lieutenant of the vessel whose guns you must have heard during the night."

The poor woman seemed to have lost all power of reply.

"I am a friend of Ernest's," continued the stranger, "and have come to prepare you to meet with him. It is now five years since his father and he were blown off to sea by a strong gale from the land. They drove before it for four days, when they were picked up by an armed vessel cruising in the North Sea, and which soon after sailed for the coast of Spanish America. The poor old man sank under the fatigues he had undergone; though Ernest, better able from his youth to endure hardship, was little affected by them. He accompanied us on our Spanish expedition—indeed, he had no choice, for we touched at no British port after meeting with him; and through good fortune, and what his companions call merit, he has risen to be the second man aboard; and has now brought home with him gold enough from the Spaniards to make his old mother comfortable. He saw your light yester evening, and steered by it to the roadstead, blessing you all the way. Tell me, for he anxiously wished me to inquire of you, whether Helen Henry is yet unmarried?"

"It is Ernest—it is Ernest himself!" exclaimed the maiden, as she started from the widow's bed. In a moment after he had locked her in his arms.

It was ill before evening with old Eachen Macinla. The fatigues of the previous day, the grief and horror of the following night, had prostrated his energies bodily and mental; and he now lay tossing in a waste apartment of the storehouse in the delirium of fever. The bodies of his two sons occupied the floor below. He muttered unceasingly in his ravings, of William and Ernest Beth. They were standing beside him, he said, and every time he attempted to pray for his poor boys and himself, the stern old man laid his cold swollen hand on his lips.

"Why trouble me?" he exclaimed. "Why stare with your white dead eyes on me? Away, old man! the little black shells are sticking in your grey hairs; away to your place! Was it I who raised the wind or the sea?—was it I—was it I? Aha!—no—no—you were asleep—you were fast asleep, and could not see me cut the *swing;* and, besides, it was only a piece of rope. Keep away—touch me not! I am a freeman, and will plead for my life. Please your honour, I did not murder these two men; *I only cut the rope that fastened their boat to the land.* Ha! ha! ha! he has ordered them away, and they have both left me unskaithed." At this moment Ernest Beth entered the apartment, and approached the bed. The miserable old man raised himself on his elbow, and, regarding him with a horrid stare, shrieked out—" Here is Ernest Beth come for me a second time!" and, sinking back on the pillow, instantly expired.

N

CHAPTER XIII.

"The silent earth
Of what it holds shall speak, and every grave
Be as a volume, shut, yet capable
Of yielding its contents to ear and eye."—WORDSWORTH.

IN the woods to the east of Cromarty, occupying the summit
of a green insulated eminence, is the ancient burying-ground
and chapel of St. Regulus. Bounding the south there is a deep
narrow ravine, through which there runs a small tinkling
streamlet, whose voice, scarcely heard during the droughts of
summer, becomes hoarser and louder towards the close of
autumn. The sides of the eminence are covered with wood,
which, overtopping the summit, forms a wall of foliage that
encloses the burying-ground except on the east, where a little
opening affords a view of the northern Sutor over the tops of
trees which have not climbed high enough to complete the
fence. In this burying-ground the dead of a few of the more
ancient families of the town and parish are still interred; but
by far the greater part of it is occupied by nameless tenants,
whose descendants are unknown, and whose bones have
mouldered undisturbed for centuries. The surface is covered
by a short yellow moss, which is gradually encroaching on the
low flat stones of the dead, blotting out the unheeded memorials
which tell us that the inhabitants of this solitary spot were
once men, and that they are now dust—that they lived, and
that they died, and that they shall live again.

Nearly about the middle of the burying-ground there is a low
flat stone, over which time is silently drawing the green veil of

oblivion.   It bears date 1690, and testifies, in a rude inscription, that it covers the remains of Paul Feddes and his son
John, with those of their respective wives.   Concerning Paul,
tradition is silent; of John Feddes, his son, an interesting
anecdote is still preserved.   Some time early in the eighteenth
century, or rather perhaps about the close of the seventeenth,
he became enamoured of Jean Gallie, one of the wealthiest and
most beautiful young women of her day in this part of the
country.   The attachment was not mutual; for Jean's affections
were already fixed on a young man, who, both in fortune and
elegance of manners, was superior, beyond comparison, to the
tall red-haired boatman, whose chief merit lay in a kind brave
heart, a clear head, and a strong arm.   John, though by no
means a dissipated man, had been accustomed to regard money
as merely the price of independence, and he had sacrificed but
little to the graces.   His love-suit succeeded as might have been
expected; the advances he made were treated with contempt;
and the day was fixed on which his mistress was to be married
to his rival.   He became profoundly melancholy; and late on
the evening which preceded the marriage-day, he was seen traversing the woods which surrounded the old castle; frequently
stopping as he went, and, by wild and singular gestures, giving
evidence of an unsettled mind.   In the morning after he was
nowhere to be found.   His disappearance, with the frightful
conjectures to which it gave rise, threw a gloom over the spirits
of the town's-folk, and affected the gaiety of the marriage party;
it was remembered, even amid the festivities of the bridal, that
John Feddes had had a kind warm heart; and it was in no
enviable frame that the bride, as her maidens conducted her to
her chamber, caught a glimpse of several twinkling lights that
were moving beneath the brow of the distant Sutor.   She could
not ask the cause of an appearance so unusual; her fears too
surely suggested that her unfortunate lover had destroyed himself, and that his friends and kinsfolk kept that night a pain-

ful vigil in searching after the body. But the search was in vain, though every copse and cavern, and the base of every precipice within several miles of the town, were visited; and though, during the succeeding winter, every wreath of sea-weed which the night-storms had rolled upon the beach, was approached with a fearful yet solicitous feeling scarcely ever associated with bunches of sea-weed before. Years passed away, and, except by a few friends, the kind enterprising boatman was forgotten.

In the meantime it was discovered, both by herself and the neighbours, that Jean Gallie was unfortunate in her husband. He had, prior to his marriage, when one of the gayest and most dashing young fellows in the village, formed habits of idleness and intemperance which he could not, or would not shake off; and Jean had to learn that a very gallant lover may prove a very indifferent husband, and that a very fine fellow may care for no one but himself. He was selfish and careless in the last degree; and unfortunately, as his carelessness was of the active kind, he engaged in extensive business, to the details of which he paid no attention, but amused himself with wild vague speculations, which, joined to his habits of intemperance, stripped him in the course of a few years of all the property which had belonged to himself and his wife. In proportion as his means decreased he became more worthless, and more selfishly bent on the gratification of his appetites; and he had squandered almost his last shilling, when, after a violent fit of intemperance, he was seized by a fever, which in a few days terminated in death. And thus, five years after the disappearance of John Feddes, Jean Gallie found herself a poor widow, with scarce any means of subsistence, and without one pleasing thought connected with the memory of her husband.

A few days after the interment, a Cromarty vessel was lying at anchor, before sunrise, near the mouth of the Spey. The master, who had been one of Feddes's most intimate friends, was seated near the stern, employed in angling for cod and ling. Between his vessel and the shore, a boat appeared in the grey light of morning, stretching along the beach under a tight, well-trimmed sail. She had passed him nearly half a mile, when the helmsman slackened the sheet, which had been close-hauled, and suddenly changing the tack, bore away right before the wind. In a few minutes the boat dashed alongside. All the crew, except the helmsman, had been lying asleep upon the beams, and now started up alarmed by the shock. "How, skipper," said one of the men, rubbing his eyes, "how, in the name of wonder, have we gone so far out of our course? What brings us here?" "You come from Cromarty," said the skipper, directing his speech to the master, who, starting at the sound from his seat, flung himself half over the gunwale to catch a glimpse of the speaker. "John Feddes," he exclaimed, "by all that is miraculous!" "You come from Cromarty, do you not?" reiterated the skipper; "Ah, Willie Mouat! is that you?"

The friends were soon seated in the snug little cabin of the vessel; and John, apparently the less curious of the two, entered, at the others' request, into a detail of the particulars of his life for the five preceding years. "You know, Mouat," he said, "how I felt and what I suffered for the last six months I was in Cromarty. Early in that period I had formed the determination of quitting my native country for ever; but I was a weak foolish fellow, and so I continued to linger, like an unhappy ghost, week after week, and month after month, hoping against hope, until the night which preceded the wedding-day of Jean Gallie. Captain Robinson was then on the coast unloading a cargo of Hollands. I had made it my business to see him; and after some little conversation, for we were old acquaintance, I broached to him my intention of leaving Scotland. It is well, said he; for friendship sake I will give you a passage to Flushing, and, if it suits your inclination, a berth in the privateer I am now fitting out for cruising along the coast of Spanish America. I find the free trade doesn't suit me ; it has no scope. I considered his proposals, and liked them hugely. There was, indeed, some risk of being knocked on the head in the cruising affair, but that weighed little with me ; I really believe that, at the time, I would as lief have run to a blow as avoided one ;—so I closed with him, and the night and hour were fixed when he should land his boat for me in the *hope* of the Sutors. The evening came, and I felt impatient to be gone. You wonder how I could leave so many excellent friends without so much as bidding them farewell. I have since wondered at it myself ; but my mind was filled, at the time, with one engrossing object, and I could think of nothing else. Positively, I was mad. I remember passing Jean's house on that evening, and catching a glimpse of her through the window. She was so engaged in preparing a piece of dress, which I suppose was to be worn on the ensuing day, that she didn't observe me. I can't tell you how I felt—indeed, I do not know ; for I have scarcely any recollection of what I did or thought until a few hours after, when I found myself aboard Robinson's lugger, spanking down the Firth. It is now five years since, and, in that time, I have both given and received some hard blows, and have been both poor and rich. Little more than a month ago, I left Flushing for Banff, where I intend taking up my abode, and where I am now on the eve of purchasing a snug little property." " Nay," said Mouat, " you must come to Cromarty." " To Cromarty ! no, that will scarcely do." " But hear me, Feddes—Jean Gallie is a widow." There was a long pause. " Well, poor young thing," said John at length with a sigh, " I should feel sorry for that ; I trust she is in easy circumstances ?" " You shall hear."

The reader has already anticipated Mouat's narrative. During the recital of the first part of it, John, who had thrown himself on the back of his chair, continued rocking backwards and forwards with the best counterfeited indifference in the world.

It was evident that Jean Gallie was nothing to him. As the story proceeded, he drew himself up leisurely, and with firmness, until he sat bolt upright, and the motion ceased. Mouat described the selfishness of Jean's husband, and his disgusting intemperance. He spoke of the confusion of his affairs. He hinted at his cruelty to Jean when he squandered all. John could act no longer—he clenched his fist and sprang from his seat. "Sit down, man," said Mouat, "and hear me out—the fellow is dead."—"And the poor widow?" said John. "Is, I believe, nearly destitute;—you have heard of the box of broad-pieces left her by her father?—she has few of them now." "Well, if she hasn't, I have; that's all. When do you sail for Cromarty?" "To-morrow, my dear fellow, and you go along with me; do you not?"

Almost any one could supply the concluding part of my narrative. Soon after John had arrived at his native town, Jean Gallie became the wife of one who, in almost every point of character, was the reverse of her first husband; and she lived long and happily with him. Here the novelist would stop; but I write from the burying-ground of St. Regulus, and the tombstone of my ancestor is at my feet. Yet why should it be told that John Feddes experienced the misery of living too long —that, in his ninetieth year, he found himself almost alone in the world? for, of his children, some had wandered into foreign parts, where they either died or forgot their father, and some he saw carried to the grave. One of his daughters remained with him, and outlived him. She was the widow of a bold enterprising man, who lay buried with his two brothers, one of whom had sailed round the world with Anson, in the depths of the ocean; and her orphan child, who, of a similar character, shared, nearly fifty years after, a similar fate, was the father of the writer.

A very few paces from the burying-ground of John Feddes, there is a large rude stone reared on two shapeless balusters,

and inscribed with a brief record of the four last generations of
the Lindsays of Cromarty—an old family now extinct.   In its
early days this family was one of the most affluent in the burgh,
and had its friendships and marriages among the aristocracy of
the country ; but it gradually sank as it became older, and, in
the year 1729, its last scion was a little ragged boy of about
ten years of age, who lived with his widow mother in one of
the rooms of a huge dilapidated house at the foot of the Chapel
hill.   Dilapidated as it was, it formed the sole remnant of all
the possessions of the Lindsays.   Andrew, for so the boy was
called, was a high-spirited, unlucky little fellow, too careless of
the school and of his book to be much a favourite with the
schoolmaster, but exceedingly popular among his playfellows,
and the projector of half the pieces of petty mischief with which
they annoyed the village.   But, about the end of the year 1731,
his character became the subject of a change, which, after un-
fixing almost all its old traits, and producing a temporary chaos,
set, at length, much better ones in their places.   He broke off
from his old companions, grew thoughtful and melancholy, and
fond of solitude, read much in his Bible, took long journeys to
hear the sermons of the more celebrated ministers of other
parishes, and became the constant and attentive auditor of the
clergyman of his own.   He felt comfortless and unhappy.   Like
the hero of that most popular of all allegories, the Pilgrim's
Progress, " he stood clothed in rags, with his face from his own
house, a book in his hand, and a burden on his back.   And
opening the book, he read thereon, and, as he read, he wept
and trembled, and, not being able to contain himself, he broke
out into a lamentable cry, saying, What shall I do ?"   Indeed,
the whole history of Andrew Lindsay, from the time of his conversion to his death, is so exact a counterpart of the journey of
Christian, from the day on which he quitted the City of Destruction until he had entered the river, that, in tracing his
course, I shall occasionally refer to the allegory ; regarding it

as the well-known chart of an imperfectly known country. All other allegories are mere mediums of instruction, and owe their chief merit to their transparency as such ; but it is not thus with the dream of Bunyan, which, through its intrinsic interest alone, has become more generally known than even the truths which are couched under it.

Some time in the year 1732, a pious Scottish clergyman who resided in England—a Mr. Davidson of Denham, in Essex, visited some of his friends who lived in Cromarty. He was crossing the Firth at this time on a Sabbath morning, to attend the celebration of the Supper in a neighbouring church, when a person pointed out to him a thoughtful-looking little boy, who sat at the other end of the boat. "It is Andrew Lindsay," said the person, "a poor young thing seeking anxiously after the truth." "I had no opportunity of conversing with him," says Mr. Davidson in his printed tract, "but I could not observe without thankfulness a poor child, on a cold morning, crossing the sea to hear the Word, without shoe or stocking, or anything to cover his head from the inclemency of the weather." He saw him again when in church—his eyes fixed steadfastly on the preacher, and the expression of his countenance varying with the tone of the discourse. Feeling much interested in him, he had no sooner returned to Cromarty than he waited upon him at his mother's, and succeeded in engaging him in a long and interesting conversation, which he has recorded at considerable length.

"How did it happen, my little fellow," said he, "that you went so far from home last week to hear sermon, when the season was so cold, and you had neither shoes nor stockings ?" The boy replied in a bashful, unassuming manner, That he was in that state of nature which is contrasted by our Saviour with that better state of grace, the denizens of which can alone inherit the kingdom of heaven. But, though conscious that such was the case, he was quite unaffected, he said, by a sense of his danger. He was anxious, therefore, to pursue those means by which such a sense might be awakened in him ; and the Word preached was one of these. For how, he continued, unless I be oppressed by the weight of the evil which rests upon me, and the woe and misery which it must necessarily entail in the future, how can I value or seek after the only Saviour ? " But what," said Mr. Davidson, " if God himself has engaged to work this affecting sense of sin in the heart ?"—" Has he so promised ?" eagerly inquired the boy. The clergyman took out his Bible, and read to him the remarkable text in John, in which our Saviour intimates the coming of the Spirit to convince the world of *sin*, of righteousness, and of judgment. Andrew's countenance brightened as he listened, and, losing his timidity in the excitement of the moment, he took the book out of Mr. Davidson's hand, and, for several minutes, contemplated the passage in silence.

" Do you ever pray ?" inquired Mr. Davidson ; Andrew shut the book, and, hanging down his head, timidly replied in the negative. " What ! not pray ! Do you go so far from home to attend sermons, and yet not bow the knee to God in prayer ?" —" Ah !" he answered, " I do bow the knee perhaps six or seven times a day, but I cannot call that praying to God—I want the spirit of prayer ; I often ask I hardly know what, and with scarce any desire to receive ; and often, when a half sense of my condition has compelled me to kneel, a vicious wandering imagination carries me away, and I rise again, not knowing what I have said."—" Oh !" rejoined the clergyman, " only persist. But tell me, was it your ordinary practice, in past years, to attend sermons as you do now ?" " No, sir, quite the reverse ; once or twice in a season, perhaps, I went to church, but I used to quit it through weariness ere the service was half completed."—" And how do you account for the change ?" " I cannot account for it ; I only know, that formerly I had no heart to go and hear of God at any time, and that now I dare not keep away." Mr. Davidson then inquired whether he had ever conversed on these matters with Mr. Gordon, the minister of the parish; but was asked with much simplicity, in return, what Mr. Gordon would think of a poor boy like him presuming to call on him? "I have many doubts and uncertainties," said he, "but I am afraid to ask any one to solve them. Once, indeed, but only once, I plucked up resolution enough to inquire of a friend how I might glorify God. He bade me obey God's commandments, for that was the way to glorify Him, and I now see the value of the advice; but I see, also, that only through faith in Jesus Christ can fallen man acquire an ability to profit by it."

"This last answer, so much above his years," says Mr. Davidson, "occasioned my asking him how he had become so intimately acquainted with these truths? He modestly answered, 'I hear Mr. Gordon preach,' as if he had said, My knowledge bears no proportion to the advantages I enjoy." And thus ended the conference; for, after exhorting him to be much in secret prayer, and to testify to the world the excellence of what he sought after, by being a diligent scholar and a dutiful son, Mr. Davidson bade him farewell. The poor little fellow was wandering, at this period, over that middle space which lies between the devoted city and the wicket gate; struggling at times in the deep mire of the slough, at times journeying beside the hanging hill. He had received, however, the roll from Evangelist, and saw the shining light of the wicket becoming clearer and brighter as he advanced.

About half a year from the time of this conversation, Mr. Davidson had again occasion to visit Cromarty; he called on Andrew, and was struck, in the moment he saw him, by a remarkable change in his appearance. Formerly, the expression of his countenance, though interesting, was profoundly melancholy; it was now lighted up by a quiet tranquil joy; and, though modest and unassuming as before, he was less timid.

He had passed the wicket. He felt he had become one of the family of God; and found a new principle implanted within him, which so operated on his affections, that he now hated the evil he had previously loved, and was enamoured of the good he had formerly rejected. Standing, as Bacon has beautifully expressed it, on the "vantage ground of truth," he could overlook the windings of the track on which he had lately journeyed, not knowing whither he went. "I see," said he to Mr. Davidson, "that the very bent of my mind was contrary to God—especially to the way of salvation by Christ—and that I could no more get rid of this disposition through any effort of my own, than I could pull the sun out of the heavens. I see, too, that not only were all my ordinary actions tainted by sin, but that even my religious duties were sins also. And yet, out of these actions and duties, was I accumulating to myself a right-eousness which I meant to barter for the favour of God; and, though he was at much pains with me in scattering the hoard in which I trusted, yet, after every fresh dispersal, would I set myself to gather anew."—When passing the wicket, he had been shot at from the castle. He was conscious that a power, detached from his mind, had been operating upon it; for, as it fluctuated on its natural balance between gaiety and depression, he had felt this power weighing it into despair as it sunk towards the lower extreme, and urging it into presumption as it ascended towards the upper. He had seen, also, the rarities at the house of the Interpreter. Religion had communicated to him the art of thinking. It first inspired him with a belief in God, and an anxious desire to know what was his character; and, as he read his Bible, and heard sermons, his mental faculties, like the wheels of a newly-completed engine, felt for the first time the impulse of a moving power, and began to revolve. It next stirred him up to stand sentinel over the various workings of his mind, and, as he stood and pondered, he became a skilful metaphysician, without so much as knowing the name

of the science.   As a last step in the process, it brought him
acquainted with those countless analogies by which the natural
world is rendered the best of all commentaries on the moral.
"I am unable," said he to his friend the clergyman, "to describe the state of my soul as I see it, but I am somewhat helped
to conceive of it by the springs which rise by the wayside, as I
pass westward from the town, along the edge of the bay.   They
contain only a scanty supply of water, and are matted over with
grass and weeds ; but even now in August, when the fierce heat
has dried up all the larger pools, that scanty supply does not
fail them.   On disentangling the weeds I see the water sparkling beneath.   It is thus, I trust, with my heart.   The life of
God is often veiled in it by the rank luxuriance of evil thoughts,
but, when a new manifestation draws these aside, I can catch
a glimpse of what they conceal.   I can hope, too, that as the
love of Christ is unchangeable, this element of life will continue
to spring up in my soul, however dry and arid the atmosphere
which surrounds it."

Bunyan has described a green pleasant valley, besprinkled
with lilies, which lies between the palace of the virgins and
the valley of the shadow of death.   "It is blessed," says he,
"with an exceedingly fertile soil, and there have been many
labouring men who have been fortunate enough to get estates
in it."   Andrew was one of these.   He was humble and unobtrusive, and but little confident in himself—a true freeman of
the valley of humiliation.   Though no longer the leader of his
school-fellows—for he had now so little influence among them,
that he could not prevail on so much as one of them to follow
him—he was much happier than before.   Leaving them at their
wild games, he would retire to his solitudes, and there hold converse with the Deity in prayer, or seek out in meditation some
of the countless parallelisms of the two great works which had
been spread out before him—Creation and the Bible.   He was
no longer a leader even to himself.   "I have been taught,"

said he, "by experience, that my heart is too stubborn a thing for my own management, and so have given it up to the management of Christ." Mr. Davidson saw him, for the last time, about the beginning of the year 1740, when he complained to him of being exposed to many sore temptations. He had met with wild beasts, and had to contend with giants—he had been astonished amid the gloom of the dark valley, and bewildered in the mists of the enchanted ground. The interesting little tract from which I have drawn nearly all the materials of my memoir, and which at the time of its first appearance passed through several editions, and was printed more recently at Edinburgh by the publishers for the Sabbath-schools, concludes with a brief notice of this conference. The rest of Andrew's story may be told in a few words. He lived virtuously and happily, supporting himself by the labour of his hands, without either seeking after wealth or attaining to it ; he bore a good name, though not a celebrated one, and lived respected, and died regretted. It is recorded on his tombstone, in an epitaph whose only merit is its truth, that " he was truly pious from a child—his whole life and conversation agreeable thereto ;" and that his death took place in 1769, in the fiftieth year of his age.

I am aware that, in thus tracing the course of my townsman, I lay myself open to a charge of fanaticism. I shall venture, however, on committing myself still further.

One night, towards the close of last autumn, I visited the old chapel of St. Regulus. The moon, nearly at full, was riding high overhead in a troubled sky, pouring its light by fits, as the clouds passed, on the grey ruins, and the mossy, tilt-like hillocks, which had been raised ages before over the beds of the sleepers. The deep, dark shadows of the tombs seemed stamped upon the sward, forming, as one might imagine, a kind of general epitaph on the dead, but inscribed, like the handwriting on the wall, in the characters of a strange tongue. A low breeze was creeping through the long withered grass at my feet ; a shower of yellow leaves came rustling, from time to time, from an old gnarled elm that shot out its branches over the burying-ground—and, after twinkling for a few seconds in their descent, silently took up their places among the rest of the departed ; the rush of the stream sounded hoarse and mournful from the bottom of the ravine, like a voice from the depths of the sepulchre ; there was a low, monotonous murmur, the mingled utterance of a thousand sounds of earth, air, and water, each one inaudible in itself ; and, at intervals, the deep, hollow roar of waves came echoing from the caves of the distant promontory, a certain presage of coming tempest. I was much impressed by the melancholy of the scene. I reckoned the tombs one by one. I pronounced the names of the tenants. I called to remembrance the various narratives of their loves and their animosities, their joys and their sorrows. I felt, and there was a painful intensity in the feeling, that the gates of death had indeed closed over them, and shut them out from the world for ever. I contrasted the many centuries which had rolled away ere they had been called into existence, and the ages which had passed since their departure, with the little brief space between—that space in which the Jordan of their hopes and fears had leaped from its source, and after winding through the cares, and toils, and disappointments of life, had fallen into the Dead Sea of the grave ; and as I mused and pondered—as the flood of thought came rushing over me—my heart seemed dying within me, for I felt that, as one of this hapless race, vanity of vanity was written on all my pursuits and all my enjoyments, and that death, as a curse, was denounced against me. But there was one tomb which I had not reckoned, one name which I had not pronounced, one story which I had not remembered. I had not thought of the tomb, the name, the story of that sleeper of hope, who had lived in the world as if he were not of the world, and had died in the full belief that because God was his friend, death could not be his enemy. My eye at length rested on the burial-ground of the Lindsays, and the feeling of deep despondency which had weighed on my spirits was dissipated as if by a charm. I saw time as the dark vestibule of eternity ;—the gate of death which separates the porch from the main building, seemed to revolve on its hinges, and light broke in as it opened ; for the hall beyond was not a place of gloom and horror, nor strewed, as I had imagined, with the bones of dead men. I felt that the sleeper below had, indeed, lived well ; the world had passed from him as from the others, but he had wisely fixed his affections, not on the transitory things of the world, but on objects as immortal as his own soul ; and as I mused on his life and his death, on the quiet and comfort of the one, and the high joy of the other, I wondered how it was that men could deem it wisdom to pursue an opposite course.—I could not, at that time, regard Lindsay as a fanatic, nor am I ashamed to confess that I have not since changed my opinion.

CHAPTER XIV.

"Around swells mony a grassy heap,
Stan's mony a sculptured stane;
An' yet in a' this peopled field
No being thinks but ane."—ANON.

THE ruins of the old chapel of St. Regulus occupy the edge
of a narrow projecting angle, in which the burying-ground terminates towards the east.   Accident and decay seem to have
wrought their worst upon them.   The greater part of the front
wall has been swallowed up piecemeal by the ravine, which,
from the continual action of the stream, and the rains, and
snows, of so many winters, has been gradually widening and
deepening, until it has at length reached the site of the building, and is now scooping out what was once the floor.   The
other walls have found enemies nearly as potent as the stream
and the seasons, in the little urchins of the town, who, for the
last two centuries, have been amusing themselves, generation
after generation, in tearing out the stones, and rolling them
down the sides of the eminence.   What is now, however, only
a broken-edged ruin, and a few shapeless mounds, was, three
hundred years ago, a picturesque-looking, high-gabled house, of
one story, perforated by a range of narrow, slit-like windows,
and roofed with ponderous grey slate.   A rude stone cross surmounted the eastern gable.   Attached to the gable which
fronted the west, there was a building roofed over like the
chapel, but much superior to it in its style of masonry.   It
was the tomb of the Urquharts.   A single tier of hewn ashlar,
with a sloping basement, and surmounted by a Gothic mould-

ing, are now almost its only remains ; but from the line of the foundation, which we can still trace on the sward, we see that it was laid out in the form of a square, with a double buttress rising at each of the angles. The area within is occupied by a mouldy half-dilapidated vault, partially filled with bones and the rubbish of the chapel.

A few yards farther away, and nearly level with the grass, there is an uncouth imitation of the human figure with the hands folded on the breast. It bears the name of the " burnt cook ;" and from time immemorial the children of the place spit on it as they pass. But though tradition bears evidence to the antiquity of the practice, it gives no account of its origin, or what perhaps might prove the same thing, of the character of the poor cook ; which we may infer, however, from the nature of the observance, to have been a bad one. I find it stated by Mr. Brady in his *Clavis Calendaria*, that as late as the last century it was customary, in some places of England, for people to spit every time they named the devil.

Viewed from the ruins, the tombstones of the burying-ground seem clustered together beneath the fence of trees which over-tops the eminence on the west. I have compared them, in some of my imaginative moods, to a covey of waterfowl sleeping beside the long rank grass and rushes of a lake. They are mostly all fashioned in that heavy grotesque style of sculpture, which, after the Reformation had pulled down both the patterns and patrons of the stone-cutter, succeeded, in this part of the country, to the lighter and more elegant style of the time of the Jameses. The centres of the stones are occupied by the rude semblances of skulls and cross-bones, dead-bells and sand-glasses, shovels and sceptres, coffins and armorial bearings ; while the inscriptions, rude and uncouth as the figures, run in continuous lines round the margins. They tell us, though with as little variety as elegance of phrase, that there is nothing certain in life except its termination ; and, taken collectively, read us a striking lesson on the vicissitudes of human affairs. For we learn from them that we have before us the burial-place of no fewer than seven landed proprietors, none of whose families now inherit their estates. One of the inscriptions, and but only one, has some little merit as a composition. It is simple and modest; and may be regarded, besides, as a specimen of the language and orthography of Cromarty in the reign of Charles II. It runs thus—

HEIR · LYES · AT · REST · AN · FAITHFVL · ONE ·
WHOM · GOD · HAITH · PLEASIT · TO · CAL · VPON ·
HIR · LYE · SHEE · LIVED · BOTH · POOR · AND · IVST ·
AND · EY · IN · GOD · SHEE · PVT · HIR · TRVST ·
GOD'S · LAWES · OBYED · TO · SIN · WAS · LEATH ·
NO · DOVBT · SHEE · DYED · ANE · HAPPIE · DATH ·

IANET · IONS^{ton} · 1679 ·

On the northern side of the burying-ground there is a low stone, sculptured like most of the others, but broken by some accident into three pieces. A few stinted shrubs of broom spread their tiny branches and bright blossoms over the figures; they are obscured, besides, by rank tufts of moss and patches of lichens; but, despite of neglect and accident, enough of the inscription remains legible to tell us that we stand on the burial-place of one John Macleod, a *merchant* of Cromarty. He kept, besides, the principal inn of the place. He had an only son, a tall, and very powerful man, who was engaged, as he himself had been in his earlier days, in the free trade, and who, for a series of years, had set the officers of the revenue at defiance. Some time late in the reign of Queen Anne, he had succeeded in landing part of a cargo among the rocks of the hill of Cromarty, and in transporting it, night after night, from the cavern in which he had first secreted it, to a vault in his father's house, which opened into the cellar. After concealing the entrance, he had seated himself beside the old man at the kitchen fire, when two revenue-officers entered the apartment, and taking their places beside a table, called for liquor. Macleod drew his bonnet hastily over his brow, and edging away from the small iron lamp which lighted the kitchen, muffled himself up in the folds of his dreadnought greatcoat. His father supplied the officers. "Where is Walter, your son ?" inquired the better-dressed of the two, a tall, thin man, equipped in a three-cornered hat, and a blue coat seamed with gold lace ; "I trust he does not still sail the Swacker." "Maybe no," said the old man dryly. "For I have just had intelligence," continued the officer, "that she was captured this morning by Captain Manton, after firing on her Majesty's flag ; and it will go pretty hard, I can tell you, with some of the crew." A third revenue-officer now entered the kitchen, and going up to the table whispered something to the others. "Please, Mr. Macleod," said the former speaker to the innkeeper, "bring us a light, and the key of your cellar." "And wherefore that ?" inquired the old man ; "show me your warrant. What would ye do wi' the key ?" "Nay, sir, no trifling ; you brought here last night three cart-loads of Geneva, and stored them up in a vault below your cellar ; the key and a light." There was no sign, however, of procuring either. "Away !" he continued, turning to the officer who had last entered ; "away for a candle and a sledge-hammer !" He was just quitting the room when the younger Macleod rose from his seat, and took his stand right between him and the door. "Look ye, gentlemen," he said in a tone of portentous coolness, "I shall take it upon me to settle this affair ; you and I have met before now, and are a little acquainted. The man who first moves out of this place in the direction of the cellar, shall never move afterwards in any direction at all." He thrust his hand, as he spoke, beneath the folds of his greatcoat, and seemed extricating some weapon from his belt. "In upon him, lads !" shouted out the tall officer, "devil though he be, he is but one ; the rest are all captured." In a moment, two of the officers had thrown themselves upon him ; the third laid hold of his father. A tremendous struggle ensued ;—the lamp was overturned and extinguished. The smuggler, with a Herculean effort, shook off both his assailants, and as they rushed in again to close with him, he dealt one of them so terrible a blow that he rolled, stunned and senseless, on the floor. The elder Macleod, a hale old man, had extricated himself at the same moment, and mistaking, in the imperfect light, his son for one of the officers, and the fallen officer for his son, he seized on the kitchen poker, and just as the champion had succeeded in mastering his other opponent, he struck at him from behind, and felled him in an instant. In less than half an hour after he was dead. The unfortunate old man did not long survive him ; for after enduring, for a few days, the horrors of mingled grief and remorse, his anguish of mind terminated in insanity, and he died in the course of the month.

For some time after, the house he had inhabited lay without a tenant, and stories were circulated among the town's-folks of it being haunted. One David Hood, a tailor of the place, was frightened almost out of his wits in passing it on a coarse winter night, when neither fire nor candle in the whole range of houses on either side, showed him that there was anybody awake in town but himself. A fearful noise seemed to proceed from one of the lower rooms, as if a party of men were engaged in some desperate struggle ;—he could hear the dashing of furniture against the floor, and the blows of the assailants ; and after a dull hollow sound twice repeated, there was a fearful shriek, and a mournful exclamation in the voice of the deceased shopkeeper, "I have murdered my son ! I have murdered my son !" The house was occupied, notwithstanding, some years after, though little to the comfort of the tenants. Often were they awakened at midnight, it is said, by noises, as if every piece of furniture in the apartment was huddled into the middle of the floor, though in the morning not a chair or table would be found displaced ; at times, too, it would seem as if some person

heavily booted was traversing the rooms overhead ; and some
of the inmates, as they lay a-bed, have seen clenched fists shaken
at them from outside the windows, and pale, threatening faces
looking in upon them through half-open doors.   There is one of
the stories which, but for a single circumstance, I should deem
more authentic, not merely than any of the others, but than most
of the class to which it belongs.   It was communicated to me by
a sensible and honest man—a man, too, of very general information.   He saw, he said, what he seriously believed to be the
apparition of the younger Macleod ; but as he was a child of
only six years at the time, his testimony may, perhaps, be more
rationally regarded as curiously illustrative of the force of imagination at a very early age, than as furnishing any legitimate
proof of the reality of such appearances.   He had a sister, a
few years older than himself, who attended some of the younger
members of the family, which tenanted, about sixty years ago,
the house once occupied by the shopkeeper.   One Sunday
forenoon, when all the inmates had gone to church except the
girl and her charge, he stole in to see her, and then amused
himself in wandering from room to room, gazing at the furniture
and the pictures.   He at length reached one of the garrets, and
was turning over a heap of old magazines in quest of the prints,
when he observed something darken the door, and looking up,
found himself in the presence of what seemed to be a very tall,
broad-shouldered man, with a pale, ghastly countenance, and
wrapt up in a brown dreadnought greatcoat.   A good deal
surprised, but not at all alarmed, for he had no thought at the
time that the appearance was other than natural, he stepped
down stairs and told his sister that there was a "muckle big
man i' the top of the house."   She immediately called in a
party of the neighbours, who, emboldened by the daylight, explored every room and closet from the garrets to the cellar, but
they saw neither the tall man nor the dreadnought greatcoat.
    The old enclosure of the burying-ground, which seems ori-

ginally to have been an earthen wall, has now sunk into a grassy mound, and on the southern and western sides some of the largest trees of the fence—a fine stately ash, fluted like a Grecian column, a huge elm roughened over with immense wens, and a low bushy larch with a bent twisted trunk, and weeping branches—spring directly out of it. At one place we see a flat tombstone lying a few yards outside the mound. The trees which shoot up on every side fling so deep a gloom over it during the summer and autumn months, that we can scarcely decipher the epitaph ; and in winter it is not unfrequently buried under a wreath of withered leaves. By dint of some little pains, however, we come to learn from the darkened and half-dilapidated inscription, that the tenant below was one Alexander Wood, a native of Cromarty, who died in the year 1690 ; and that he was interred in this place at his own especial desire. His wife and some of his children have taken up their places beside him ; thus lying apart like a family of hermits ; while his story—which, almost too wild for tradition itself, is yet as authentic as most pieces of written history—affords a curious explanation of the circumstance which directed their choice.

Wood was a man of strong passions, sparingly gifted with common sense, and exceedingly superstitious. No one could be kinder to one's friends or relatives, or more hospitable to a stranger ; but when once offended, he was implacable. He had but little in his power either as a friend or an enemy—his course through the world lying barely beyond the bleak edge of poverty. If a neighbour, however, dropped in by accident at meal-time, he would not be suffered to quit his house until he had shared with him his simple fare. There was benevolence in the very grasp of his hand and the twinkle of his eye, and in the little set speech, still preserved by tradition, in which he used to address his wife every time an old or mutilated beggar came to the door :—" Alms, gudewife," he would say ; " alms to the cripple, and the blin', and the broken-down."

When injured or insulted, however, and certainly no one could do either without being very much in the wrong, there was a toad-like malignity in his nature, that would come leaping out like the reptile from its hole, and no power on earth could shut it up again. He would sit hatching his venom for days and weeks together with a slow, tedious, unoperative kind of perseverance, that achieved nothing. He was full of anecdote; and, in all his stories, human nature was exhibited in only its brightest lights and its deepest shadows, without the slightest mixture of that medium tint which gives colour to its working, everyday suit. Whatever was bad in the better class, he transferred to the worse, and *vice versa;* and thus not even his narratives of the supernatural were less true to nature and fact than his narratives of mere men and women. And he dealt with the two classes of stories after one fashion—lending the same firm belief to both alike.

In the house adjoining the one in which he resided, there lived a stout little man, a shoemaker, famous in the village for his great wit and his very considerable knavery. His jokes were mostly practical, and some of the best of them exceedingly akin to felonies. Poor Wood could not understand his wit, but, in his simplicity of heart, he deemed him honest, and would fain have prevailed on the neighbours to think so too. He knew it, he said, by his very look. Their gardens, like their houses, lay contiguous, and were separated from each other, not by a fence, but by four undressed stones laid in a line. Year after year was the garden of Wood becoming less productive; and he had a strange misgiving, but the thing was too absurd to be spoken of, that it was growing smaller every season by the breadth of a whole row of cabbages. On the one side, however, were the back walls of his own and his neighbour's tenements; the four large stones stretched along the other; and nothing, surely, could be less likely than that either the stones or the houses should take it into their heads to rob him of his property. But the more he strove to exclude the idea the more it pressed upon him. He measured and remeasured to convince himself that it was a false one, and found that he had fallen on just the means of establishing its truth. The garden was actually growing smaller. But how? Just because it was bewitched! It was shrinking into itself under the force of some potent enchantment, like a piece of plaiding in the fulling-mill. No hypothesis could be more congenial; and he would have held by it, perhaps, until his dying day, had it not been struck down by one of those chance discoveries which destroy so many beautiful systems and spoil so much ingenious philosophy, quite in the way that Newton's apple struck down the vortices of Descartes.

He was lying a-bed one morning in spring, about day-break, when his attention was excited by a strange noise which seemed to proceed from the garden. Had he heard it two hours earlier, he would have wrapped up his head in the bedclothes and lain still; but now that the cock had crown, it could not, he concluded, be other than natural. Hastily throwing on part of his clothes, he stole warily to a back window, and saw, between him and the faint light that was beginning to peep out in the east, the figure of a man, armed with a lever, tugging at the stones. Two had already been shifted a full yard nearer the houses, and the figure was straining over a third. Wood crept stealthily out at the window, crawled on all fours to the intruder, and, tripping up his heels, laid him across his lever. It was his knavish neighbour the shoemaker. A scene of noisy contention ensued; groups of half-dressed town's-folk, looming horrible in their shirts and nightcaps through the grey of morning, came issuing through the lanes and the closes; and the combatants were dragged asunder. And well was it for the shoemaker that it happened so; for Wood, though in his sixtieth year, was strong enough, and more than angry enough, to have torn him to pieces. Now, however, that the warfare had to be carried on by words, the case was quite reversed.

" Neebours," said the shoemaker, who had the double advantage of being exceedingly plausible, and decidedly in the wrong, "I'm desperately ill-used this morning—desperately ill-used ;— he would baith rob and murder me. I lang jaloused, ye see, that my wee bit o' a yard was growing littler and littler ilka season ; and, though no very ready to suspect folk, I just thought I would keep watch, and see wha was shifting the mark-stanes. Weel, and I did ;—late and early did I watch for mair now than a fortnight; and wha did I see this morning through the back winnock but auld Sandy Wood there in his verra sark—Oh, it's no him that has ony thought o' his end!—poking the stones wi' a lang kebar, intil the very heart o' my grun'? See," said he, pointing to the one that had not yet been moved, " see if he hasna shifted it a lang ell; and only notice the craft o' the bodie in tirring up the yird about the lave, as if they had been a' moved frae my side. Weel, I came out and challenged him, as wha widna?—says I, Sawney my man, that's no honest; I'll no bear that; and nae mair had I time to say, when up he flew at me like a wull-cat, and if it wasna for yoursels I daresay he would hae throttled me. Look how I am bleedin';— and only look till him—look till the cankart deceitful bodie, if he has one word to put in for himsel'."

There was truth in, at least, this last assertion; for poor Wood, mute with rage and astonishment, stood listening, in utter helplessness, to the astounding charge of the shoemaker, —almost the very charge he himself had to prefer. Twice did he spring forward to grapple with him, but the neighbours held him back, and every time he essayed to speak, his words —massed and tangled together, like wreaths of sea-weed in a hurricane—stuck in his throat. He continued to rage for three days after, and when the eruption had at length subsided, all his former resentments were found to be swallowed up, like the lesser craters of a volcano, in the gulf of one immense hatred.

His house, as has been said, lay contiguous to the house of the shoemaker, and he could not avoid seeing him, every time he went out and came in—a circumstance which he at first deemed rather gratifying than otherwise. It prevented his hatred from becoming vapid by setting it a-working at least ten times a day, as a musket would a barrel of ale if discharged into the bunghole. Its frequency, however, at length sickened him, and he had employed a mason to build a stone wall, which, by stretching from side to side of the close, was to shut up the view, when he sickened in right earnest, and at the end of a few days found himself a-dying. Still, however, he was possessed by his one engrossing resentment. It mingled with all his thoughts of the past and the future ; and not only was he to carry it with him to the world to which he was going, but also to leave it behind him as a legacy to his children. Among his many other beliefs, there was a superstition, handed down from the times of the monks, that at the day of final doom all the people of the sheriffdom were to be judged on the moor of Navity ; and both the judgment and the scene of it he had indissolubly associated with the shoemaker and the four stones. Experience had taught him the importance of securing a first hearing for his story ; for was his neighbour, he concluded, to be beforehand with him, he would have as slight a chance of being righted at Navity as in his own garden. After brooding over the matter for a whole day, he called his friends and children round his bed, and raised himself on his elbow to address them.

"I'm wearing awa, bairns and neebours," he said, "and it vexes me sair that that wretched bodie should see me going afore him. Mind, Jock, that ye'll build the dike, and make it heigh, heigh, and stobbie on the top; and oh! keep him out o' my lykewake, for should he but step in at the door, I'll rise, Jock, frae the verra straiking-board, and do murder! Dinna let him so muckle as look on my coffin. I have been pondering a' this day about the meeting at Navity, and the march-stanes; and I'll tell you, Jock, how we'll match him. Bury me ayont the saint's dike on the Navity side, and dinna lay me deep. Ye ken the bonny green hillock, spreckled o'er wi' gowans and puddock-flowers—bury me there, Jock; and yoursel', and the auld wife, may just, when your hour comes, tak up your places beside me. We'll a' get up at the first tout—the ane helping the other; and I'se wad a' I'm worth i' the warld, we'll be half-way up at Navity afore the shochlan, short-legged bodie wins o'er the dike." Such was the dying injunction of Sandy Wood: and his tombstone still remains to testify that it was religiously attended to. An Englishman who came to reside in the parish, nearly an age after, and to whom the story must have been imparted in a rather imperfect manner, was shocked by what he deemed his unfair policy. The litigants, he said, should start together; he was certain it would be so in England where a fair field was all that would be given to St. Dunstan himself though he fought with the devil. And that it might be so here, he buried the tombstone of Wood in an immense heap of clay and gravel. It would keep him down, he said, until the little fellow would have clambered over the wall. The town's-folk, however, who were better acquainted with the merits of the case, shovelled the heap aside; and it now forms two little hillocks which overtop the stone, and which, from the nature of the soil, are still more scantily covered with verdure than any part of the surrounding bank.

CHAPTER XV.

" Oh ! I do ponder with most strange delight
On the calm slumbers of the dead-man's night."
HENRY K. WHITE.

WE have lingered long in the solitary burying-ground of St. Regulus ; the sun hastens to its setting ; and the slanting beam of red light that comes pouring in through an opening amid the trees, catches but the extreme tops of the loftier monuments, and the higher pinnacles of the ruin beyond.   There is a little bird chirping among the graves ; we may hear the hum of the bee as it speeds homeward, and the low soothing murmur of the stream in the dell below ; all else is stillness and solitude in this field of the dead.

There are times when, amid scenes such as the present, one can almost forget the possible, and wish that the silence were less deep.   The most contemplative of modern poets, in giving voice to a similar wish, has sublimed it into poetry.   " Would," he says of his churchyard among the hills, in the stanza I have already employed as a motto,

" Would that the silent earth
Of what it holds could speak, and every grave
Be as a volume, shut, yet capable
Of yielding its contents to ear and eye."

The dead of a thousand years are sleeping at our feet ; the poor peasant serf of ten centuries ago, whom the neighbouring baron could have hung up at his cottage door, with the intelligent mechanic of yesterday, who took so deep an interest in the emancipation of the negroes. What strange stories of the past, what striking illustrations of the destiny and nature of man, how important a chronicle of the progress of society, would this solitary spot present us with, were it not that, like the mysterious volume in the Apocalypse, no man can open the book or unloose the seals thereof! There are recollections associated with some of the more recent graves, of interest enough to show us how curious a record the history of the whole would have furnished.

It is now well-nigh thirty years since Willie Watson returned, after an absence of nearly a quarter of a century, to the neighbouring town. He had been employed as a ladies' shoemaker in some of the districts of the south; but no one at home had heard of Willie in the interval, and there was little known regarding him at his return, except that when he had quitted town so many years before, he was a neat-handed industrious workman, and what the elderly people called a quiet decent lad. And he was now, though somewhat in the wane of life, even a more thorough master of his trade than before. He was quiet and unobtrusive, too, as ever, and a great reader of serious books. And so the better sort of the people were beginning to draw to Willie by a kind of natural sympathy; some of them had learned to saunter into his workshop in the long evenings, and some had grown bold enough to engage him in serious conversation when they met with him in his solitary walks; when out came the astounding fact—and important as it may seem, the simple-minded mechanic had taken no pains to conceal it—that, during his residence in the south country, he had laid down Presbyterianism, and become the member of a Baptist church. There was a sudden revulsion of feeling towards him, and all the people of the town began to speak of Willie Watson as " a poor lost lad."

The " poor lost lad," however, was unquestionably a very excellent workman; and as he made neater shoes than anybody else, the ladies of the place could see no great harm in wearing them. He was singularly industrious, too, and indulged in no extraordinary expense, except when he now and then bought a good book, or a few flower-seeds for his garden. He was withal a single man, with only himself, and an elderly sister who lived with him, to provide for ; and, what between the regularity of his gains on the one hand, and the moderation of his desires on the other, Willie, for a person of his condition, was in easy circumstances. It was found that all the children in the neighbourhood had taken a wonderful fancy to his shop. Willie was fond of telling them good little stories out of the Bible, and of explaining to them the prints which he had pasted on the walls. Above all, he was anxiously bent on teaching them to read. Some of their parents were poor, and some of them were careless ; and he saw that, unless they learned their letters from him, there was little chance of their ever learning them at all. Willie in a small way, and to a very small congregation, was a kind of missionary ; and what between his stories and his pictures, and his flowers and his apples, his labours were wonderfully successful. Never yet was school or church half so delightful to the little men and women of the place as the workshop of Willie Watson, " the poor lost lad."

Years of scarcity came on ; taxes were high, and crops not abundant ; and the soldiery abroad, whom the country had employed to fight against Bonaparte, had got an appetite at their work, and were consuming a good deal of meat and corn. The price of food rose tremendously ; and many of the town's-people, who were working for very little, were not in every case secure of that little when the work was done. Willie's small congregation began to find that the times were exceedingly bad ; there were no more morning *pieces* among them, and the porridge was less than enough. It was observed, however, that in the midst of their distresses Willie got in a large stock of meal, and that his sister began to bake as if she were making ready for a wedding. The children were wonderfully interested in the work, and watched it to the end; when, lo! to their great and joyous surprise, Willie divided the whole baking among them. Every member of the congregation got a cake; there were some who had little brothers and sisters at home who got two; and from that day forward, till times got better, none of Willie's young people lacked their morning *piece*. The neighbours marvelled at Willie; and all agreed that there was something strangely puzzling in the character of the " poor lost lad."

I have alluded to Willie's garden. Never was there a little bit of ground better occupied; it looked like a piece of rich needlework. He had got wonderful flowers too—flesh-coloured carnations streaked with red, and double roses of a rich golden yellow. Even the commoner varieties—auriculas and anemones, and the party-coloured polyanthus—grew better with Willie than with anybody else. A Dutchman might have envied him his tulips, as they stood row beyond row on their elevated beds, like so many soldiers on a redoubt; and there was one mild dropping season in which two of these beautiful flowers, each perfect in its kind, and of different colours, too, sprang apparently from the same stem. The neighbours talked of them as they would have talked of the Siamese Twins; but Willie, though it lessened the wonder, was at pains to show them that the flowers sprang from different roots, and that what seemed to be their common stem, was in reality but a green hollow sheath formed by one of the leaves. Proud as Willie was of his flowers, and with all his humility he could not help being a little proud of them, he was yet conscientiously determined to have no miracle among them, unless, indeed, the miracle should chance to be a true one. It was no fault of Willie's that all his neighbours had not as fine gardens as himself; he gave them slips of his best flowers, flesh-coloured carnation, yellow rose, and all; he grafted their trees for them too,

THE POOR LOST LAD.   225

and taught them the exact time for raising their tulip-roots, and the best mode of preserving them. Nay, more than all this, he devoted whole hours at times to give the finishing touches to their parterres and borders, just in the way a drawing-master lays in the last shadings, and imparts the finer touches, to the landscapes of his favourite pupils. All seemed impressed by the unselfish kindliness of his disposition; and all agreed that there could not be a warmer-hearted or more obliging neighbour than Willie Watson, " the poor lost lad."

Everything earthly must have its last day. Willie was rather an elderly than an old man, and the childlike simplicity of his tastes and habits made people think of him as younger than he really was; but his constitution, never a strong one, was gradually failing; he lost strength and appetite; and at length there came a morning in which he could no longer open his shop. He continued to creep out at noon, however, for a few days after, to enjoy himself among his flowers, with only the Bible for his companion; but in a few days more he had declined so much lower, that the effort proved too much for him, and he took to his bed. The neighbours came flocking in; all had begun to take an interest in poor Willie; and now they had learned he was dying, and the feeling had deepened immensely with the intelligence. They found him lying in his neat little room, with a table bearing the one beloved volume drawn in beside his bed. He was the same quiet placid crea- ture he had ever been; grateful for the slightest kindness, and with a heart full of love for all—full to overflowing. He said nothing about the Kirk, and nothing about the Baptists, but earnestly did he urge his visitors to be good men and women, and to be availing themselves of every opportunity of doing good. The volume on the table, he said, would best teach them how. As for himself, he had not a single anxiety; the great Being had been kind to him during all the long time he had been in the world, and He was now kindly calling him out

P

of it. Whatever He did to him was good, and for his good, and why then should he be anxious or afraid? The hearts of Willie's visitors were touched, and they could no longer speak or think of him as " the poor lost lad."

A few short weeks went by, and Willie had gone the way of all flesh. There was silence in his shop, and his flowers opened their breasts to the sun, and bent their heads to the bee and the butterfly, with no one to take note of their beauty, or to sympathize in the delight of the little winged creatures that seemed so happy among them. There was many a wistful eye cast at the closed door and melancholy shutters by the members of Willie's congregation, and they could all point out his grave. Yonder it lies, in the red light of the setting sun, with a carpeting of soft yellow moss spread over it. This little recess contains, doubtless, to use Wordsworth's figure, many a curious and many an instructive volume, and all we lack is the ability of deciphering the characters; but a better or more practical treatise on toleration than that humble grave, it cannot contain. The point has often been argued in this part of the country—argued by men with long beards, who preached bad grammar in behalf of Johanna Southcote, and by men who spoke middling good sense for other purposes, and shaved once a day. But of all the arguments ever promulgated, those which told with best effect on the town's-people were the life and death of Willie Watson, " the poor lost lad."

We have perused the grave of the " poor lost lad," and it turns out to be a treatise on toleration. The grave beside it may be regarded as a ballad—a short plaintive ballad—moulded in as common a form of invention, if I may so express myself, as any, even the simplest, of those old artless compositions which have welled out from time to time from among the people. Indeed, so simple is the story of it, that we might almost deem it an imitation, were we not assured that all the volumes of this solitary recess are originals from beginning to end.

It was forty years last March since the Champion man-of-war entered the bay below, with her *ancient* suspended half-way over the deck. Old seamen among the town's-folk, acquainted with that language of signs and symbols in which fleets converse when they meet at sea, said that either the captain or one of his officers was dead; and the town's-people, interested in the intelligence, came out by scores to gaze on the gallant vessel as she bore up slowly and majestically in the calm, towards the distant roadstead. The sails were furled, and the anchors cast; and as the huge hull swung round to the tide, three boats crowded with men were seen to shoot off from her side, and a strain of melancholy music came floating over the waves to the shore. A lighter shallop, with only a few rowers, pulled far a-head of the others, and as she reached the beach, the shovels and pickaxes, for which the crew relinquished their oars, revealed to the spectators more unequivocally than even the half-hoisted ensign or the music, the sad nature of their errand. The other boats approached with muffled and melancholy stroke, and the music waxed louder and more mournful. They reached the shore; the men formed at the water's edge round a coffin covered by a flag, and bearing a sword a-top, and then passed slowly amid the assembled crowds to the burying-ground of St. Regulus. Arms glittered to the sun. The echoes of the tombs and of the deep precipitous dell below were awakened awhile by unwonted music, and then by the sharp rattle of musketry; the smoke went curling among the trees, or lingered in a blue haze amid the dingier recesses of the hollow; the coffin was covered over: a few of the officers remained behind the others; and there was one of the number, a tall handsome young man, who burst out, as he was turning away, into an uncontrollable fit of weeping. At length the whole pageant passed, and there remained behind only a darkened little hillock, with whose history no one was acquainted, but which was known for many years after as the " officer's grave."

Twenty years went by, and the grave came to be little thought of, when a townsman, on going up one evening to the burying-ground, saw a lady in deep mourning sitting weeping beside it, and a tall handsome gentleman in middle life, the same individual who had been so much affected at the funeral, standing, as if waiting for her, a little apart. They were brother and sister. The storms of twenty seasons had passed over the little mossy hillock. The deep snows had pressed upon it in winter; the dead vegetation of succeeding summers and autumns had accumulated around it, and it had gradually flattened to nearly the level of the soil. It had become an old grave; but the grief, that for the first time was now venting itself over it, had remained fresh as at first. There are cases, though rare, in which sorrow does not yield to time. A mother loses her child just as its mind has begun to open, and it has learned to lay hold of her heart by those singularly endearing signs of infantine affection and regard, which show us how the sympathies of our nature, which serve to bind us to the species, are awakened to perform their labour of love with even the first dawn of intelligence. Little missed by any one else, or at least soon to be forgotten, it passes away; but there is one who seems destined to remember it all the more vividly just because it *has* passed. To her, death serves as a sort of *mordant* to fix the otherwise flying colours in which its portraiture had been drawn on her heart. Time is working out around her his thousand thousand metamorphoses. The young are growing up to maturity, the old dropping into their graves; but the infant of her affections ever remains an infant—her charge in middle life, when all her other children have left her and gone out into the world, and, amid the weakness of decay and decrepitude, the child of her old age. There arises, however, a more enduring sorrow than even that of the mother, when, in the midst of hopes all but gratified, and wishes on the eve of fulfilment, the ties of the softer passion are rudely dissevered by death. Feelings, evanescent in their nature, and restricted to one class of circumstances and one stage of life, are uneradicably fixed through the event in the mind of the survivor. Youth first passes away, then the term of robust and active life, and last of all, the cold and melancholy winter of old age; but through every succeeding change, until the final close, the bereaved lover remains a lover still. Death has fixed the engrossing passion in its tenderest attitude by a sort of petrifying process; and we are reminded by the fact of those delicate leaves and florets of former creations, which a common fate would have consigned to the usual decay, but which were converted, when they died by some sudden catastrophe, into a solid marble that endures for ever. The lady who wept this evening beside the " officer's grave," was indulging in a hopeless, enduring passion of the character described ; but all that now remains of her story forms but a mere outline for the imagination to fill up at pleasure. Her lover had been the sole heir of an ancient and affluent family; the lady herself belonged to rather a humbler sphere. He had fixed his affections upon her when almost a boy, and had succeeded in engaging hers in turn; but his parents, who saw nothing desirable in a connexion which was to add to neither the wealth nor the honours of the family, interfered, and he was sent to sea; where a disappointed attachment, preying on a naturally delicate constitution, soon converted their fears for his marriage into regret for his death. Did I not say truly that the " officer's grave " was a simple little ballad, moulded in one of the commonest forms of invention ?

Let us peruse one other grave ere we quit the burying-ground —the grave of Morrison the painter. It treats of morals, like that of " the poor lost lad," but it enforces them after a different mode. We shall find it in the strangers' corner, beside the graves of the two foreign seamen, whose bodies were cast upon the beach after a storm.

Morrison, some sixty or seventy years ago, was a tall, thin, genteel-looking young man, who travelled the country as a portrait and miniature painter. The profession was new at the time to the north of Scotland; and the people thought highly of an artist who made likenesses that could be recognised. But they could not think more highly of him than Morrison did of himself. He was one of the class who mistake the imitative faculty for genius, and the ambition of rising in a genteel profession for that energy of talent whose efforts, with no higher object often than the mere pleasure of exertion, buoy up the possessor to his proper level among men. There was a time when Morrison's pictures might be seen in almost every house—in little turf cottages even among halfpenny prints and broadsheet ballads; nor were instances wanting of their finding place among the paintings of a higher school:—some proprietor of the district retained an eccentric piper or gamekeeper in his establishment, or, like the baron of a former age, kept a fool, and Morrison had been employed to confer on all that was droll or picturesque in his appearance, the immortality of colour and canvas. Like the painter in the fable who pleased everybody, he drew, in his serious portraits, all his men after one model, and all his women after another; but, unlike the painter, he copied from neither Apollo nor Venus. His gentlemen had sloping shoulders and long necks, and looked exceeding grave and formidable; his ladies, on the contrary, were sweet simpering creatures, with waists almost tapering to a point, and cheeks and lips of as bright a crimson as that of the bunch of roses which they bore in their hands.

I have said that Morrison thought more highly of his genius than even his countryfolk. As the member of a highly liberal profession, too, he naturally enough took rank as a gentleman. Geniuses were eccentric in those days, and gentlemen not very moral; and Morrison, in his double capacity of genius and gentleman, was skilful enough to catch the eccentricity of the one class and the immorality of the other. He raked a little, and drank a great deal; and when in his cups said and did things which were thought very extraordinary indeed. But though all acknowledged his genius, he was less successful in establishing his gentility. There was, indeed, but one standard of gentility in the country at the time, and fate had precluded the painter from coming up to it; no one was deemed a gentleman whose ancestors had not been useless to the community for at least five generations. It must be confessed, too, that some of Morrison's schemes for establishing his claims were but ill laid. On one occasion he attended an auction of valuable furniture in the neighbouring town, and though a wanderer at the time, as he had been all his life long, and miserably poor to boot, he deemed it essential to the maintenance of his character, that, as all the other gentlemen present were bidding with spirit, he should now and then give a spirited bid too. He warmed gradually as the sale proceeded, offered liberally for beds and carpets, and made a dead set on a valuable pianoforte. The purchasers were sadly annoyed; and the auctioneer, who was a bit of a wag, and laboured to put down the painter by sheer force of wit, found that he had met with as accomplished a wit as himself. Morrison lost the piano, and then fell in love with a moveable wooden house, which had served as a sort of meat preserve, and was secured by a strong lock. " You had better examine it inside, Mr. Morrison," said the auctioneer; " in fact, the whole merit of the thing lies inside." Morrison went in, and the auctioneer shut and locked the door. There could not be a more grievous outrage on the feelings of a gentleman; but though the poor man went bouncing against the cruel walls of his prison like an incarcerated monkey, and grinned with uncontrollable wrath at all and sundry through its little wire-woven window, pity or succour was there none; he was kept in close durance for four long hours till the sale terminated, and found his claim to gentility not in the least strengthened when he got out.

After living, as he best could, for about forty years, the painter took to himself a wife. No woman should ever have thought of marriage in connexion with such a person as Morrison, nor should Morrison have ever thought of marriage in connexion with such a person as himself. But so it was—for ladies are proverbially courageous in such matters, and Morrison could bid as dauntlessly for a wife as for a pianoforte—that he determined on marrying, and succeeded in finding a woman bold enough to accept of him for her husband. She was a rather respectable sort of person, who had lived for many years as housekeeper in a gentleman's family, and had saved some money. They took lodgings in the neighbouring town ; Morrison showed as much spirit, and got as often drunk as before ; and in little more than a twelvemonth they came to be in want. They lingered on, however, in miserable poverty for a few months longer, and then quitted the place, leaving behind them all Mrs. Morrison's well-saved wardrobe under arrestment for debt. The large trunk which contained it lay unopened till about five years after the poor woman had been laid in her grave, the victim of her miserable marriage ; and the contents formed a strange comment on her history. *There* were fine silk gowns, sadly marred by mildew, and richly flowered petticoats eaten by the moths. *There,* too, were there pretty little heads of the virgin and the apostles, and beads and a crucifix of some value ; the loss of which, as the poor owner had been a zealous Roman Catholic, had affected her more than the loss of all the rest. And there, also, like the Babylonish garment among the goods of Achan, there was a packet of Morrison's letters, full of flames and darts, and all those little commonplaces of love which are used by men clever on a small scale, who think highly of their own parts, and have no true affection for any one but themselves.

It has been told me by an acquaintance, who resided for some time in one of our northern towns, that when hurrying to his lodgings on a wet and very disagreeable winter evening, his curiosity was attracted by a red glare of light which he saw issuing through the unglazed window and partially uncovered rafters of a deserted hovel by the wayside. He went up to it, and found the place occupied by two miserable-looking wretches, a man and woman, who were shivering over a smouldering fire of damp straw. These were Morrison and his wife, neither of them wholly sober; for the woman had ere now broken down in character as well as in circumstances. They had neither food nor money; the rain was dropping upon them through the roof, and the winter wind fluttering through their rags; and yet, as if there was too little in all this to make them unhappy enough, they were adding to their miseries by mutual recriminations. The woman, as I have said, soon sank under the hardships of a life so entirely wretched; her unlucky partner survived until the infirmities of extreme old age were added to his other miseries. It is not easy to conceive how any one who passed such a life as Morrison should have lived for the greater part of a century; and yet so it was, that, when he visited the neighbouring town for the last time, he was in his eighty-fifth year. And never, certainly, was the place visited by a more squalid, miserable-looking creature; he resembled rather a corpse set a-walking than a living man. He was still, however, Morrison the painter, feebly eccentric, and meanly proud: even when compelled to beg, which was often, he could not forget that he was an artist and a gentleman. In his younger days he had skill enough to make likenesses that could be recognised; the things he now made scarcely resembled human creatures at all; but he went about pressing his services on every one who had children and spare sixpences, till he had at length well-nigh filled the town with pictures of little boys and girls, which, in every case, the little boys and girls got to themselves. On one occasion he went into the shop of one of the town traders, and insisted on furnishing the trader with the picture of one of his daughters, a little laughing *blonde*, who was playing in front of the counter.

He produced his colours, and began the drawing ; but the girl, after wondering at him till his work was about half finished, escaped into the street, and one of her sisters, a sober-eyed *brunette*, who had heard of the strange old man who was " making pictures," came running in, and took her place. The painter held fast the intruder, and continued his drawing. " Hold, hold, Mr. Morrison, that is another little girl you have got !" said the trader ; " that is but the sister of the first." " Heaven bless the dear sweet creature !" said Morrison, still plying the pencil, " they are so very like that there can be no mistake."

The closing scene to poor Morrison came at last. He left his bed one day after an illness of nearly a week, and crawled out into the street to beg. A gentleman in passing dropped him a few coppers, and Morrison felt indignant that any one should have offered an artist less than silver. But on second thoughts he corrected himself. " Heaven help me !" he ejaculated, " I have been a fool all life long, and I am not wise yet !" He crept onwards along the pavement to the house of a gentleman whom he had known thirty years before. " I am dying," he said, " and I am desirous that you should see my body laid decently under ground ; I shall be dead in less than a week." The gentleman promised to attend the funeral ; Morrison crept back to his lodgings, and was dead in less than a day. Yonder he lies in the strangers' corner ; the parish furnished the shroud and the coffin, and the gentleman whom he had invited to his burial carried his head to the grave, and paid the sexton. There are few real stories consistently gloomy throughout. Nature delights in strange compounds of the *bizarre* and the serious ; and Morrison's story, like some of the old English dramas that terminate unfortunately, has a mixture of the comic in it. And yet, notwithstanding its lighter touches, I question whether we shall be able to find a deeper tragedy among all the volumes of the churchyard.

CHAPTER XVI.

" Like a timeless birth, the womb of fate
Bore a new death of unrecorded date,
And doubtful name."—MONTGOMERY.

IN the history of every community there are periods of comparative quiet, when the great machine of society performs all its various movements so smoothly and regularly, that there is nothing to remind us of its being in motion. And who has not remarked that when an unlooked-for accident sets it a-jarring, by breaking up some minor wheel or axis, there follows a whole series of disasters—pressing the one upon the other, with stroke after stroke. We live, perhaps, in some quiet village, and see our neighbours, the inhabitants, moving noiselessly around us—the young rising up to maturity, the old descending slowly to the grave. Death for a long series of years drafts out his usual number of conscripts from among only the weak and the aged ; and there is no irregular impressment of the young and vigorous in the way of accident. Anon, however, there succeeds a series of disasters. One of the villagers topples over a precipice, one is engulfed in a morass, one is torn to pieces by the wheels of an engine, one perishes in fording a river, one falls by the hand of an enemy, one dies by his own. And then in a few months, perhaps, the old order of things is again established, and all goes on regularly as before. In the phenomena of even the inanimate world we see marks of a similar economy. Whoever has mused for a single half hour by the side of a waterfall, must have remarked that, without any apparent change in the volume of the stream, the waters descend at one time louder and more furious, at another gentler and more subdued. Whoever has listened to the howlings of the night wind, must have heard it sinking at intervals into long hollow pauses, and then rising and sweeping onwards, gust after gust. Whoever has stood on the sea-shore during a tempest, must have observed that the waves roll towards their iron barrier in alternate series of greater and lesser—now fretting ineffectually against it, now thundering irresistibly over. But between the irregularities of the inanimate world, and those of the rational, there exists one striking difference. We may assign natural causes for the alternate rises and falls of the winds and waters ; but it is not thus in most instances with those ebbs and flows, gusts and pauses, which occur in the world of man. They set our reasonings at defiance, and we can refer them to only the will of Deity. We can only say regarding them, that the climax is a favourite figure in the book of Providence ;—that God speaks to us in His dispensations, and, in the more eloquent turns of His discourse, piles up instance upon instance with sublime and impressive profusion.

To the people of Scotland the whole of the seventeenth century was occupied by one continuous series of suffering and disaster. And though we can assign causes for every one of the evils which compose the series, just as we can assign causes for every single accident which befalls the villagers, or for the repeated attacks and intervening pauses of the hurricane, it is a rather different matter to account for the series itself. In flinging a die we may chance on any one certain number as readily as on any other; but it would be a rare occurrence, indeed, should the same number turn up some eight or ten times together. And is there nothing singular in the fact, that, for a whole century, a nation should have been invariably unfortunate in every change with which it was visited, and have met with only disaster in all its undertakings? There turned up an unlucky number at every cast of the die. Even when the shout of the persecutor, and the groans of his victim, had ceased to echo among our rocks and caverns, the very elements arrayed themselves against the people, and wasting famine and exterminating pestilence did the work of the priest and the tyrant. I am acquainted with no writer who has described this last infliction of the series so graphically, and with such power, as Peter Walker in his Life of Cargill. Other contemporary historians looked down on this part of their theme from the high places of society;—they were the soldiers of a well-victualled garrison, situated in the midst of a wasted country, and sympathized but little in the misery that approached them no nearer than the outer gate. But it was not thus with the poor Pedlar;—he was barred out among the sufferers, and exposed to the evils which he so feelingly describes.

One night in the month of August 1694, a cold east wind, accompanied by a dense sulphurous fog, passed over the country, and the half-filled corn was struck with mildew. It shrank and whitened in the sun, till the fields seemed as if sprinkled with flour, and where the fog had remained longest—for in some places it stood up like a chain of hills during the greater part of the night—the more disastrous were its effects. From this unfortunate year, till the year 1701, the land seemed as if struck with barrenness, and such was the change on the climate, that the seasons of summer and winter were cold and gloomy in nearly the same degree. The wonted heat of the sun was withholden, the very cattle became stunted and meagre, the moors and thickets were nearly divested of their feathered inhabitants, and scarcely a fly or any other insect was to be seen even in the beginning of autumn. November and December, and in some places January and February, became the months of harvest; and labouring people contracted diseases which terminated in death, when employed in cutting down the corn among ice and snow. Of the scanty produce of the fields, much was left to rot on the ground, and much of what was carried home proved unfit for the sustenance of either man or beast. There is a tradition that a farmer of Cromarty employed his children, during the whole winter of 1694, in picking out the sounder grains of corn from a blasted heap, the sole product of his farm, to serve for seed in the ensuing spring.

In the meantime the country began to groan under famine. The little portions of meal which were brought to market were invariably disposed of at exorbitant prices, before half the people were supplied; "and then," says Walker, "there would ensue a screaming and clapping of hands among the women." " How shall we go home," he has heard them exclaim, "and see our children dying of hunger?—they have had no food for these two days and we have nothing to give them." There was many " a black and pale face in Scotland;" and many of the labouring poor, ashamed to beg, and too honest to steal, shut themselves up in their comfortless houses, to sit with their eyes fixed on the floor till their very sight failed them. The savings of the careful and industrious were soon dissipated; and many who were in easy circumstances when the scarcity came on, had sunk into abject poverty ere it passed away. Human nature is a sad thing when subjected to the test of circumstances so trying. As the famine increased, people came to be so wrapped up in their own sufferings, that "wives thought not of their husbands, nor husbands of their wives, parents of their children, nor children of their parents." " And their staff of bread," says the Pedlar, "was so utterly broken, that when they ate they were neither satisfied nor nourished. They could think of nothing but food, and being wholly unconcerned whether they went to heaven or hell, the success of the gospel came to a stand."

The pestilence which accompanied this terrible visitation broke out in November 1694, when many of the people were seized by " strange fevers, and sore fluxes of a most infectious nature," which defied the utmost power of medicine. " For the oldest physicians," says Walker, "had never seen the like before, and could make no help." In the parish of West Calder, out of nine hundred "examinable persons" three hundred were swept away; and in Livingston, in a little village called the Craigs, inhabited by only six or eight families, there were thirty corpses in the space of a few days. In the parish of Resolis whole villages were depopulated, and the foundations of the houses, for they were never afterwards inhabited, can still be pointed out by old men of the place. So violent were the effects of the disease, that people, who in the evening were in apparent health, would be found lying dead in their houses next morning, "the head resting on the hand, and the face and arms not unfrequently gnawed by the rats." The living were wearied with burying the dead; bodies were drawn on sledges to the place of interment, and many got neither coffin nor winding-sheet. "I was one of four," says the Pedlar, "who carried the corpse of a young woman a mile of way; and when we came to the grave, an honest poor man came and said—'You must go and help me to bury my son; he has lain dead these two days.' We went, and had two miles to carry the corpse, many neighbours looking on us, but none coming to assist." "I was credibly informed," he continues, "that in the north, two sisters, on a Monday morning, were found carrying their brother on a barrow with bearing-ropes, resting themselves many times, and none offering to help them." There is a tradition that in one of the villages of Resolis the sole survivor was an idiot, whose mother had been, of all its more sane inhabitants, the last victim to the disease. He waited beside the corpse for several days, and then taking it up on his shoulders carried it to a neighbouring village, and left it standing upright beside a garden wall.

Such were the sufferings of the people of Scotland in the seventeenth century, and such the phenomena of character which the sufferings elicited. We ourselves have seen nearly

the same process repeated in the nineteenth, and with nearly
the same results. The study of mind cannot be prosecuted in
quite the same manner as the study of matter. We cannot
subject human character, like an earth or metal, to the test
of experiments which may be varied or repeated at pleasure;
on the contrary, many of its most interesting traits are developed only by causes over which we have no control. But
may we not regard the whole world as an immense laboratory,
in which the Deity is the grand chemist, and His dispensations
of Providence a course of experiments? We are admitted into
this laboratory, both as subjects to be acted upon and as spectators; and, though we cannot in either capacity materially
alter the course of the exhibition, we may acquire much whole-
some knowledge by registering the circumstances of each process, and its various results.

In the year 1817 a new and terrible pestilence broke out in
a densely-peopled district of Hindostan. During the twelve
succeeding years it was " going to and fro, and walking up and
down," in that immense tract of country which intervenes
between British India and the Russian dominions in Europe. It
passed from province to province, and city to city. Multitudes,
" which no man could number," stood waiting its approach in
anxiety and terror; a few solitary mourners gazed at it from
behind. It journeyed by the highways, and strewed them with
carcases. It coursed along the rivers, and vessels were seen
drifting in the current with their dead. It overtook the caravan
in the desert, and the merchant fell from his camel. It followed
armies to the field of battle, struck down their standards, and
broke up their array. It scaled the great wall of China, forded
the Tigris and the Euphrates, threaded with the mountaineer the
passes of the frozen Caucasus, and traversed with the mariner
the wide expanse of the Indian Ocean. Vainly was it deprecated with the rites of every religion, exorcised in the name
every god. The Brahmin saw it rolling onwards, more terrible

than the car of Juggernaut, and sought refuge in his temple; but the wheel passed over him, and he died. The wild Tartar raised his war-cry to scare it away, and then, rushing into a darkened corner of his hut, prostrated himself before his idol, and expired. The dervish ascended the highest tower of his mosque to call upon Allah and the prophet; but it grappled with him ere he had half repeated his prayer, and he toppled over the battlements. The priest unlocked his relics, and then, grasping his crucifix, hied to the bedside of the dying; but, as he doled out the consolations of his faith, the pest seized on his vitals, and he sunk howling where he had kneeled. And alas for the philosopher! silent and listless he awaited its coming; and had the fountains of the great deep been broken up, and the proud waves come rolling, as of old, over wide-extended continents, foaming around the summit of the hills, and prostrating with equal ease the grass of the field and the oak of the forest, he could not have met the inundation with a less effective resistance. It swept away in its desolating progress a hundred millions of the human species.

In the spring of 1831 the disease entered the Russian dominions, and in a few brief months, after devastating the inland provinces, began to ravage the shores of the Baltic. The harbours, as is usual in the summer season, were crowded with vessels from every port of Britain: and the infection spread among the seamen. To guard against its introduction into this country, a rigid system of quarantine was established by the Government; and the Bay of Cromarty was one of the places appointed for the reception of vessels until their term of restriction should have expired. The whole eastern coast of Britain could not have afforded a better station; as, from the security and great extent of the bay, entire fleets can lie in it safe from every tempest, and at a distance of more than two miles from any shore.

On a calm and beautiful evening in the month of July 1831,

Q

a little fleet of square-rigged vessels were espied in the offing, slowly advancing towards the bay. They were borne onwards by the tide, which, when flowing, rushes with much impetuosity through the narrow opening, and, as they passed under the northern Sutor, there was seen from the shore, relieved by the dark cliffs which frowned over them, a pale yellow flag dropping from the mast-head of each. As they advanced farther on, the tide began to recede. The foremost was towed by her boats to the common anchoring-ground; and the burden of a Danish song, in which all the rowers joined, was heard echoing over the waves with a cadence so melancholy, that, associating in the minds of the town's-people with ideas of death and disease, it seemed a coronach of lamentation poured out over the dead and the expiring. The other vessels threw out their anchors opposite the town;—groups of people, their countenances shaded by anxiety, sauntered along the beach; and children ran about, shouting at the full pitch of their voices that the ships of the plague had got up as far as the ferry. As the evening darkened, little glimmering lights, like stars of the third magnitude, twinkled on the mast-heads from whence the yellow flags had lately depended; and never did astrologer experience greater dismay when gazing at the two comets, the fiery and the pale, which preceded those years of pestilence and conflagration that wasted the capital of England, than did some of the people of Cromarty when gazing at these lights.

Day after day vessels from the Baltic came sailing up the bay, and the fears of the people, exposed to so continual a friction, began to wear out. The first terror, however, had been communicated to the nearer parishes, and from them to the more remote; and so on it went, escorted by a train of vagabond stories, that, like felons flying from justice, assumed new aspects at every stage. The whole country talked of nothing but Cholera and the Quarantine port. Such of the shopkeepers of Cromarty as were most in the good graces of the countrywomen who came to town laden with the produce of the dairy and hen-cot, and return with their little parcels of the luxuries of the grocer, experienced a marked falling away in their trade. Occasionally, however, a few of the more courageous housewives might be seen creeping warily along our streets; but, in coming in by the road which passes along the edge of the bay, they invariably struck up the hill if the wind blew from off the quarantine vessels, and, winding by a circuitous route among the fields and cottages, entered the town on the opposite side. A lad who ran errands to a neighbouring burgh, found that few of the inhabitants were so desperately devoted to business as to incur the risk of receiving the messages he brought them; and, from the inconvenient distance at which he was held by even the less cautious, he entertained serious thoughts of providing himself with a speaking-trumpet. Our poor fishermen, too, fared but badly in the little villages of the Firth where they went to sell their fish. It was asserted on the very best authority, by the villagers, that dead bodies were flung out every day over the sides of the quarantine vessels, and might be seen, bloated by the water and tanned yellow by disease, drifting along the surface of the bay. Who could eat fish in such circumstances? There was one person, indeed, who remarked to them, that he might perhaps venture on eating a haddock or whiting; but no man in his senses, he said, would venture on eating a cod. He himself had once found a bunch of furze in the stomach of a fish of this species, and what might not that throat contrive to swallow that had swallowed a bunch of furze? The very fishermen themselves added to the general terror by their wild stories. They were rowing homewards one morning, they said, in the grey uncertain light which precedes sunrise, along the rough edge of the northern Sutor, when, after doubling one of the rocky promontories which jut into the sea from beneath the crags of the hill, they saw a gigantic figure, wholly attired in white, winding slowly along the beach.

It was much taller than any man, or as Cowley would perhaps have described it, than the shadow of any man in the evening; and at intervals, after gliding round the base of some inaccessible cliff, it would remain stationary for a few seconds, as if gazing wistfully upon the sea. No one who believed this apparition to be other than a wreath of vapour, entertained at the time the slightest doubt of its portending the visitation of some terrible pestilence, which was to desolate the country.

About eighty or a hundred years ago the port of Cromarty was occupied, as in 1831, by a fleet performing quarantine. Of course none of the town's-people recollected the circumstance; but a whole host of traditions connected with it, which had been imparted to them by their fathers, and had lain asleep in the recesses of some of their memories for a full half century, were awakened at this time, and sent wandering over the town, like so many ghosts. Some one had heard it told that a crew of Cromarty fishermen had, either in ignorance or contempt of the quarantine laws, boarded one of the vessels on this occasion; and that aboard they were compelled to remain for six tedious weeks, exposed to the double, but very unequally appreciated hardship of getting a great deal to drink and very little to eat. Another vessel had, it was said, entered the bay deeply laden; but every morning, for the time she remained there, she was seen to sit lighter on the water, and when she quitted it on her return to Flushing, she had scarcely ballast enough aboard to render the voyage practicable. Gin and tobacco were rife in Cromarty for twelve months thereafter. A third vessel carried with her into the bay the disease to guard against which the quarantine had been established; and opposite the place where the fleet lately lay, there are a few little mounds on a patch of level sward, still known to children of the town as the Dutch-men's graves. About fifty years ago, when the present harbour of Cromarty was in building, a poor half-witted man, one of the labourers employed in quarrying stone, was told one day by some of his companions, that a considerable sum of money had been deposited in this place with the bodies. In the evening he stayed on some pretext in the quarry until the other workmen had gone home, and then repairing to the graves, with his shovel and pickaxe he laid one of them open ; but, instead of the expected treasure, he found only human bones and wasted fragments of woollen cloth. Next morning he was seized by a putrid fever, and died a few days after. Miss Seward tells a similar story in one of her letters ; but in the case of the Cromarty labourer no person suffered from his imprudence except himself ; whereas, in the one narrated by Miss Seward, a malignant disease was introduced into a village near which the graves were opened, which swept away seventy of the inhabitants.

In a central part of the churchyard of Nigg there is a rude undressed stone, near which the sexton never ventures to open a grave. A wild apocryphal tradition connects the erection of this stone with the times of the quarantine fleet. The plague, as the story goes, was brought to the place by one of the vessels, and was slowly flying along the ground, disengaged from every vehicle of infection, in the shape of a little yellow cloud. The whole country was alarmed, and groups of people were to be seen on every eminence, watching with anxious horror the progress of the little cloud. They were relieved, however, from their fears and the plague by an ingenious man of Nigg, who, having provided himself with an immense bag of linen, fashioned somewhat in the manner of a fowler's net, cautiously approached the yellow cloud, and, with a skill which could have owed nothing to previous practice, succeeded in enclosing the whole of it in the bag. He then secured it by wrapping it up carefully, fold after fold, and fastening it down with pin after pin ; and as the linen was gradually changing, as if under the hands of the dyer, from white to yellow, he consigned it to the churchyard, where it has slept ever since. But to our narrative.

The cholera was at length introduced into Britain, and shortly after into Ireland; not, however, at any of the quarantine ports, but at places where scarcely any precautions had been taken to exclude it, or any danger apprehended; much in the manner that a beleaguered garrison is sometimes surprised at some unnoticed bastion, or untented angle, after the main points of attack have withstood the utmost efforts of the besiegers. It had previously been remarked that the disease traversed the various countries which it visited, at nearly the same pace with the inhabitants. In Persia, where there is little trade, and neither roads nor canals to facilitate intercourse, it was a whole year in passing over a distance of somewhat less than three hundred leagues; while among the more active people of Russia, it performed a journey of seven hundred in less than six months. In Britain it travelled through the interior with the celerity of the mail, and voyaged along the coasts with the speed of the trading vessels; and in a few weeks after its first appearance, it was ravaging the metropolis of England, and the southern shores of the Firth of Forth. It was introduced by some south country fishermen into the town of Wick, and a village of Sutherlandshire, in the month of July 1832; and from the latter place in the following August, into the fishing villages of the peninsula of Easter Ross. It visited Inverness, Nairn, Avoch, Dingwall, Urquhart, and Rosemarkie, a few weeks after.

I shall pass hurriedly over the sad story of its ravages. Were I to dwell on it to the extent of my information, and I know only a little of the whole, the reader might think I was misanthropically accumulating into one gloomy heap all that is terrible in the judgments of God, and all that is mean and feeble in the character of man. The pangs of the rack, the boot, the thumbscrew—all that the Dominican or the savage has inflicted on the heretic or the white man, were realized in the tortures of this dreadful disease. Utter debility, intense thirst, excruciating cramps of the limbs, and an unimpaired intellect, were its chief characteristics. And the last was not the least terrible. Amid the ruins of the body, from which it was so soon to part, the melancholy spirit looked back upon the past with regret, and on the future with terror. Or even if the sufferer amid his fierce pain " laid hold on the hope that faileth not;" with what feelings must he have looked around the deserted cottage, when the friends in whom he had trusted proved unfaithful—or, more melancholy still, on the affectionate wife or the dutiful child struck down by the bedside in agonies as mortal as his own.

In the villages of Ross the disease assumed a more terrible aspect than it had yet presented in any other part of Britain. In the little village of Portmahomack one-fifth of the inhabitants were swept away; in the still smaller village of Inver, one-half. So abject was the poverty of the people, that in some instances there was not a candle in any house in a whole village; and when the disease seized on the inmates in the night-time, they had to grapple in darkness with its fierce agonies and mortal terrors, and their friends, in the vain attempt to assist them, had to grope round their beds. The infection spread with frightful rapidity. At Inver, though the population did not much exceed a hundred persons, eleven bodies were committed to the earth, without shroud or coffin, in one day; in two days after they had buried nineteen more. Many of the survivors fled from the village, and took shelter, some in the woods, some among the hollows of an extensive tract of sand-hills. But the pest followed them to their hiding-places, and they expired in the open air. Whole families were found lying dead on their cottage floor. In one instance, an infant, the only survivor, lay grovelling on the body of its mother—the sole mourner in a charnel-house of the pestilence. Rows of cottages, entirely divested of their inhabitants, were set on fire and burned to the ground. The horrors of the times of Peter Walker were more than realized. Two young persons, a lad and his sister,

were seen digging a grave for their father in the churchyard of
Nigg; and then carrying the corpse to it on a cart, no one ven-
turing to assist them. The body of a man who died in a cottage beside the ferry of Cromarty, was borne to a hole, hurriedly
scooped out of a neighbouring sand-bank, by his brother and
his wife. During the whole of the preceding day, the unfortunate woman had been seen from the opposite shore, flitting around
the cottage like an unhappy ghost; during the whole of the preceding night had she watched alone by the dead. The coffin lay
beside the door; the corpse in the middle of the apartment.—
Never shall I forget the scene which I witnessed from the old
chapel of St. Regulus on the evening of the following Sabbath.

It was one of those lovely evenings which we so naturally
associate with ideas of human enjoyment; when, from some
sloping eminence, we look over the sunlit woods, fields, and
cottages, of a wide extent of country, and dream that the inhabitants are as happy as the scene is beautiful. The sky was
without a cloud, and the sea without a wrinkle. The rocks
and sandhills on the opposite shore lay glistening in the sun,
each with its deep patch of shadow resting by its side; and the
effect of the whole, compared with the aspect which it had presented a few hours before, was as if it had been raised on its
groundwork of sea and sky from the low to the high relief of
the sculptor. There were boats drawn up on the beach, and a
line of houses behind; but where were the inhabitants? No
smoke rose from the chimneys; the doors and windows were
fast closed; not one solitary lounger sauntered about the harbour
or the shore; the inanity of death and desertion pervaded the
whole scene. Suddenly, however, the eye caught a little dark
speck moving hurriedly along the road which leads to the ferry.
It was a man on horseback. He reached the cottages of the
boatmen, and flung himself from his horse; but no one came at
his call to row him across. He unloosed a skiff from her moorings, and set himself to tug at the oar. The skiff flew athwart

the bay. The watchmen stationed on the shore of Cromarty moved down to prevent her landing. There was a loud cry passed from man to man; a medical gentleman came running to the beach, he leapt into the skiff, and laying hold of an oar as if he were a common boatman, she again shot across the bay. A case of cholera had just occurred in the parish of Nigg. I never before felt so strongly the force of contrast. There is a wild poem of the present age which presents the reader with a terrible picture of a cloak of utter darkness spread over the earth, and the whole race of man perishing beneath its folds, like insects of autumn in the chills of a night of October. There is another modern poem, less wild, but not less sublime, in which we see, as in a mirror of a magician, the sun dying in the heavens, and the evening of an eternal night closing around the last of our species. I trust I am able in some degree to appreciate the merits of both; and yet, since witnessing the scene which I have so feebly attempted to describe, I am led to think that the earth, if wholly divested of its inhabitants, would present a more melancholy aspect, should it still retain its fertility and beauty, than if wrapped up in a pall of darkness, surrounded by dead planets and extinguished suns.

CHAPTER XVII.

"He sat upon a rock and bobbed for whale."—KENRICK.

ON the fourth Tuesday of November every year, there is a kind of market held at Cromarty, which for the last eighty years has been gradually dwindling in importance, and is now attended by only the children of the place, and a few elderly people, who supply them with toys and sweetmeats. Early in the last century, however, it was one of the most considerable in this part of the country; and the circumstance of its gradual decline is curiously connected with the great change which has taken place since that period in the manners and habits of the people. It flourished as long as the Highlander legislated for himself and his neighbour on the good old principle so happily described by the poet,[1] and sunk into decay when he had flung down his broadsword, and become amenable to the laws of the kingdom. The town of Cromarty, as may be seen by consulting the map, is situated on the extremity of a narrow promontory, skirted on three of its sides by the sea, and bordered on the fourth by the barren uninhabited waste described in a previous chapter. And though these are insurmountable defects of situation for a market of the present day, which ought always to be held in some central point of the interior that commands a wide circumference of country, about a century ago they were positive advantages. It was an important circumstance that

[1] For why? because the good old rule
Sufficeth them, the simple plan,
That they should take who have the power,
And they should keep who can.

the merchants who attended the fair could convey their goods to it by sea, without passing through any part of the Highlands; and the extent of moor which separated it by so broad a line from the seats of even the nearer clans, afforded them no slight protection when they had arrived at it. For further security the fair was held directly beneath the walls of the old castle, in the gorge of a deep wooded ravine, which now forms part of the pleasure-grounds of Cromarty House.

The progress of this market, from what it was once to what it is at present, was strongly indicative of several other curious changes which were taking place in the country. The first achievement of commerce is the establishment of a market. In a semi-barbarous age the trader journeys from one district to another, and finds only, in a whole kingdom, that demand for his merchandise which, when in an after period civilisation has introduced her artificial wants among the people, may be found in a single province. So late as the year 1730, one solitary shopkeeper more than supplied the people of Cromarty with their few, everyday necessaries, of foreign manufacture or produce; I say more than supplied them, for in summer and autumn he travelled the country as a pedlar. For their occasional luxuries and finery they trusted to the traders of the fair. Times changed, however, and the shopkeeper wholly supplanted the travelling merchant; but the fair continued to be frequented till a later period by another class of traders, who dealt in various articles, the produce and manufacture of the country. Among these were a set of dealers who sold a kind of rude harness for horses and oxen, made of ropes of hair and twisted birch; a second set who dealt in a kind of conical-shaped carts made of basket-work; and a third who supplied the house-builders of the period with split lath, made of moss-fir, for thatched roofs and partitions. In time, however, the harness-maker, cart-wright, and house-carpenter of modern times, dealt by these artists as the shopkeeper had done by the market-trader. The broguer, or maker of Highland shoes, kept the field in spite of the regular shoemaker half a century later, and disappeared only about five years ago. The dealer in home-grown lint frequented it until last season; but the low wages, and sixteen-hour-per-day employment of the south country weaver, were gradually undermining his trade, and the steam-loom seems to have given it its deathblow.

Prior to the Revolution, and as late as the reign of Queen Anne, Cromarty drove a considerable trade in herrings. About the middle of July every year, immense bodies of this fish came swimming up the Moray Firth; and after they had spawned on a range of banks not more than eight miles from the town, quitted it for the main sea in the beginning of September. In the better fishing seasons they filled the bays and creeks of the coast, swimming in some instances as high as the ferries of Fowlis and Ardersier. There is a tradition that, shortly after the Union, a shoal of many hundred barrels, pursued by a body of whales and porpoises, were stranded in a little bay of Cromarty, a few hundred yards to the east of the town. The beach was covered with them to the depth of several feet, and salt and casks failed the packers when only an inconsiderable part of the shoal was cured. The residue was carried away for manure by the neighbouring farmers; and so great was the quantity used in this way, and the stench they caused so offensive, that it was feared disease would have ensued. The season in which this event took place is still spoken of as the "har'st of the Herring-drove."

About thirty years ago some masons, in digging a foundation in the eastern extremity of the town, discovered the site of a packing-yard of this period; and threw out vast quantities of scales which glittered as bright as if they had been stripped from the fish only a few weeks before. Near the same place, there stood about twenty years earlier a little grotesque building two storeys in height, and with only a single room on each floor.

The lower was dark and damp, and had the appearance of a cellar or storehouse ; the upper was lighted on three sides, and finished in a style which, at the period of its erection, must have led to a high estimate of the taste of the builder. A rich cornice, designed doubtless on the notion of Ramsay, that good herrings and good claret are very suitable companions, curiously united bunches of grapes with clusters of herrings, and divided the walls from the ceiling. The walls were neatly panelled, the centre of the ceiling was occupied by a massy circular patera, round which a shoal of neatly relieved herrings were swimming in a sea of plaster. This building was the place of business of Urquhart of Greenhill, a rich herring merchant and landed proprietor, and a descendant of the old Urquharts of Cromarty. But it was destined long to survive the cause of its erection.

In a fishing season late in this period, two men of the place, who, like most of the other inhabitants, were both tradesfolks and fishermen, were engaged one morning in discussing the merits of an anker of Hollands which had been landed from a Dutch lugger a few evenings before. They nodded to each other across the table with increasing heartiness and good-will, until at length their heads almost met ; and as quaich after quaich was alternately emptied and replenished, they began to find that the contents of the anker were best nearest the bottom. They were interrupted, however, before they had fully ascertained the fact, by the woman of the house tapping at the window, and calling them out to see something extraordinary ; and, on going to the door, they saw a plump of whales blowing, and tumbling, and pursuing one another, in a long line up the bay. A sudden thought struck one of the men : "It would be gran' fun, Charlie man," said he, addressing his companion, " to hook ane o' yon chiels on Nannie Fizzle's crook." " Ay, if we had but bait," rejoined the other ; " but here's a gay fresh codling on Nannie's hake, an' the yawl lies on the tap o' the fu' sea." The crook——a chain about six feet in length, with a hook at one end, and a large ring at the other, and which, when in its proper place, hung in Nannie's chimney to suspend her pots over the fire—was accordingly baited with the cod, and fastened to a rope; and the two men, tumbling into their yawl rowed out to the *cossmee*.   Like the giant of the epigram they sat bobbing for whale, but the plump had gone high up the Firth; and, too impatient to wait its return, they hollowed to a friend to row out his skiff for them; and leaving their own at anchor, with the crook hanging over the stern, they returned to Nannie Fizzle's, where they soon forgot both the yawl and the whales.

They were not long, however, in being reminded of both.   A person came bellowing to the window, " Charlie, Willie, the yawl! the yawl!" and, on staggering out, they saw the unfortunate yawl darting down the Firth with twice the velocity of a king's cutter in a fresh breeze.   Ever and anon she would dance, and wheel, and plunge, and then shoot off in a straight line.   Wonderful to relate! one of the whales had swallowed the crook; the little skiff was launched and manned; but the Hollands had done its work; one of the poor fellows tumbled over the thaft, the other snapped his oar;—all was confusion. Luckily, however, the rope fastened to the crook broke at the ring; and the yawl, after gradually losing way, began to drift towards the shore.   The adventure was bruited all over the town; and every one laughed at the whale-fishers except Nannie Fizzle, who was inconsolable for the loss of her crook.

It was rumoured a few weeks after that the carcass of a whale had been cast ashore somewhere in the Firth of Beauly, near Redcastle, and the two fishermen set off together to the place, in the hope of identifying the carcass with the fish in which they had enfeoffed themselves at the expense of Nannie Fizzle. The day of the journey chanced to be also that of a Redcastle market; and, as they approached the place, they were encountered by parties of Highlanders hurrying to the fair.   Most of them had heard of the huge fish, but none of them of the crook.

When the Cromarty men came up to the carcass, they found it surrounded by half the people of the fair, who were gazing, and wondering, and pacing it from head to tail, and poking at it with sticks and broadswords. "It is our property every inch," said one of the men, coming forward to the fish; "we hooked it three weeks ago on the *cossmee*, but it broke off; and we have now come here to take possession. It carried away our tackle, a chain and a hook. Lend me your dirk, honest man," he continued, addressing a Highlander; "we shall cut out hook and chain, and make good our claim." "O ay! nae doubt," said the Highlander, as he obligingly handed him the weapon; "but och! it's no me that would like to eat her, for she maun be a filthy meat." The crowd pressed round to witness the dissection, which ended in the Cromarty man pulling out the crook from among the entrails, and holding it up in triumph "Did I no tell you?" he exclaimed; "the fish is ours beyond dispute." "Then," said a smart-looking little pedlar, who had just joined the throng, "ye have made the best o' this day's market. I'se warrant your fishing worth a' the plaiding sold to-day." The Highlanders stared. "For what is it worth?" asked a tacksman of the place. "Oh, look there! look there!" replied the pedlar, tapping the blubber with his elwand, "ulzie clear as usquebaugh. I'se be bound it's as richly worth four hunder punds Scots as ony booth at the fair." This piece of mischievous information entirely altered the circumstances of the case as it regarded the two fishermen; for the tacksman laid claim to the fish on his own behalf and the laird's, and, as he could back his arguments by a full score of broadswords, the men were at length fain to content themselves with being permitted to carry away with them Nannie Fizzle's crook. I am afraid it is such of our naturalists as are best acquainted with the habits of the *cetacea* that will be most disposed to question the truth of the tradition just related. But, however doubtful its foundation, a tradition it is.

The mishap of the whale-fishers was followed by a much greater mishap—the total failure of the herring fishery. The herring is one of the most eccentric little fishes that frequents our seas. For many years together it visits regularly in its season some particular firth or bay;—fishing villages spring up on the shores, harbours are built for the reception of vessels; and the fisherman and merchant calculate on their usual quantum of fish, with as much confidence as the farmer on his average quantum of grain. At length, however, there comes a season, as mild and pleasant as any that have preceded it, in which the herring does not visit the firth. On each evening, the fisherman casts out his nets on the accustomed bank, on each morning he draws them in again, but with all the meshes as brown and open as when he flung them out; in the following season he is equally unsuccessful; and, ere the shoal returns to its accustomed haunts, the harbour has become a ruin, and the village a heap of green mounds. It happened thus, late in the reign of Queen Anne, with the herring trade of the Moray Firth. After a busy and successful fishing, the shoal as usual left the Firth in a single night; preparations were made for the ensuing season; the season came, but not the herrings; and for more than half a century from this time Cromarty derived scarcely any benefit from its herring fishery.

My town's-folk in this age—an age in which every extraordinary effect was coupled with a supernatural cause—were too ingenious to account for the failure of the trade by a simple reference to the natural history of the herring; and two stories relating to it still survive, which show them to have been strangely acute in rendering a reason, and not a little credulous in forming a belief. Great quantities of fish had been caught and brought ashore on a Saturday, and the packers continued to work during the night; yet on the Sunday morning much still remained to be done. The weather was sultry, and the fish were becoming soft; and the merchants, unwilling to

lose them, urged on the work throughout the Sabbath. Towards evening the minister of the parish visited the packers;
and, as they had been prevented from attending church, he
made them a short serious address. They soon, however, became impatient; the diligent began to work, the mischievous
to pelt him with filth; and the good man abruptly concluded
his exhortation by praying that the besom of judgment would
come and sweep every herring out of the Firth. On the following Monday the boats went to sea as usual, but returned
empty; on the Tuesday they were not more successful, and it
was concluded that the shoal had gone off for the season; but
it proved not for the season merely; for another and another
season came, and still no herrings were caught. In short, the
prayer, as the story goes, was so fully answered, that none of
the unlucky packers who had insulted the minister witnessed
the return of the shoal.

The other story accounts for its flight in a different and
somewhat conflicting manner. Tradition, who, as I have already
shown, is even a more credulous naturalist than historian,
affirms that herrings have a strong antipathy to human blood,
especially when spilt in a quarrel. On the last day of the
fishing, the nets belonging to two boats became entangled; the
crew that first hauled applied the knife to their neighbours'
baulks and meshes, and, with little trouble or damage to themselves, succeeded in unravelling their own. A quarrel was the
consequence; and one of the ancient modes of naval warfare,
the only one eligible in their circumstances, was resorted to—
they fought leaning over the gunwales of their respective boats.
Blood was spilt, unfortunately spilt in the sea; the affronted
herrings took their departure, and for more than half a century
were not the cause, in even the remotest degree, of any quarrel
which took place on the Moray Firth or its shores. One of the
combatants, who distinguished himself either by doing or suffering in this unlucky fray, was known ever after by the name of

Andrew *Bleed;* and there are men still living who remember to have seen him.

The failure of the herring trade was followed by that of Urquhart of Greenhill.   He is said to have been a shrewd industrious man, of great force of character, and admirably fitted by nature and habit, had he lived in better times, to have restored the dilapidated fortunes of his house.   During the reign of William he was adding ship to ship, and field to field, until about the year 1700, when he was possessed of nearly one-half the lands of the parish, and of five large vessels.   But it was his lot to speculate in an unfortunate age ; and having, with almost all the other merchants of Scotland, suffered severely from the Union, the failure of the herring fishery completed his ruin.   He sank by inches ; striving to the last, with a proud heart and a bitter spirit, against the evils which assailed him. All his ships were at length either knocked down by the hammer of the auctioneer, or broken up by the maul of the carpenter, except one ; and that one, the Swallow of Cromartie, when returning homewards from some port of the Continent, was driven ashore in a violent night-storm on the rocky coast of Cadboll, and beaten to pieces before morning.   It was with difficulty the crew was saved.   One of them, a raw young fellow, a much better herdsman than sailor, escaped to his friends, full of the wild scenes he had just witnessed, and set himself to relate to them the particulars of his voyage ;—it was his first and his last.   Smooth water and easy sailing may be delineated in common language ; he warmed, however, as the narrative proceeded. He described the gathering of the tempest, the darkening of the night, the dashing of the waves, the howling of the winds, and the rolling of the vessel ; but being unfortunately no master of climax, language failed him in the concluding scene, where there were rocks, and breakers, and midnight darkness, and a huge ship wallowing in foam, like a wounded boar in the toils of the hunters.   "Oh !" exclaimed the sailor herdsman, "I can think o' nae likening to that puir ship, and the awfu' crags and awfu' jaws, except the nowt i' the byre, when they break their fastenings i' the mirk night, and rout and gore, and rout and gore, till the roof-tree shakes wi' the brattle." The people of the present age may not think much of the comparison ; but it was deemed a piece of very tolerable humour in Cromarty in the good year 1715. Greenhill's remark, when informed of the disaster, had more of philosophy in it. "Aweel," said he, taking a deliberate pinch of snuff, and then handing the box to his informant, " I have lang warstled wi' the warld, and fain would I have got on the tap o't ; but I may be just as weel as I am. Diel haet can harm me now, if the laird o' Cadboll, honest man, doesna put me to the law for dinting the Swallow against his march-stanes."

One other passage relating to the Greenhill branch of the family of the Urquharts, ere I take leave of it for the time. It has produced, in a lady of Aberdeenshire, one of the most pleasing poetesses of our age and country—not, however, one of the most celebrated. Her exquisite little pieces, combining with singular felicity the simplicity and pathos of the old ballad with the refinement and elegance of our classical poets, have been flung as carelessly into the world as the rich plumes of the birds of the tropics on the plains and forests of the south. But they have not lain altogether unnoticed. The nameless little foundlings have been picked out from among the crowd, and introduced into the best company on the score of merit alone.—The genealogist was of a different spirit from his relative ; he would have inscribed his name on the face of the sun could he have but climbed to it ;—but may not there be something to regret in even the more amiable extreme ? The prophecies of that sibyl who committed her writings to the loose leaves of the forest, were lost to the world on the first slight breeze. I present the reader with a pleasing little poem of this descendant of the Urquharts, in which, though perhaps not

one of the most finished of her pieces, he will find something
better than mere finish. It may not be quite new to him,
having found its way into Macdiarmid's Scrap-Book, and several
other collections of merit; but he may peruse it with fresh interest, as the production of a relative of Sir Thomas, who seems
to have inherited all his genius, undebased by any mixture of
his eccentricity.

ON HEARING A LIVELY PIECE OF MUSIC,

"THE WATERLOO WALTZ."

A moment pause, ye British fair,
  While pleasure's phantom ye pursue,
And say if dance and sprightly air
  Suit with the name of Waterloo.
    Dearly bought the victory,
    Chasten'd should the triumph be;
    'Midst the laurels she has won,
    Britain weeps for many a son.

Veil'd in clouds the morning rose,
  Nature seem'd to mourn the day
Which consign'd before its close
  Thousands to their kindred clay.
    How unfit for courtly ball,
    Or the giddy festival,
    Was the grim and ghastly view
    Ere evening closed on Waterloo.

See the Highland warrior rushing,
  First in danger, on the foe,
Till the life-blood, stemless gushing,
  Lays the plaided hero low.
    His native pipe's heart-thrilling sound,
    'Mid war's infernal concert drown'd,
    Cannot soothe his last adieu,
    Nor wake his sleep on Waterloo.

Crashing o'er the cuirassier,
  See the foaming charger flying,
Trampling in his wild career,
  All alike, the dead and dying.
    See the bullets pierce his side,
    See, amid a crimson tide,
    Helmet, horse, and rider too,
    Roll on bloody Waterloo.

Shall sights like these the dance inspire,
 Or wake the jocund notes of mirth ?
Oh, shiver'd be the recreant lyre
 That gave the base idea birth !
  Other sounds, I ween, were there,
  Other music rent the air,
  Other *Waltz* the warriors knew,
  When they closed at Waterloo.

Forbear, till time with lenient hand
 Has heal'd the wounds of recent sorrow,
And let the picture distant stand,
 The softening hue of years to borrow.
  When our race has pass'd away,
  Hands unborn may wake the lay,
  And give to joy alone the view
  Of victory at Waterloo.

About the time of the Rebellion, or a little after, the trade of the place began to recover itself much through the influence of a vigorous-minded man, a merchant of the period. Urquhart of Greenhill had sunk with the sinking trade of the country; his townsman, William Forsyth, enjoyed the advantage of being born at least forty years later, and rose as it revived. The nature of the business which the latter pursued may be regarded as illustrating, not inaptly, the condition of society in the north of Scotland at the time. It was of a miscellaneous character, as became the state of a country so poor and so thinly peopled, and in which, as there was scarce any division of labour, one merchant had to perform the part of many. He supplied the proprietors with teas, wines, and spiceries; with broad-cloths, glass, Delft ware, Flemish tiles, and pieces of japanned cabinet-work; he furnished the blacksmith with iron from Sweden, the carpenter with tar and spars from Norway, and the farmer with flax-seed from Holland. He found, too, in other countries, markets for the produce of our own. The exports of the north of Scotland, at this period, were mostly malt, wool, and salmon. Almost all rents were paid in kind or in labour—the proprietors retaining in their hands a portion of their estates, termed demesnes or *mains*, which was cultivated mostly by their tacksmen or feuars as part of their proper service. Each proprietor, too, had his storehouse or girnal—a tall narrow building, the strong-box of the time—which, at the Martinmas of every year, used to be filled from gable to gable with the grain-rents paid him by his tenants, and the produce of his own farm. His surplus cattle found their way south under charge of the drovers of the period; but it proved a more difficult matter to dispose to advantage of his surplus corn, mostly barley, until some one, more fertile in speculation than the others, originated the scheme of converting it into malt, and exporting it into England and Flanders. And to so great an extent was this trade carried on about the middle of the last century, that in the town of Inverness the English under Cumberland found almost every second building a malt-barn.

It is quite according to the nature of the herrings to resume their visits as suddenly and unexpectedly as they have broken them off, though not until after a lapse of so many seasons, that the fishermen have ceased to watch for their appearance in their old haunts, or to provide the tackle necessary for their capture; and in this way a number of years are sometimes suffered to pass after the return of the fish, ere the old trade is re-established. It was a main object with William Forsyth to guard against any such waste of opportunity on the part of his town's-people; and representing the case to the more intelligent gentlemen of the district, and some of the wealthier merchants of Inverness, he succeeded in forming them, for the encouragement of the herring fishery, into a society, which provided a yearly premium of twenty merks Scots for the first barrel of herrings caught every season in the Moray Firth. The sum was small; but as money at the time was greatly more valuable than now, it proved a sufficient inducement to the fishermen and tradespeople of the place to fit out, about the beginning of autumn every year, a few boats that swept over the various fishing banks for the herrings; and there were not many seasons in which some one crew or other did not catch enough to entitle them to the premium. At length, however, their tackle wore out, and Mr. Forsyth, in pursuance of his scheme, provided himself, at some little expense, with a complete *drift* of nets, which were carried to sea each season by a crew of boatmen, and the search kept up. His exertions, however, could only merit success, without securing it. The fish returned for a few seasons in considerable bodies, and the fishermen procuring nets, several thousand barrels were caught ; but they soon deserted the Firth as entirely as before. It was at the period of this second return that the " Herring Fishery," according to Goldsmith, " employed all Grub Street ;" and " formed the topic of every coffee-house, and the burden of every ballad." The sober English of the times of George II. had got sanguine on the subject, and hope had broken out into poetry. They were " to drag up oceans of gold from the bottom of the sea, and to supply all Europe with herrings on their own terms ;" but their expectations outran the capabilities of the speculation ; " they fished up very little gold" that the essayist " ever heard of, nor did they furnish the world with herrings." Their herring fishery turned out in short to be a mere herring fishery, and not even that for any considerable length of time.

Sir John Sinclair marks the autumn of the year 1770 as a season in which the herring fishery of Caithness suddenly doubled its amount. " From that time," he adds, " the fishery gradually increased for a few years, but afterwards fell off again, and did not revive with spirit until the year 1788." During the short period in which it was plied with success, it was prosecuted by several crews of Cromarty fishermen ; and their first visit to the coast of this northern county, I find connected with a curious anecdote of the class whose extreme singularity gives in some measure evidence of their truth. Invention generally loves a beaten track—it has its rules and its formulas, beyond which it rarely ventures to expatiate ; but the course of real events is narrowed by no such contracting barrier ; the range of possibility is by far too extensive to be fully occupied by the anticipative powers of imagination; and hence it is that true stories are often stranger than fictions, and that their very strangeness, and their dissimilarity from all the models of literary plot and fable, guarantee in some measure their character as authentic.

The hill of Cromarty is skirted, as I have said, by dizzy precipices, some of them more than a hundred yards in height; and one of these, for the last hundred and fifty years, has borne the name of the Caithness-man's Leap. The sheer descent is broken by projecting shelves, covered with a rank vegetation, and furrowed by deep sloping hollows, filled at the bottom with long strips of loose *débris*, which, when set in motion by the light foot of the goat, falls rattling in continuous streams on the beach. The upper part of the precipice is scooped out by a narrow and perilous pathway, which, rising slantways from the shore, along the face of the neighbouring precipices, makes an abrupt turn on the upper edge of the "leap," and then gains the top. Immediately above, on a sloping acclivity, covered for the last century by a thick wood, there was a little field, the furrows of which can still be distinctly traced among the trees, and which, about the time of the Revolution, was tenanted by a wild young fellow, quite as conversant with his fowling-piece as with his plough. He was no favourite with such of the neighbouring proprietors as most resembled himself; the game-laws in Scotland were not quite so stringent at that period as they are now, but game had its value; and sheriffs and barons, addicted to hunting and the chase, who had dungeons in their castles, and gibbets on their Gallow Hills, neither lacked the will nor the power to protect it. And so the tacksman of the the little field found poaching no safe employment; but the dangers he incurred had only the effect common in such cases, of imparting to his character a sort of Irish-like recklessness— a carelessness both of his own life and the lives of others. He had laid down his little field with peas, and was seriously annoyed, when they began to ripen, by the town's boys—mischievous little fellows—who, when on their fishing excursions, would land in a little rocky bay, immediately below the pathway, and ascending the cliffs, carry away his property by armfuls at a time. The old northern pirates were scarcely more obnoxious to the early inhabitants of Scotland than the embryo fishermen to the man of the gun: nay, the man of the gun was himself scarcely more obnoxious to the proprietors. There was no possibility of laying hold of the intruders; a few minutes were sufficient on the first alarm, to bring them from the top to the bottom of the cliffs—a few strokes of the oar set them beyond all reach of pursuit—and he saw that, unless he succeeded in terrifying them into honesty with his gun, they might go on robbing him with impunity until they had left nothing behind them to rob. Matters were in this state when a Caithness boat, laden with timber, moored one morning in the bay below, and one of the crew, a young fellow of eighteen, after climbing the pathway on an excursion of discovery, found out the field of peas. The farmer, on this unlucky morning, had been rated and collared by the laird for shooting a hare, and, very angry, and armed with the gun as usual, he came up to his field, and found the Caithness-man employed in leisurely filling his pockets. He presented his piece and drew the trigger, but the powder flashed in the pan. "The circumstance of being shot," says the ingenious author of Cyril Thornton, "produces a considerable confusion in a man's ideas." The ideas of the Caithness-man became confused in circumstances one degree less trying; for starting away with the headlong speed of a hare roused out of her form, instead of following the windings of the path, he shot right over the precipice at the abrupt angle. Downwards he went from shelf to shelf—now tearing away with him a huge bush of ivy—now darting along a stream of *débris*—now making somersets in mid-air over the perpendicular walls of rock which alternate with the shelving terraces. The fear of the gun precluded every other fear; he reached the beach un-harmed, except by a few slight sprains and a few scratches, and bolting up, tumbled himself into the boat, and dived for shelter under the folds of the sail. The farmer had pursued him to the top of the rock, and had turned the angle just in time to see him dash over; when, horror-struck at so terrible an accident, for he had intended only to shoot the man, he flung away his gun and ran home. Years and generations passed away; the good King William was succeeded by the good Queen Anne, and Anne by the three Georges, successively; the farmer and all his contemporaries passed to the churchyard—his very fields were lost in the thickets of a deep wood;—the story of the Caithness-man had become traditional—elderly men said it had happened in their grandfather's days, and pointing out to the "leap," they adverted to the name which the rock still continued to bear, as proofs that the incident had really occurred —incredible as it might seem that a human creature could possibly have survived such a fall. Ninety years had elapsed from the time, ere the Cromarty fishermen set out on their Caithness expedition. In the first year of the enterprise one of their fleet was storm-bound in a rocky bay, and the crew found shelter in a neighbouring cottage. There was a spectral-looking old man seated in a corner beside the fire. On learning they had come from Cromarty, he seemed to shake off the apathy of extreme age, and began to converse with them; and they were astonished to learn from his narrative that they had before them the hero of the "leap," at that time in his hundred and eighth year.

CHAPTER XVIII.

"He whom my restless gratitude has sought
So long in vain."—THOMSON.

EARLY in the month of April 1734, three Cromarty boatmen, connected with the custom-house, were journeying along the miserable road which at this period winded between the capital of the Highlands and that of the kingdom. They had already travelled since morning more than thirty miles through the wild highlands of Inverness-shire, and were now toiling along the steep side of an uninhabited valley of Badenoch. A dark sluggish morass, with a surface as level as a sheet of water, occupied the bottom of the valley; a few scattered tufts of withered grass were mottled over it, but the unsolid, sooty-coloured spaces between were as bare of vegetation as banks of sea-mud left by the receding tide. On either hand, a series of dreary mountains thrust up their jagged and naked summits into the middle sky. A scanty covering of heath was thrown over their bases, except where the frequent streams of loose *débris* which had fallen from above, were spread over them; but higher up, the heath altogether disappeared, and the eye rested on what seemed an endless file of bare gloomy cliffs, partially covered with snow.

The evening, for day was fast drawing to a close, was as melancholy as the scene. A dense volume of grey cloud hung over the valley like a ceiling, and seemed descending along the cliffs. There was scarcely any wind, but at times a wreath of vapour would come rolling into a lower region of the valley, as if shot out from the volume above; and the chill bleak air was filled with small specks of snow, so light and fleecy that they seemed scarcely to descend, but, when caught by the half perceptible breeze, went sailing past the boatmen in long horizontal lines. It was evident there impended over them one of those terrible snow-storms which sometimes overwhelm the hapless traveller in these solitudes ; and the house in which they were to pass the night was still nearly ten miles away.

The gloom of evening, deepened by the coming storm, was closing around them as they entered one of the wildest recesses of the valley, an immense precipitous hollow scooped out of the side of one of the hills ; the wind began to howl through the cliffs, and the thickening flakes of snow to beat against their faces. "It will be a terrible night, lads, in the Moray Firth," said the foremost traveller, a broad-shouldered, deep-chested, strong-looking man, of about five feet eight ; "I would ill like to hae to beat up through the drift along the rough shores o' Cadboll. It was in just such a night as this, ten year ago, that old Walter Hogg went down in the Red Sally."—"It will be as terrible a night, I'm feared, just where we are, in the black strath o' Badenoch," said one of the men behind, who seemed much fatigued ; "I wish we were a' safe i' the clachan." —"Hoot, man," said Sandy Wright, the first speaker, "it canna now be muckle mair than sax miles afore us, an' we'll hae the tail of the gloamin' for half an hour yet. But, gude safe us ! what's that ?" he exclaimed, pointing to a little figure that seemed sitting by the side of the road, about twenty yards before him ; "it's surely a fairy !" The figure rose from its seat, and came up, staggering apparently from extreme weakness, to meet them. It was a boy scarcely more than ten years of age. "O my puir boy !" said Sandy Wright, "what can hae taken ye here in a night like this ?"—"I was going to Edinburgh to my friends," replied the boy, "for my mother died and left me among the *freme*; but I'm tired, and canna walk farther ; and I'll be lost, I'm feared, in the *yowndrift*."—

"That ye winna, my puir bairn," said the boatman, "if I can help it; gi'es a haud o' your han'," grasping, as he spoke, the extended hand of the boy; "dinna tine heart, an' lean on me as muckle's ye can." But the poor little fellow was already exhausted, and, after a vain attempt to proceed, the boatman had to carry him on his back. The storm burst out in all its fury; and the travellers, half suffocated, and more than half blinded, had to grope onwards along the rough road, still more roughened by the snow-wreaths that were gathering over it. They stopped at every fiercer blast, and turned their backs to the storm to recover breath; and every few yards they advanced, they had to stoop to the earth to ascertain the direction of their path, by catching the outline of the nearer objects between them and the sky. After many a stumble and fall, however, and many a groan and exclamation from the two boatmen behind, who were well-nigh worn out, they all reached the clachan in safety about two hours after nightfall.

The inmates were seated round an immense peat fire, placed, according to the custom of the country, in the middle of the floor. They made way for the travellers; and Sandy Wright, drawing his seat nearer the fire, began to chafe the hands and feet of the boy, who was almost insensible from cold and fatigue. "Bring us a mutchkin o' brandy here," said the boatman, "to drive out the cauld frae our hearts; an', as supper canna be ready for a while yet, get me a piece bread for the boy. He has had a narrow escape, puir little fellow; an' maybe there's some that would miss him, lanerly as he seems. Only hear how the win' roars on the gable, an' rattles at the winnocks and the door. It's an awfu' night in the Moray Firth."

"It's no gude," continued the boatman, as he tendered a half glass of the brandy and a cake of bread to his *protégé*, "it's no gude to be ill-set to boys. My own loon, Willie, that's the liftenant now, taught me a lesson o' that. He was a wild roytous laddie, fu' o' droll mischief, an' desperately fond o' doos an' rabbits. He had a doo's nest out in the Crookburn Wood ; but he was muckle in the dread o' fighting Rob Moffat, the gamekeeper ; an', on the day it was ripe for harrying, what did he do but set himself to watch Rob, at his house at the Mains ? He saw him setting off to the hill, as he thought, wi' his gun an' his twa dogs ; an' then awa sneaks he to the burn, thinking himsel' out o' Rob's danger. He could climb like a cat, an' so up he clamb to the nest ; an' then wi' his bonnet in his teeth, an' the twa doos in his bonnet, he drapped down frae branch to branch. But, as ill luck would hae it, the first thing he met at the bottom was muckle Rob. The cankered wretch raged like a madman, an' laying hold on the twa birds by the feet, he dawded them about Willie's face till they were baith massacred. It was an ill-hearted cruel thing ; an', had I been there, I would hae tauld him sae on the deafest side o' his head, lang though he be. Willie cam' hame wi' his chafts a' swelled an' bluidy, an' the greet, puir chield, in his throat, for he was as muckle vexed as hurt. He was but a thin slip o' a callant at the time ; but he had a high spirit, an', just out o' the healey, awa he went in young Captain Robinson's lugger, an' didna come near the place, though he sent his mither pennies now an' then by the Campvere traders, for about five years. Weel, back he cam' at last, a stalwart young fallow o' sax feet, wi' a grip that would spin the bluid out at the craps o' a chield's fingers ; an' we were a' glad to see him ! 'Mither,' said he, ' is fighting Rob Moffat at the Mains yet ?' ' O ay !' quo' she. ' Weel, then, I think I'll call on him in the morning,' says he, ' an' clear aff an old score wi' him ;' an' his brow grew black as he spoke. We baith kent what was working wi' him ; an', after bedtime, his mither, puir body, gaed up a' the length o' the Mains to warn Rob to keep out o' the way. An' weel did he do that ; for, for the three weeks that Willie stayed at hame wi' us, not a bit o' Rob was to be seen at either kirk or market.—Puir Willie ! he has got fighting enough sinsyne."

Sandy Wright shared with the boy his supper and his bed; and, on setting out on the following morning, he brought him along with him, despite the remonstrances of the other boatmen, who dreaded his proving an incumbrance. The story of the little fellow, though simple, was very affecting. His mother, a poor widow, had lived for the five preceding years in the vicinity of Inverness, supporting herself and her boy by her skill as a seamstress. As early as his sixth year he had shown a predilection for reading; and, with the anxious solicitude of a Scottish mother, she had wrought late and early to keep him at school. But her efforts were above her strength, and, after a sore struggle of nearly four years, she at length sank under them. "Oh!" said the boy to his companion, "often would she stop in the middle of her work, and lay her hand on her breast, and then she would ask me what would I do when she would be dead—and we would both greet. Her fingers grew white and sma', and she couldna sit up at nights as before; but her cheeks were redder and bonnier than ever, and I thought that she surely wouldna die;—she has told me that she wasna eighteen years older than mysel'. Often, often when I waukened in the morning, she would be greetin' at my bedside; and I mind one day, when I brought home the first prize from school, that she drew me till her, an' told me wi' the tear in her ee, that the day would come, when her head would be low, that my father's gran' friends, who were ashamed o' her because she was poor, would be proud that I was connected wi' them. She soon couldna hold up her head at all, and if it wasna for a neighbour woman, who hadna muckle to spare, we would have starved. I couldna go to the school, for I needed to stay and watch by her bedside, and do things in the house; and it vexed her more that she was keeping me from my learning, than that hersel' was sae ill. But I used to read chapters to her out of the Bible. One day when she was very sick, two neighbour women came in, and she called me to her and told me, that when she would be dead I would need to go to Edinburgh, for I had no friends anywhere else. Her own friends were there, she said, but they were poor, and couldna do muckle for me ; and my father's friends were there too, and they were gran' and rich, though they wadna own her. She told me no to be feared by the way, for that Providence kent every bit o't, and He would make folk to be kind to me ; and then she kissed me, and grat, and bade me go to the school. When I came out she was lying wi' a white cloth on her face, and the bed was all white. She was dead ; and I could do nothing but greet a' that night ; and she was dead still. I'm now travelling to Edinburgh, as she bade me, and folk are kind to me just as she said ; and I have letters to show me the way to my mother's friends when I reach the town ; for I can read write." Such was the narrative of the poor boy.

Throughout the whole of the journey, Sandy Wright was as a father to him. He shared with him his meals and his bed, and usually for the last half dozen miles of every stage, he carried him on his back. On reaching the Queensferry, however, the boatman found that his money was wellnigh expended. I must just try and get him across, thought he, without paying the fare. The boat had reached the middle of the ferry, when one of the ferrymen, a large gruff-looking fellow, began to collect the freight. He passed along the passengers one after one, and made a dead stand at the boy. " Oh !" said Sandy Wright, who sat by him, " dinna stop at the boy ;—it's a puir orphan ; see, here's my groat." The ferryman still held out his hand. " It's a puir orphan," reiterated the boatman ; " we found him bewildered, on the bursting out o' the last storm, in a dismal habitless glen o' Badenoch, an' we've ta'en him wi' us a' the way, for he's going to seek his friends at Edinburgh ; surely ye'll no grudge him a passage?" The ferryman, without deigning him a reply, plucked off the boy's bonnet ; the boatman instantly twitched it out of his hand. " Hoot, hoot,

hoot!" he exclaimed, "the puir fatherless and motherless boy! —ye'll no do that?" "Take tent, my man," he added, for the ferryman seemed doggedly resolved on exacting the hire; "take tent; we little ken what may come o' oursel's yet, forbye our bairns." "By ——, boatman, or whatever ye be," said the ferryman, "I'll hae either the fare or the fare's worth, though it should be his jacket;" and he again laid hold on the boy, who began to cry. Sandy Wright rose from his seat in a towering passion. "Look ye, my man," said he, as he seized the fellow by the collar with a grasp that would have pulled a bull to the ground, "little hauds me from pitching ye out owre the gunwale. Only crook a finger on the poor thing, an' I'll knock ye down, man, though ye were as muckle as a bullock. Shame! shame ye for a man!—ye hae nae mair natural feeling than a sealchie's bubble."[1] The cry of shame! shame! was echoed from the other passengers, and the surly ferryman gave up the point.

"An' now, my boy," said the boatman as they reached the West Port, "I hae business to do at the Customhouse, an' some money to get; but I maun first try and find out your friends for ye. Look at the letters and tell me the street where they put up." The boy untied his little bundle, which contained a few shirts and stockings, a parcel of papers, and a small box.—"What's a' the papers about?" inquired the boatman; "an' what hae ye in the wee box?" "My mither," said the boy, "bade me be sure to keep the papers, for they tell of her marriage to my father; and the box hauds her ring. She could have got money for it when she was sick and no able to work, but she would sooner starve, she said, than part wi' it; and I widna like to part wi' it, either, to ony bodie but yoursel'—but if ye would take it?" He opened the box and passed it to his companion. It contained a valuable diamond ring. "No, no, my boy," said the boatman, "that widna do;

---

[1] Sea-nettle.

S

the ring's a bonny ring, an' something bye ordinar, though I be no judge; but, blessings on your heart! tak ye care o't, an' part wi't on no account to ony bodie;—Hae ye found out the direction?" The boy named some place in the vicinity of the Cowgate, and in a few minutes they were both walking up the Grassmarket.

" O, yonder's my aunt!" exclaimed the boy, pointing to a young woman who was coming down the street; " yonder's my mither's sister;" and away he sprang to meet her. She immediately recognised and welcomed him; and he introduced the boatman to her as the kind friend who had rescued him from the snow-storm and the ferryman. She related in a few words the story of the boy's parents. His father had been a dissipated young man of good family, whose follies had separated him from his friends; and the difference he had rendered irreconcilable by marrying a low-born but industrious and virtuous young woman, who, despite of her birth, was deserving of a better husband. In a few years he had sunk into indigence and contempt; and in the midst of a wretchedness which would have been still more complete had it not been for the efforts of his wife, he was seized by a fever, of which he died. " Two of his brothers," said the woman, " who are gentlemen of the law, were lately inquiring about the boy, and will, I hope, interest themselves in his behalf." In this hope the boatman cordially acquiesced. " An' now, my boy," said he, as he bade him farewell, " I have just one groat left yet; —here it is; better in your pocket than wi' the gruff carle at the ferry. It's an honest groat, anyhow; an' I'm sure I wish it luck."

Eighteen years elapsed before Sandy Wright again visited Edinburgh. He had quitted it a robust, powerful man of forty-seven, and he returned to it a greyheaded old man of sixty-five. His humble fortunes, too, were sadly in the wane. His son William, a gallant young fellow, who had risen in a few years,

on the score of merit alone, from the forecastle to a lieutenancy,
had headed, under Admiral Vernon, some desperate enterprise,
from which he never returned : and the boatman himself, when
on the eve of retiring on a small pension from his long service
in the Customhouse, was dismissed without a shilling, on the
charge of having connived at the escape of a smuggler.   He
was slightly acquainted with one of the inferior clerks in the
Edinburgh Customhouse; and in the slender hope that this
person might use his influence in his behalf, and that that
influence might prove powerful enough to get him reinstated,
he had now travelled from Cromarty to Edinburgh, a weary
journey of nearly two hundred miles.   He had visited the clerk,
who had given him scarcely any encouragement, and he was
now waiting for him in a street near Brown Square, where he
had promised to meet him in less than half an hour.   But more
than two hours had elapsed; and Sandy Wright, fatigued and
melancholy, was sauntering slowly along the street, musing on
his altered circumstances, when a gentleman, who had passed
him with the quick hurried step of a person engaged in business, stopped abruptly a few yards away, and returning at a
much slower pace, eyed him steadfastly as he repassed.   He
again came forward and stood.   " Are you not Mr. Wright ?"
he inquired.   " My name, sir, is Sandy Wright," said the
boatman, touching his bonnet.   The face of the stranger glowed
with pleasure, and grasping him by the hand, " Oh, my good
kind friend, Sandy Wright!" he exclaimed, " often, often, have
I inquired after you, but no one could tell me where you
resided, or whether you were living or dead.   Come along with
me—my house is in the next square.   What! not remember
me; ah, but it will be ill with me when I cease to remember
you!   I am Hamilton, an advocate—but you will scarcely
know me as that."

The boatman accompanied him to an elegant house in Brown
Square, and was ushered into a splendid apartment, where

there sat a Madonna-looking young lady engaged in reading.
" Who of all the world have I found," said the advocate to the
lady, " but good Sandy Wright, the kind brave man who
rescued me when perishing in the snow, and who was so true a
friend to me when I had no friend besides." The lady welcomed the boatman with one of her most fascinating smiles,
and held out her hand. " How happy I am," she said, " that
we should have met with you! Often has Mr. Hamilton told
me of your kindness to him, and regretted that he should have
no opportunity of acknowledging it." The boatman made one
of his best bows, but he had no words for so fine a lady.

The advocate inquired kindly after his concerns, and was told
of his dismissal from the Customhouse. " I'll vouch !" he exclaimed, " it was for nothing an honest man should be ashamed
of." " Oh ! only a slight matter, Mr. Hamilton," said the
boatman ; " an' troth I couldna' weel do other than what I
did though I should hae to do't o'er again. Captain Robinson
o' the Free Trade was on the coast o' Cadboll last har'st, about
the time o' the *Equinoxal*, unlading a cargo o' Hollands, whan
on cam' the storm o' the season, an' he had to run for Cromarty
to avoid shipwreck. His loading was mostly out, except a few
orra kegs that might just make his lugger seizable if folk gied
a wee owre strict. If he could but show, however, that he had
been at the Isle o' Man, an' had been forced into the Firth by
mere stress o' weather frae his even course to Flushing, it would
set him clear out o' our danger. I had a strong liking to the
Captain, for he had been unco kind to my poor Willie, that's
dead now ; an' when he tauld our officer that he had been at
Man, an' the officer asked for proof, I contrived to slide twa
Manks baubees intil his han', an' he held them out just in a
careless way, as if he had plenty mair proof besides. Weel,
this did, an' the puir chield wan off ; but hardly was he down
the Firth when out cam' the haill story. Him they coudna
harm, but me they could ; an' after muckle ill words, (an' I

had to bear them a', for I'm an auld failed man now,) instead o' getting retired on a pension for my forty years' service, I was turned aff without a shilling. I have an acquaintance in the Customhouse here, Mr. Scrabster the clerk ; an' I came up ance errand to Edinburgh in the hope that he might do something for me ; but he's no verra able I'm thinking, an' I'm feared no verra willing ; an' so, Mr. Hamilton, I just canna help it. My day, o' coorse o' nature, canna be verra lang, an' Providence, that has aye carried me through as yet, winna surely let me stick now."—" Ah no, my poor friend !" said the advocate. " Make up your mind, however, to stay for a few weeks with Helen and me, and I'll try in the meantime what my little influence may be able to do for you at the Customhouse."

A fortnight passed away very agreeably to the boatman. Mrs. Hamilton, a fascinating young creature of very superior mental endowments, was delighted with his character and his stories :—the latter opened to her a new chapter in her favourite volume—the book of human life ; and the advocate, a man of high talent and a benevolent heart, seemed to regard him with the feelings of an affectionate son. At length, however, he began to weary sadly of what he termed the life of a gentle-man, and to sigh after his little smoky cottage, and " the puir auld wife." " Just remain with us one week longer," said the advocate, " and I shall learn in that time the result of my application. You are not now quite so active a man as when you carried me ten miles through the snow, and frightened the tall ferryman, and so I shall secure for you a passage in one of the Leith traders." In a few days after, when the boatman was in the middle of one of his most interesting stories, and Mrs. Hamilton hugely delighted, the advocate entered the apart-ment, his eyes beaming with pleasure, and a packet in his hand. " This is from London," he said, as he handed it to the lady ; " it intimates to us, that ' Alexander Wright, Customhouse boatman,' is to retire from the service on a pension of twenty

pounds per annum."—But why dwell longer on the story ?
Sandy Wright parted from his kind friends, and returned to
Cromarty, where he died in the spring of 1769, in the eighty-
second year of his age. "Folk hae aye to learn," he used to
say, "an', for my own pairt, I was a saxty-year-auld scholar
afore I kent the meaning o' the verse, ' Cast thy bread on the
waters, and thou shalt find it after many days.' "

CHAPTER XIX.

" I'll give thee a wind."—SHAKSPERE.

FOR about thirty years after the failure of the herring fishery, the population of the town of Cromarty gradually decreased. Many of the young men became sailors and went into foreign parts, from whence few of them returned. One of their number, poor Lieutenant Wright, the boatman's son, served in the unfortunate expedition of Vernon, and left his bones under the walls of Carthagena; another, after sailing round the world with Anson, died on his passage homewards when within sight of the white cliffs of England; a third was barbarously murdered on the high seas by the notorious Captain James Lowrie. Such of the town's-people as had made choice of the common mechanical professions, plied their respective trades in the fishing towns of the north of Scotland; and I have seen, among old family papers, letters of these emigrants written from Lerwick, Kirkwall, and Stornoway. As the population gradually decreased in this way, house after house became tenantless and fell into decay; until the main street was skirted by roof-less tenements, and the town's cross, which bears date 1578, was bounded by a stone wall on the one side, and a hawthorn hedge on the other.

The domestic economy of the people, who still continue to inhabit the town, differed considerably from what it had been when their circumstances were more prosperous. There was now no just division of labour among them—working people of all the different denominations encroaching each on the bounds

of the others' profession.   Fishermen wrought as labourers,
tradesfolk as fishermen, and both as farmers.   In the latter
part of spring every year, and the two first months of summer,
the town's-people spent their evenings in angling with rods and
hand-lines in their boats, or from the rocks at the entrance of
the bay ; towards the end of July they formed themselves into
parties of eight or ten, and sailed to Tarbat Ness, a fishing
station of the Moray Firth, where they remained for several
weeks catching and storing up fish for winter.   At night they
converted their sails into tents, ranged in the manner of an
encampment, at the edge of the little bay where they moored
their boats.

The long low promontory of Tarbat Ness forms the north-
eastern extremity of Ross-shire.   Etymologists derive its name
from the practice which prevailed among mariners in this
country during the infancy of navigation, of drawing their light
shallops across the necks of such promontories instead of sailing
round them.   On a moor of this headland there may be traced
the vestiges of an encampment, which some have deemed
Roman, and others Danish ; and there is a cave among the
low rocks by which it is skirted, which, according to tradition,
communicates with another cave on the coast of Caithness.
The scenery of Tarbat Ness is of that character which Addison
regarded as the most sublime ; but it has something more to
recommend it than a mere expansiveness, like that predicated
by the poet, in which no object, tree, house, or mountain, con-
tracts the view of the vast arch of heaven or the huge circle of
earth.   Instead of a low plain bounded by the sky, there is
here a wide expanse of ocean encircling a narrow headland—
brown, sterile, solitary, edged with rock, and studded with
fragments of stone.   On the one hand, the mountains of Sutherland are seen rising out of the sea like a volume of blue clouds ;
on the other, at a still greater distance, the hills of Moray stretch
along the horizon in a long undulating strip, so faintly defined

in the outline that they seem almost to mingle with the firmament. Instead, however, of contracting the prospect, they serve
but to enhance, by their diminished bulk, the immense space in
which they are included. Space—wide, interminable space—
in which he who contemplates it finds himself lost, and is
oppressed by a sense of his own littleness, is at all times the
circumstance to which the prospect owes most of its power;
but it is only during the storms of winter, when the firmament
in all its vastness seems converted into a hall of the tempest,
and the earth in all its extent into a gymnasium for contending
elements, that the scene assumes its full sublimity and grandeur.
On the north a chain of alternate currents and whirlpools howl,
toss, and rage, as if wrestling with the hurricane; on the east
the huge waves of the German ocean come rolling against the
rocky barrier, encircling it with a broad line of foam, and join
their voices of thunder to the roar of current and whirlpool;
cloud after cloud sweeps along the brown promontory, flinging
on it their burdens as they pass; the sea-gull shrieks over it as
he beats his wings against the gale; the distant hills seem
blotted from the landscape; occasionally a solitary bark, half
enveloped in cloud and spray, with its dark sails furled to the
yards, and its topmasts lowered to the deck, comes drifting
over the foam; and the mariner, anxious, afraid, and lashed to
the helm, looks wistfully over the waves for the headlands of
the distant haven.

A party of Cromarty tradesfolk, who had prosecuted the fishing on the promontory of Tarbat Ness for part of the summer
and autumn of 1738, had been less successful that season than
most of their neighbours, and had lingered for several days on
the station after the tents of the encampment had been struck,
and the boats had sailed for home. At length, however, a day
was fixed for their return, but when it arrived the wind had
set in strongly from the south-west; and, instead of prosecuting their voyage, they were compelled to haul up their boat to

the site of the encampment. The storm continued for more than two weeks, accompanied by heavy showers, which extinguished their fire, and so saturated the cover of their tent, that the water dropped on their faces as they lay folded in the straw and blankets with which they had covered the floor. Their provisions too, except the salted fish, which they had secured in barrels, began to fail them ; and they became exceedingly anxious for a change of wind. But the storm seemed to mock at their anxiety ; night after night were they awakened by the rain pattering against the sail, and when they raised its edge every succeeding morning, they saw the sea whitened by the gale, and clouds laden with water rolling heavily from the south-west.

Not more than a mile from the tent there stood an inhabited cottage. The solitary tenant, an elderly woman, still known to tradition as *Stine Bheag o' Tarbat*, was famous at this time as one in league with Satan, and much consulted by seafaring men when windbound in any of the neighbouring ports. And her history, as related by her neighbours, formed, like the histories of all the other witches of Scotland, a strange medley of the very terrible and the very ludicrous. A shipmaster, who had unwittingly offended her, had moored his vessel one evening within the rocky bay of Portmahomack, a haven of Tarbat ; but on going on deck next morning, he found that the vessel had been conveyed during the night over the rocks and the beach, a broad strip of meadow, two corn-fields, and a large moor, into a deep muddy ditch ; and there would she have lain till now had he not found means to conciliate the witch, who on the following night transported her to her former moorings. With all this power, however, it so happened, that only a few weeks after a farmer of the parish, whom she had long annoyed in the shape of a black beetle, succeeded in laying hold of her as she hummed round his bonnet, and confined her for four days in his snuff-box.

Shortly before the arrival of the Cromarty men, a small sloop had been weather-bound for a few days in a neighbouring port; and the master applied to *Stine* for a wind. Part of his cargo consisted of foreign spirits; and on taking leave of the witch he brought with him two empty bottles, which he promised to fill, and send to her by the ship-boy. It was evening, however, before he reached the vessel; the boy would not venture on carrying the bottles by night to the witch's cottage; and on the following morning they were forgotten in the hurry of sailing. The wind blew directly off the land, from what the master deemed the very best point of the compass; the vessel scudded down the Firth before it under a tight sail; it freshened as the land receded, and the mainsail was lowered reef after reef, until as the evening was darkening it had increased into a hurricane. The master stood by the helm, and in casting an anxious glance at the binnacle, to ascertain his course, his eye caught the two bottles of *Stine Bheag*. " Ah, witch!" he muttered, " I must get rid of thee;" and taking up one of the bottles he raised his arm to throw it over the side, when he was interrupted by a hoarse croaking above-head, and on looking up saw two ravens hovering round the vane. The bottle was replaced. An immense wave came rolling behind in the wake of the vessel; it neared; it struck the stern, and, rushing over the deck, washed everything before it, spars, coops, cordage; but only the bottles were carried overboard. In the moment they rose to the surface the ravens darted upon them like sea-gulls on a shoal of coal-fish; and the master, as the vessel swept along, could see them bearing the bottles away. The hurricane gradually subsided into a moderate breeze, and the rest of the voyage was neither rough nor unprosperous; but the ship-master, it was said, religiously determined never again to purchase a wind. And the Cromarty men, who had heard the story, were so much of the master's opinion, that it was not until the second week after the wind had set so stiffly into the south-west, and when all their provisions were expended, that they resolved on risking **a** visit to *Stine Bheag.*

One of them, a tall robust young fellow, named Macglashan, accompanied by two others, after collecting all the placks and boddles of the party—little pieces of copper coin, with the head of Charles II. on one side, and the Scotch thistle on the other—set out for the hovel of the witch. It was situated on the shore of a little sandy bay, which opens into the Dornoch Firth, and formed one of a range, four in number, three of which were now deserted. The roof of one had fallen in; the two others, with their doors ajar, the casements of the windows bleached white by the sea winds, and with wreaths of chickweed mantling over the sloping sides, and depending from the eaves, seemed very dwellings of desolation. From the door and window of the inhabited hovel, which joined to the one which had fallen, and which in appearance was as ruinous and weather-beaten as either of the other two, there issued dense volumes of smoke, accompanied by a heavy oppressive scent, occasioned apparently by the combustion of some marine vegetable. The range had been inhabited about ten years before by a crew of fishermen and their families; one of them the husband, another the son of *Stine Bheag.* The son had unluckily chanced to come upon her when she was engaged in some of her orgies, and telling his father of what he had seen, they deliberated together, it was said, on delating her as a witch before the presbytery of Tain; but ere they came to a full determination they unluckily went to sea. *Stine* was not idle;—there arose a terrible hurricane, and the boat was driven on a quicksand, where she was swallowed up with all her crew. The widows, disturbed by supernatural sights and noises, deserted their cottages soon after, and *Stine Bheag* became the sole tenant of the range.

Macglashan walked up to the door, which hung half open, and tapped against it, but the sound was lost in a loud crackling noise, resembling a ceaseless discharge of pocket pistols, which proceeded from the interior. He tapped a second time, but the crackling continued, and, despairing of making himself heard, he stooped and entered. The apartment was so filled with smoke, that for the first few seconds he could only distinguish a red glare of light upon the hearth, and a small patch of sky, which appeared of a rusty-brown colour through the dense volume which issued out at the window. The hag sat on a low stool beside the wall, and fronting the fire, into which at intervals she flung handfuls of dried sea-weed, of that kind (*Fucus nodosus*) which consists of chains of little brown bladders filled with air, and which is used in the making of kelp. As the bladders, one after one, expanded and burst with the heat, she continued to mutter a Gaelic rhyme. The thick smoke circled round her as she bent over the fire, and when the flame shot up through the eddies, Macglashan could see her long sharp features, but when it sunk her eyes were alone visible. Her grizzled hair escaped from a red coif, and fell over her shoulders, round which there was wrapped a square of red tartan, held on by a large silver brooch. The imagination of a poet could scarcely have invested one of the ancient sibyls with more circumstances of the wild and terrible, or have placed her in a scene of a character so suited to her own. "Sad weather this," said Macglashan;—the hag started at the unexpected address, and rising up gazed at the intruder with a mixed expression of anger and surprise. " I come," he continued, " from *the point*, where I and my companions have been windbound for the last fortnight, and half starved with cold and hunger to boot. Could you not favour us with a breeze that would serve for Cromarty?" Without waiting a reply, he thrust into her hand the joint contribution of the crew. She spread out her palm to the light, looked at the coins, then at Macglashan, and shook her head. " For *that!!*" she said contemptuously. He shook his head in turn. " Bad times, mother, bad times; not a rap more among us; but we will not forget you should we once reach

home." " Then send one of your companions," said the witch,
" for your lugged water-stoup."—" Ay, an' so you know of
them, and of the stoup," muttered Macglashan ;—" Jock, Sandy,
this way, lads." The two men entered the apartment. " Run,
Sandy," continued the young fellow, "for the muckle stoup," and
drawing in a huge settle of plank which stood in the middle
of the floor, he seated himself, all unbidden, before the fire of
Stine Bheag.

The place was darkened, as I have said, with smoke, but at
intervals the flames glanced on the naked walls of turf and
stone, and on a few rude implements of housewifery which were
ranged along the sides, together with other utensils of a more
questionable form and appearance. A huge wooden trough,
filled with water, from whence there proceeded a splashing
bubbling noise, as if it were filled with live fish, occupied one
of the corners; and was sentinelled by a black cat, that sat
purring on a stool beside it, and that on every louder splash
rose from her seat, and stretched her neck over the water. A
bundle of dried herbs; a table bearing the skeleton of some
animal, round part of which a kind of red clay had been
moulded, as if by a statuary; a staff, with the tail of a fish
fastened to one end, and the wings of a raven to the other;
and a large earthen vessel, like that in which Hercules sailed
to release Prometheus, with a white napkin tied in the manner
of a sail to a stick, which served for a mast, were ranged along
the wall. As Macglashan surveyed the apartment, *Stine* seemed
lost in a reverie, with her head bent, and her eyes fixed upon
the fire; but as if struck by a sudden thought, she started into
a more erect posture, and regarded him with a malignant scowl,
clutched her hands into a hill of dried weed, and flung a fresh
heap on the fire, which for a few seconds seemed extinguished.
" Od, mother," said the young fellow, nothing appalled by the
darkness, "ye lead a terrible lonely life of it here; were I in
your place I would die of sheer longing in less than a fortnight."

"Lonely," muttered the hag, who seemed in no communicative mood; "how ken ye that?" As she spoke the croak of a raven was heard from the chimney, accompanied by the flutter of wings. "Ug, ug!" ejaculated Macglashan's companion; "let's out, Mac, and see what's keeping Sandy." "Nay, here he comes," said the other; and as he spoke Sandy entered with the stoup. "And now," said *Stine*, rising and laying hold of it, "ye maun out, an' bide at the rock yonder till I call."

Macglashan and his companions waited for nearly half an hour; night was fast falling, and the ruinous cottages, as the twilight darkened round them, assumed a more dismal appearance. From the window of the inhabited one there glimmered a dull red light, which was repeatedly eclipsed, as if by the shadows of persons passing between the window and the fire. At length the door opened, and the sharp harsh voice of *Stine Bheag* was heard calling from the entrance. Macglashan stepped up to her, and received the stoup, stoppled with a bunch of straw. "Set off," said she, as she delivered it, "on the first blink of to-morrow; but as ye love life, touch not the wisp till ye reach Cromarty." Macglashan promised a strict observance of the injunction, and, taking his leave, set out with his companions for the tent.

The wind lowered during the night, and when early next morning Macglashan raised the edge of the sail, the wide extent of the Moray Firth presented a surface as glassy as that of a mirror; though it still heaved in long ridges, on which the reflection of the red light that preceded sunrise, danced and flickered like sheets of flame. He roused his companions; the tent was struck, the boat launched, the thwarts manned; and before the sun had risen, the whole party were toiling at the oar. A light breeze from the north-east began to ruffle the surface of the water; it increased into a brisk gale, and the boat, with both her sails set, was soon scudding before it. The ancient towers of Balone, the still more ancient towers of Cad-

boll, Hilton with its ruinous chapel, and Shandwick with its
sculptured obelisk, neared and then receded, as she swept along
the shore; and the sun was yet low in the sky, when, after
passing the steep overhanging precipices of the hill of Nigg, she
opened the bay of Cromarty. "What in the name of wonder,"
asked one of the crew, "can *Stine Bheag* hae put in the stoup?"
"Rax it this way," said another; "we would better be ony
gate than in Cromarty should the minister come to hear of it;
I'm thinking Mac had as weel fling out the wisp here as on the
shore."—"Think you so?" said Macglashan, "then send the
stoup this way." He drew out the stopple, and flung it over
his head into the sea; but in the next moment, when half-a-
dozen necks were stretched out to pry into the vessel, which
proved empty, the man stationed at the bows roared out, "For
heaven's sake, lads, mind your haulyards! lower, lower, a squall
from the land! we shall back-fill and go down like a mussel-
shell." The crew clustered round the sails, and had succeeded
in lowering them, when the squall struck the boat ahead with
the fury of a tornado, and almost forced her out of the water.
The thwarts were manned, but ere the rowers had bent to the
first stroke, the oars were wrested out of their hands by the
force of the hurricane. The bay around them was agitated as
if beaten by rods; the wind howled in one continuous gust,
without pause or intermission; and a cloud of spray which arose
from the waves, like a sheet of drift from a field of snow, swept
over them in so dense a volume, that it hid the land and
darkened the heavens. As the boat drifted before the tempest,
the bay receded, the cliffs, the villages, the castles, were passed
in hasty succession, and before noon the crew had landed at
Tarbat Ness, where they found *Stine Bheag* sitting on the shore,
as if waiting their arrival.

"Donnart deevils, what tak's ye here?" was the first saluta-
tion of the witch. "Ah, mother, that cursed wisp!" groaned
out Macglashan. "Wisp!—Look ye, my frack young man,

your weird may have hemp in it, an' sae ye may tempt salt water when ye like; but a' the ither drookit bodies there have nae such protection. An' now ye may tak' the road, for here maun your boat gizzen till the drift o' Januar be heapit oure her gunwale." "Ah, mother!" said Macglashan, "what could we do on the road? and home were but a cold home without either our fish or our winbread. Od, it were better for us to plenish the old bothies at the bay, and go and live wi' yoursel'; but ye must just try and put another wisp in the stoup." To this she at length consented; and on the following morning the party arrived in Cromarty without any new adventure. The one detailed did not become history until many years after, when it was related by Macglashan. He was probably well enough acquainted with the tenth book of Homer's Odyssey to know of that ill-improved gift bestowed on Ulysses by old king Æolus, when

> "The adverse winds in leathern bags he braced,
> Compressed their force, and locked each struggling blast,
> Securely fettered by a silver thong."

T

CHAPTER XX.

" Implore his aid, for Proteus only knows
The secret cause and cure of all thy woes,
But first the wily wizard must be caught,
For unconstrain'd he nothing tells for naught,
Nor is with prayers, or bribes, or flattery bought,
Surprise him first, and with strong fetters bind."—GEORGICS.

OF all the old mythologic existences of Scotland—half earth, half air—there was none with whom the people of Cromarty were better acquainted than with the mermaid. Thirty years have not yet gone by since she has been seen by moonlight sitting on a stone in the sea, a little to the east of the town ; and scarce a winter passed, forty years earlier, in which she was not heard singing among the rocks, or seen braiding up her long yellow tresses on the shore. Like her contemporaries the river-wraiths and fairies—like the nymphs and deities, too, of the Greeks and Romans—she was deemed scarcely less material than the favoured individuals of our own species, who, in the grey of the morning or at the close of evening, had marked her sitting on some desert promontory, or frolicking amid the waves of some solitary arm of the sea. But it is not so generally known, that though in some respects less potent even than men—than at least the very strong and very courageous—she had a power through her connexion with the invisible world over human affairs, and could control and remodel even the decrees of destiny. A robust, fearless man might treat her, it is said, as Ulysses did Circe, or Diomedes Venus ; but then, more potent than these goddesses, she could render all his future undertakings either successful or unfortunate, or, if a seafaring man, could either bury him in the waves or protect him from their fury. It is said, too, that like the Proteus of classical mythology (and the coincidence, if merely such, is at least a curious one), she never exerted this power in a good direction except when compelled to it. She avoided in the daytime shores frequented by man, and when disturbed by him in her retreats, escaped into her native element ; but if he succeeded in seizing and overpowering her, she always purchased her release by granting him any three wishes he might form, connected with either his own fortunes or those of his friends. Her strength, however, was superior to that of most men ; and, if victorious in the struggle, she carried the unfortunate assailant with her into the sea.

It is now nearly a hundred and twenty years since honest John Reid, the Cromarty shipmaster, was positively the most unhappy man in the place. He was shrewd, sensible, calculating, good-humoured, in comparatively easy circumstances, and at this time in his thirtieth year. The early part of his life had been spent abroad ; he had voyaged over the wide Pacific, and traded to China and both the Indies; and to such purpose— for he was quite the sort of person one would most like to have for one's grandfather—that in about fourteen years after sailing from Cromarty a poor ship-boy, he had returned to it with money enough to purchase a fine large sloop, with which he engaged in the lucrative trade carrying on at this period between Holland and the northern ports of Scotland. His good luck still followed him ; nor was he of the class who are ingenious in discovering imaginary misfortunes. What is more, too, he was of so cool a temperament, that when nature rendered him capable of the softer passion at all, it seemed as if she had done so by way of after-thought, and contrary to her original intention. And yet, John Reid, with all his cool prudence, and his good humour and good fortune to boot, was one of the unhappiest men in the place—and all this because he had been just paying his addresses to one of its prettiest girls.

He had first seen Helen Stuart when indulging in a solitary walk on the hill of Cromarty, shortly after his return from the Indies. Helen was fully twelve years younger than himself, slightly but elegantly formed, with small regular features, and a complexion in which the purest white was blended with the most exquisite red. Never before had the sailor seen a creature half so lovely; he thought of her all the evening after, and dreamed of her all the night. But there was no corresponding impression on the other side; the maiden merely remembered that she had met in the wood with the newly-arrived shipmaster and described him to one of her companions as a strongly-built man of barely the middle size, broad-shouldered and deep-chested, with a set of irregular, good-humoured features, over which a tropical sun had cast its tinge of the deepest bronze. Helen was a village heiress, with a good deal of the pride of beauty in her composition, and a very little of the pride of wealth, and, with what was perhaps as unfavourable to the newly-formed passion of Reid as either, a romantic attachment to that most perfect man of the imagination, the *maid's husband*—a prince in disguise, the Admirable Crichton in a revised edition, or the hero of an old ballad.

This dangerous, though shadowy rival of the true lover, who assumes in almost every feminine mind a shape of its own, was in the present instance handsome as Helen herself, with just such a complexion and such eyes and hair; and, excelling all men in fine clothes, fine speeches, and fine manners, he excelled them in parts, and wealth, and courage too. What had the robust, sunburned sailor of thirty to cast into the opposite scale? Besides, Helen, though she had often dreamed of courtship, had never seriously thought of marriage; and so, partly for the sake of her ideal suitor, partly through a girlish unwillingness to grapple with the realities of life, the real suitor was rejected.

Grave natures, says Bacon, are ever the most constant in their attachments. Weeks and months passed away, and still there was an uneasy void in the mind of the sailor, which neither business nor amusement could fill——a something which differed from grief, without affecting him less painfully. He could think and dream of only Helen Stuart. Her image followed him into Holland among the phlegmatic Dutchmen, who never break their hearts for the sake of a mistress, and watched beside him for many a long hour at the helm. He ever saw her as as he had first seen her on the hill; there were trees in the background, and the warm mellow flush of a setting sun, while in front there tripped lightly along a sylph-looking creature, with bright happy eyes, and cheeks glowing with crimson.

He had returned from one of his voyages late in April, and had risen, when May-day arrived, ere the first peep of daylight, in the hope of again meeting Helen among the woods of the hill. Were he but to see her, barely see her, he could be happy, he thought, for months to come; and he knew she would be gathering May-dew this morning, with all her companions, on the green slopes of Drieminory. Morning rose upon him as he sauntered eastward along the edge of the bay; the stars sunk one by one into the blue; and on reaching a piece of rocky beach that stretches along the brow of the hill, the sun rose all red and glorious out of the Firth, and flung a broad pathway of flame across the waters to the shore. The rocks, the hill, the little wavelets which came toppling against the beach, were tinged with the orange light of morning; and yet, from the earliness of the hour, and the secluded character of the scene, a portion of terror might well have mingled with one's quieter feelings of admiration when in the vicinity of a place so famous for the wild and the wonderful as the Dropping-Cave. But of the cave more anon. Darkness and solitude are twin sisters, and foster nearly the same emotions; but they failed this morning to awaken a single fear in the mind of the shipmaster, sailor as he was, and acquainted, too, with every story of the cave. He could think of only Helen Stuart.

An insulated pile of rock, roughened with moss and lichens, which stands out of the beach like an old ruinous castle, surmounted by hanging bartisans and broken turrets, conceals the cave itself, and the skerries abreast of it, from the traveller who approaches them from the west. It screened them this morning from the view of the shipmaster, as, stepping lightly along the rough stones, full of impossible wishes and imaginings, he heard the low notes of a song. He looked round to ascertain whether a boat might not be passing, or a shepherd seated on the hill; but he could see only a huge overgrown seal that had raised its head over the waves, and seemed listening to the music with its face towards the east. On turning, however, the edge of the cliff, he saw the musician, apparently a young girl, who seemed bathing among the cliffs, and who was now sitting half on the rock, half in the water, on one of the outer skerries, opposite the cave. Her long yellow hair fell in luxuriant profusion on her snowy shoulders, and as she raised herself higher on the cliff, the sun shone on the parts below her waist with such dazzling brightness, that the sailor raised his hands to his eyes, and a shifting speck of light, like the reflection of a mirror, went dancing over the shaded roughnesses of the opposite precipice. Her face was turned towards the cave, and the notes of her song seemed at times to be answered from it in a chorus, faint and low indeed, but which could not, he thought, be wholly produced by echo.

Reid was too well acquainted with the beliefs of the age not to know that he looked upon the mermaid. And were he less a lover than he was, he would have done nothing more. But, aware of her strange power over the destinies of men, he only thought that now or never was his opportunity for gaining the hand of Helen. "Would that there were some of my lads here to see fair play!" he muttered, as, creeping amid the crags, and availing himself of every brake that afforded the slightest cover, he stole towards the shelf on which the creature was seated. She turned round in the moment he had gained it; the last note of her song lengthened into a shriek; and with an expression of mingled terror and surprise, which clouded a set of the loveliest features, she attempted to fling herself into the water; but in the moment of the attempt, the brawny arms of the shipmaster were locked round her waist. Her arms clasped his shoulders in turn, and with a strength scarcely inferior to that exerted by the snake of India when struggling with the tiger, she strove to drag him to the edge of the rock; but though his iron sinews quivered under her grasp like the beams of his vessel when straining beneath a press of canvas, he thought of Helen Stuart, and bore her down by main force in the opposite direction. A fainter and a still fainter struggle ensued, and she then lay passive against the cliff. Never had Reid seen aught so beautiful—and he was convinced of it, lover as he was—as the half-fish half-woman creature that now lay prostrate before him.

"Man, what with me?" she said, in a tone of voice which, though sweet as the song of a bird, had something so unnatural in it that it made his blood run cold. "Wishes three," he replied, in the prescribed formula of the demonologist, and then proceeded to state them. His father, a sailor like himself, had been drowned many years before; and the first wish suggested to him by the circumstance was, that neither he himself nor any of his friends should perish by the sea. The second—for he feared lest Helen, so lady-looking a person, and an heiress to boot, might yet find herself the wife of a poor man—was, that he should be uninterruptedly fortunate in all his undertakings. The third wish he never communicated to any one except the mermaid, and yet no one ever failed to guess it. "Quit, and have," replied the creature. Reid slackened his hold; and pressing her tail against the rock until it curled to her waist, and raising her hands, the palms pressed together, and the edge to her face, she sprang into the sea. The spray dashed to the sun ; the white shoulders and silvery tail gleamed for a moment through the green depths of the water. A slight ripple splashed against the beach, and when it subsided, every trace of the mermaid had vanished. Reid wiped his brow, and ascending by one of the slopes of the hill towards the well-known resorts of his town's-women—not the less inclined to hope from the result of his strange contest—he found Helen Stuart seated with one of her companions, a common acquaintance, on the grassy knoll over the Lover's Leap. The charm, thought he, already begins to work.

He bowed to Helen, and addressed her companion. " The man of all the world," said the latter, " whom we most wished to see. Helen has been telling me one of the strangest dreams ; and it is not half an hour yet since we both thought we were going to see it realized ; but you must assist us in reading it. She had just fallen asleep last night, when she found herself on the green slope covered with primroses and cuckoo-flowers, that lies, you know, to the west of the Dropping-Cave ; and there she was employed, she thought, as we have been this morning, in gathering May-dew. But the grass and bushes seemed dry and parched, and she had gathered only a few drops, when, on hearing some one singing among the rocks beside the cave, she looked that way, and saw you sleeping on the beach, and the singer, a beautiful lady, watching beside you. She turned again to the bushes, but all was dry ; and she was quite unhappy that she could get no dew, and unhappy, too, lest the strange lady should suffer you to sleep till you were covered by the tide ; when suddenly you stood beside her, and began to assist her in shaking the bushes. She looked for the lady, and saw her far out among the skerries, floating on the water like a white sea-gull ; and as she looked and wondered, she heard a shower of drops which you had shaken down, tinkling against the bottom of the pitcher. And only think of the prettiness of the fancy !—the drops were all drops of pure gold, and filled the pitcher to the brim. So far the dream. But this is not all. We both passed the green primrose slope just as the sun was rising, and—can you believe it ?—we heard from among the rocks the identical song which Helen heard in her dream. It was like nothing else I ever listened to ; and now here are you to fill our pitchers with gold, like the genie of a fairy tale."

"And so you have really heard music from among the rocks?" said Reid. "Well, but I have more than heard it—I have seen and conversed with the musician ; the strange unearthly lady of Helen's dream. I have visited every quarter of the globe, and sailed over almost every ocean, but never saw the mermaid before."

"Seen the mermaid !" exclaimed Helen.

"Seen and conversed with the mermaid !" said her companion ; "Heaven forbid ! The last time she appeared at the Dropping-Cave was only a few days before the terrible storm in which you lost your father. Take care you repeat not her words— for they thrive ill who carry tales from the other world to this."

" But I am the creature's master," said the sailor, "and need not be so wary."

He told his story ; how he had first seen the mysterious creature sitting in the sea, and breathing exquisite music, as she combed down her long yellow tresses ; how he had stolen warily among the crags, with a heart palpitating betwixt dread and eagerness ; and how, after so fearful a struggle, she had lain passive against the cliff. Helen listened with feelings of wonder and admiration, dashed with terror ; and in returning home, though the morning was far advanced, and the Dropping-Cave a great way below, she leaned for support and protection on the arm of the sailor—a freedom which no one would have remarked as over great at May-day next year, for the sailor had ere then become her husband. For nearly a century after, the family was a rising one; but it is now extinct. Helen, for the last seventy years, has been sleeping under a slab of blue marble within the broken walls of the Chapel of St. Regulus; her only daughter, the wife of Sir George Mackenzie of Cromarty, lies in one of the burying-grounds of Inverness, with a shield of I know not how many quarterings over her grave; and it is not yet twenty years since her grandson, the last of the family, died in London, bequeathing to one of his Cromarty relatives several small pieces of property, and a legacy of many thousand pounds.

There is on the northern side of the Firth of Cromarty, a shallow arm of the sea several miles in length, which dries during stream tides throughout almost its entire extent, and bears the name of the sands of Nigg. Like the sands of the Solway, it has been a frequent scene of accidents. Skirting a populous tract of country on both sides, it lies much in the way of travellers; and the fords, which shift during land floods and high winds, are often attempted at night, and occasionally at improper times of the tide. A narrow river-like channel in the middle, fed by the streams which discharge themselves into the estuary from the interior, and which never wholly dries, bears the name of "The Pot," and was infamous during even the present century for its death-lights and its wraiths, and for the strange mysterious noises which used to come sounding from its depths to either shore previous to "a drowning." Little more than half a century ago, a farmer of the district who had turned aside to see an acquaintance, an old man who lived on the northern shore of the sands of Nigg, found him leaning over the fence of his little garden, apparently so lost in thought that he seemed unconscious of his presence. "What ails you, Donald?" inquired the visitor. "There will be a drowning to-day in the Pot," replied Donald. "A drowning in the Pot!—what makes you say so?" "Do you hear nothing?" "No'o—and yet I rather think I do;—there are faint sounds as of a continual knocking—are there not?—so very faint, that they seem rather within the ear, than without; and yet they surely come from the Pot;—knock, knock, knock—what can it mean?" "That knocking," said the old man, "has been sounding in my ears all this morning. I have never known a life lost on the sands but that knocking has gone before." As he spoke, a horseman was seen riding furiously along the road which skirts the opposite shore of the estuary. On reaching the usual ford, though the rise of the tide had rendered it impracticable for more than an hour before, he spurred his horse across the beach and entered the water. "Surely," said the old man to his friend, "that madman is not taking the ford, and the sea nearly at full?" "Ay, but he is though," said the other; "if the distance does not deceive me, it is Macculloch the corn-agent, in hot haste for the Tain market. See how he spurs through the shallows; and see, he has now reached the Pot, and the water deepens—he goes deeper, and deeper, and deeper. Merciful heavens! he is gone!" Horse and rider had sunk into one of the hollows. The horse rose to the surface a moment after, and swam to the shore; but the rider had disappeared for ever. A story of nearly the same part of the country connects the mysterious knocking with the mermaid.

In the immediate neighbourhood of the Old Abbey of Fearn, famous for its abbot, Patrick Hamilton, our first Protestant martyr, there stood, rather more than ninety years ago, a little turf cottage, inhabited by a widow, whose husband, a farmer of the parish, had died suddenly in the fields about ten years before. The poor woman had been within doors with her only child, a little girl of seven years of age, at the time; and when, without previous preparation, she had opened the door on a hurried summons, and seen the corpse of her husband on the threshold, her mind was totally unhinged by the shock. For the ten following years she went wandering about like a ghost, scarce conscious apparently of anything; no one ever heard her speak, or saw her listen ; and save that she retained a few of the mechanical neatnesses of her earlier years—which, standing out alone on a groundwork of vacuity, seemed akin to the instincts of the inferior animals—her life appeared to be nearly as much a blank as that of the large elm-tree which stretched its branches over her cottage. Her husband's farm, shortly after his death, had been put into the hands of a relation of the family, a narrow sordid man who had made no generous use, it was thought, of the power which the imbecility of the poor woman and the youth of her daughter gave him over their affairs ; it was at least certain that he became comparatively wealthy, and they very poor ; and in the autumn of 1742, the daughter, now a pretty girl of seventeen, had to leave her mother on the care of a neighbour, and to engage as a reaper with a farmer in the neighbouring parish of Tarbat. She had gone with a heavy heart to work for the first time among strangers, but her youth and beauty, added to a quiet timidity of manner, that showed how conscious she was of having no one to protect her, had made her friends ; and now that harvest was over, she was returning home, proud of her slender earnings, and full of hope and happiness. It was early on a Sabbath morning, and her path winded along the southern bank of Loch-Slin, where the parish of Tarbat borders on that of Fearn.

Loch-Slin is a dark sluggish sheet of water, bordered on every side by thick tangled hedges of reeds and rushes ; nor has the surrounding scenery much to recommend it. It is comparatively tame—tamer perhaps for the last thirty years than at any former period ; for the plough has been busy among its green undulating slopes, and many of its more picturesque thickets of alders and willows have disappeared. It possesses, however, its few points of interest ; and its appearance at this time in the quiet of the Sabbath morning, was one of extreme seclusion. The tall old castle of Loch-Slin, broken and weather-worn, and pregnant with associations of the remote past, stood up over it

like some necromancer beside his mirror ; and the maiden, as
she tripped homewards along the little blind pathway that went
winding along the quiet shore—now in a hollow, anon on a
height—could see the red image of the ruins heightened by the
flush of the newly-risen sun, reflected on the calm surface that
still lay dark and grey under the shadow of the eastern bank.
All was still as death, when her ear suddenly caught a low
indistinct sound as of a continuous knocking, which heightened
as she went, until it was at length echoed back from the old
walls ; and which, had she heard it on a week morning, she
would have at once set down as that of the knocking of clothes
at a washing.   But who, she thought, can be " knocking claes"
on the Sabbath ?   She turned a projecting angle of the bank,
and saw, not ten yards away, what seemed to be a tall female
standing in the water immediately beyond the line of flags and
rushes which fringed the shore, and engaged apparently in knocking clothes on a stone, with the sort of bludgeon still used in
the north country for the purpose.   The maiden hurried past,
convinced that the creature before her could be none other than
the mermaid of Loch-Slin ; but in the midst of her terror she
was possessed enough to remark that the beautiful goblin seemed
to ply its work with a malignant pleasure, and that on a grass
plot directly opposite where it stood, there were spread out as
if to dry, more than thirty smocks and shirts, all horribly
dabbled with blood.   As the poor girl entered her mother's
cottage, the excitement that had borne her up in her flight
suddenly failed, and she sunk insensible upon the floor.   For
a moment the mother seemed roused by the circumstance, but
as her daughter recovered, she again relapsed into her accustomed
apathy.

The spirits of the maiden were much flurried, and there was
one to whom she would have fain communicated her strange
story, and sought relief in his society from the terror that made
her heart still palpitate against her side.   But her young cousin

(the son of her unkind relation, the farmer), with whom she had so often herded on the same knoll, and wrought on the same harvest-furrow, had set out for a neighbouring farm, on his way to church, and so there was no probability of her seeing him before evening. She sickened at the gloom of her mother's cottage, where the scowling features of the mermaid seemed imprinted on every darker recess; and, taking her mother by the hand, she walked out with her to the fields. It was now about an hour after noon, and the sun in his strength was looking down in the calm on the bare stubbly campaign, and the old abbey in the midst, with its steep roof of lichened stone, and its rows of massy buttresses. The maiden could hear the higher notes of the congregational psalm as they came floating along the slope from the building, when—fearful catastrophe!—sudden as the explosion of a powder magazine, or the shock of an earthquake, there was a tremendous crash heard, accompanied by a terrific cry; a dense cloud of dust enveloped the ancient abbey, and when it cleared away, it was seen that the ponderous stone roof of the building had sunk in. " O wretched day!" exclaimed the widow, mysteriously restored by the violence of one shock to that full command of her faculties which she had lost by another, and starting at once from the deathlike apathy of years, " O wretched day! the church has fallen, and the whole congregation are buried in the ruins. Fearful calamity!—a parish destroyed at a blow. Dear, dear child, let us haste and see whether something cannot be done—whether some may not be left." The maiden followed her mother to the scene of the accident in distraction and terror.

As they approached the churchyard gate they met two young women covered with blood, who were running shrieking along the road, and shortly after an elderly man so much injured, that he was creeping for support along the wall. " Go on," he said to the widow, who had stopped to assist him; " I have gotten my life as a ransom, but there are hundreds perishing yonder."

They entered the churchyard; two-thirds of the roof had fallen, and nearly half the people were buried in the ruins; and they could see through the shattered windows men all covered with blood and dust, yelling, like maniacs, and tearing up the stones and slates that were heaped over their wives and children. As the sufferers were carried out one by one, and laid on the flat tombstones of the churchyard, the widow, so strangely restored to the energies of her better years, busied herself in stanching their wounds, or restoring them to animation; and her daughter, gathering heart, strove to assist her. A young man came staggering from among the ruins, his face suffused with blood, and bearing a dead body on his shoulders, when, laying down his charge beside them, he sunk over it in a swoon. It was the young cousin of the maiden, and the mutilated corpse which he carried was that of his father. She sobbed over him in an agony of grief and terror; but the exertions of the widow, who wonderfully retained her self-possession, soon recovered him to consciousness, though in so weak a state from exhaustion and loss of blood, that some time elapsed ere he was able to quit the burying-ground, leaning on the arm of his cousin. Thirty-six persons were killed on the spot, and many more were so dreadfully injured that they never recovered. The tombstones were spread over with dead bodies, some of them so fearfully gashed and mangled that they could scarce be recognised, and the paths that wended throughout the churchyard literally ran with blood. It was not until the maiden had reached her mother's cottage, and the heart-rending clamour had begun to fall more faintly on the ear, that she thought of the mysterious washing of Loch-Slin, with its bloody shirts, and felt that she could understand it.

There were lights that evening in many a cottage, and mourners beside many a bed. The widow and her daughter watched beside the bed of their young relative, and though the struggle for life was protracted and doubtful, the strength of his constitution at length prevailed, and he rose, pale and thin, and taller than before, with a scar across his left temple. But ere the first spring had passed, with its balmy mornings and clear sunshine days, he had recovered his former bloom, and more than his former strength. The widow retained the powers so wonderfully restored to her; for the dislocation of faculty effected by one shock had been completely reset by another, and the whole intellect refitted. She had, however, her season of grief to pass through, as if her husband had died only a few days before; and when the relations of the lately perished came to weep over the newly-formed graves that rose so thickly in the burying-place, and around which the grass and hemlock stalks still bore the stain of blood, the widow might be seen seated by a grave covered with moss and daisies, and sunk so low that it was with difficulty its place could be traced on the sward. Of the ten previous years she retained only a few doubtful recollections, resembling those of a single night spent in broken and feverish dreams. At length, however, her grief subsided; and though there were louder and gayer guests at the bridal of her daughter and her young cousin, which took place about two years after the washing of the mermaid, there were none more sincerely happy on that occasion than the widow.

CHAPTER XXI.

" They said they were an hungry; sigh'd forth proverbs—
That hunger broke stone walls; that dogs must eat;
That meat was made for mouths; that the gods sent not
Corn for the rich men only:—With these shreds
They vented their complainings."—CORIOLANUS.

THE autumn and winter of the year 1740 were, like the black
years which succeeded the Revolution, long remembered all over
Scotland, and more especially to the north of the Grampians.
One evening late in the summer of this year, crops of rich
promise were waving on every field, and the farmer anticipated
an early harvest; next morning, a chill dense fog had settled
on the whole country, and when it cleared up, the half-filled
ears drooped on their stalks, and the long-pointed leaves slanted
towards the soil, as if scathed by fire.   The sun looked out
with accustomed heat and brilliancy, and a light breeze from
the south rolled away every lingering wreath of vapour; there
succeeded pleasant days and mild evenings: but the hope of
the season was blasted; the sun only bleached and shrivelled
the produce of the fields, and the breeze rustled through unpro-
ductive straw.   Harvest came on, but it brought with it little
of the labour and none of the joy of other harvests.   The
husbandman, instead of carousing with his reapers, brooded in
the recesses of his cottage over the ruin which awaited him;
and the poor craftsman, though he had already secured his
ordinary store of fish, launched his boat a second time to provide against the impending famine.

Towards the close of autumn not an ounce of meal was to be

U

had in the market; and the housewives of Cromarty began to discover that the appetites of their children had become appallingly voracious. The poor things could not be made to understand why they were getting so much less to eat than usual, and the monotonous cry of "Bread, mammy, bread!" was to be heard in every house. Groups of the inhabitants might be seen on the beach below the town watching the receding tide, in the expectation of picking up a few shell-fish; and the shelves and ledges of the hill were well-nigh stripped by them of their dulce and tangle: but with all their industry they throve but ill. Their eyes receded, and their cheekbones stuck out; they became sallow, and lank of jaw, and melancholy; and their talk was all about the price of corn, bad times, and a failing trade. Poor people! it was well for both themselves and the Government, that politics had not yet come into fashion; for had they lived and been subjected to such misery eighty years later, they would have become Radicals to a man: they would have set themselves to reform the State; and, as they were very hungry, no moderate reform would have served.

The winter was neither severe nor protracted, but to the people of Cromarty it was a season of much suffering; and with the first month of spring there came down upon them whole shoals of beggars from the upper part of the country, to implore the assistance which they were, alas! unable to render them, and to share with them in the spoils of the sea. The unfortunate paupers, mostly elderly men and women, were so modest and unobtrusive, so unlike common beggars in their costume, which in most instances was entire and neat, and so much more miserable in aspect, for they were wasted by famine, that the hearts of the people of the town bled for them. It is recorded of a farmer of the parish, whose crops did not suffer quite so much as those of his neighbours, that he prepared every morning a pot of gruel, and dealt it out by measure to the famishing strangers—giving to each the full of a small ladle. There was a widow gentlewoman, too, of the town, who imparted to them much of her little, and yet, like the widow of Zarephath, found enough in what remained. On a morning of this spring, she saw a thin volume of smoke rising from beside the wall of a corn-yard, which long before had been emptied of its last stack; and approaching it, she found that it proceeded from a little fire, surrounded by four old women, who were anxiously watching a small pot suspended over the fire by a pin fixed in the wall. Curiosity induced her to raise the lid; and as she stretched out her hand the women looked up imploringly in her face. The little pot she found about half filled with fish entrails, which had been picked up on dunghills and the shore; her heart smote her, and hastening home for a cake of bread, she divided it among the women. And never till her dying day did she forget the look which they gave her when, breaking the cake, she doled out a portion to each.

Towards the end of the month of February, when the sufferings of the people seemed almost to have reached their acme, a Mr. Gordon, one of the most considerable merchants of the town, set out to the country, armed with a warrant from the Sheriff, and backed by a small party in quest of meal. The old laws of the sheriffdom, though still unrepealed, were well-nigh exploded, but what was lacking in authority was made up by force; and so, when Mr. Gordon entered their houses to ransack the girnals and meal-chests, there were many attempts made at concealment, but none at open resistance. The magistrate found one ingenious gudewife buried in a mountainous heap of bedclothes; the gudeman, it was said, had gone for the *howdie;* but one of the party mistrusting the story, raised the edge of a blanket, and lo! two sacks were discovered lying quietly by her side. She was known ever after by the name of " the pocks' mither." The meal procured by the party was carefully portioned out, a quantity deemed sufficient for the farmer and his household being left with him, and the remainder, which was paid for by Mr. Gordon, was carried to town, and sold out to the people in pounds and half-pounds.

In the midst of the general distress, a small sloop from the village of Gourac entered the Firth, to take in a lading of meal, which, by dint of grievous pinching and hoarding, had been scraped together by some of the farmers of Easter Ross. The vessel was the property of a Mr. Matthew Simpson, who acted as skipper and supercargo; and she lay on the sands of Nigg, the creek or inlet to which, in the foregoing chapter, I have had occasion to refer. Twice every twenty-four hours was she stranded on the bottom of the inlet, and the wicker carts, laden with sacks, could be seen from the shore of Cromarty driving up to her side;—it was evident, too, that she floated heavier every tide; and many were the execrations vented by the half-starved town's-people against Simpson and the farmers. Plans innumerable were formed among them for seizing on the vessel and disposing of her cargo; but their schemes fell to the ground, for there was none of them bold or skilful enough to take the lead in such an enterprise; and, in all such emergencies, a party without a leader is a body without a soul. Meanwhile the sloop left the creek deeply laden, and threw out her anchors opposite the town, where she lay waiting a fair wind.

Towards the evening of the 9th of April 1741, a shopkeeper of Cromarty was half sitting, half reclining, on his counter, humming a tune, and beating time with his ellwand on the point cf his shoe. He was a spruce, dapper, little personage, of great flexibility of countenance, full of trick and intrigue, and much noted among his simple town's-folk for a lawyer-like ingenuity. He was, withal, a man of considerable courage when contemplating a distant danger, but somewhat of a coward when it came near. His various correspondents addressed him by the name of Mr. Alexander Ross—the town's-people called him Silken Sawney. On an opposite angle of the counter sat Donald Sandison, a tall, robust, red-haired man, who wrought in wood, but whose shop, from the miserable depression of trade, had been shut up for the last two months. He had resided at Edinburgh about five years before; and when there, with another man at Cromarty named Bain, had the satisfaction of escorting the notorious Porteous from the Tolbooth to the Grass-market; and had been much edified, for he was in at the death, by the earnest remonstrances and dying ejaculations of that worthy. A few days afterwards, however, he found his services to the commonwealth on this occasion so ill appreciated, that he deemed it prudent to quit the metropolis for the place of his nativity. No one had ever heard him boast of the exploit; but Bain, who was a tailor, was not so prudent, and so the story came out.

" Weel, Sandison, what are we gaun to do wi' the meal ship ?" said the shopkeeper, laying down his ellwand, and sitting up erect.

" Do wi' the ship ?" replied the mechanic, scratching his head with a half-perplexed, half-humorous expression ; " man, I dinna weel ken. It's bad enough to see a' yon meal going down the Firth, an' folk at hame dying o' hunger !"

" But, Sandison," rejoined the wily shopkeeper, " if it does a' go down the Firth, I'm just thinking it will be nobodie's wyte but your ain."

" How that, man ?" rejoined Sandison.

" I'll tell you how that, an' in your ain words too. Whig as ye are, ye say that all men are no born alike. Some come intil the world to do just what they're bid, an' go just where they're bid, and say just what they hear their neebours saying ; while ithers, again, come into it to think baith for themsels an' the folk round them.—Is that no your own sentiment ?"

" Weel, an' is it no true ?"

" Ay, an' I'll gie you a proof o't. What takes the town's-folk to your shop when any thrawart matter comes in their way that they canna redd up o' themselves ? And why do they ask your advice before entering into a law-plea ? or whether they should try the fishing ? or whether the strange minister gied a gude discoorse; you're no a lawyer, nor a boatman, nor a divine. Why do they call for you to lay a tulzie when you're no a magistrate ? and why do folk that quarrel wi' everybody else, take care an' no quarrel wi' you ? Just because they ken that you were born wi' a bigger mind an' a bolder heart than themsels— born a gentleman, as it were, in spite o' your hamely birth an' your serge coat; an' now that the puir folk are starving, an' a shipful o' meal going down the Firth, you slink awa from your proper natural office o' leader, an' just let them starve on."

"Sawney," said the mechanic, "ye have such a natural turn for flattery, that ye fleech without hope o' fee or bountith. But even allowing that I am a clever enough chiel to make an onslaught on the shipman's meal (a man wi' mair wit, I'm fear'd, would be hungrier than ony o' us afore he would think o't), I may hesitate a wee in going first in the ploy. I have a wife an' twa bairnies. Were there naething to fear but the stroke o' a cutlass, or the flash o' a musket, I widna muckle hesitate, maybe; but the law's a rather bad thing in these quiet times; an' I daresay 'twould be better to want cravat an' nightcap a' thegither than to hae the ane o' brown hemp an' the ither o' white cotton."

"Hoot, man, ye're thinking o' Jock Porteous—we can surely get the meal without hanging onybodie. Hunger breaks through stone walls, an' our apology will be written on the verra face o' the affair. Besides, we're no going to steal the meal; we're only going to sell it out on behalf o' the inhabitants, as Mr. Gordon did the meal o' the parish. An' as for risk—gang ye first, and here's my hand I'll go second :—if I had only your brow, I would willingly go first mysel."

But why record the whole dialogue ? Sandison, though characteristically wary, was, in reality, little averse from the scheme : he entered into it; and, after fully digesting it with the wily shopkeeper, set out to impart it to some of the bolder townsmen.

" Now haud ye in readiness," said he to the man of silk as he quitted his shop ; " I shall call ye up at midnight."

The hour of midnight arrived, and a party of about thirty men, their faces blackened, and their persons enveloped, some in women's cloaks, some in their own proper vestments turned inside out, marched down the lane which, passing the shop-keeper's door, led to the beach. They were headed by a tall active-looking man, wrapped up in a seaman's greatcoat. No one, in the uncertain gloom of midnight, could have identified his sooty features with those of the peaceable mechanic San-dison ; but there was light enough to show the but-ends of two pistols stuck in the leathern belt which clasped his middle, and that there hung by his side an enormous basket-hilted broad-sword. Stopping short at the domicile of the shopkeeper, he tapped gently against a window ;—no one made answer. He tapped again. " Wha's there ?" exclaimed a shrill female voice from within. " Sawney, man, Sawney, wauken up !"—" Oh, Sawney's frae hame !" rejoined the voice ; " there came an express for him ance errand, just i' the gloamin', an' he's awa to the sheriffdom to see his sick mither."—" Daidlin' deceitfu' body !" exclaimed Sandison ; " wha could hae reckoned on this ! But it were shame, lads, to turn back now that we hae gane sae far ; an' besides, if ill comes o' the venture, he canna escape. An' now, shaw yoursels to be men, an' keep as free frae fear or anger as if ye were in the parish kirk. Launch down the yawls ane by ane, and dinna let their keels skreigh alang the stanes ; an' be sure an' put in the spile plugs, that we mayna swamp by the way. Let ilk rower muffle his oar wi' his neckcloth, just i' the clamp ; an', for gudesake, skaith nane o' the crew. Willie, dinna forget the nails an' the hammer ; Bernard, man, bring up the rear." The cool resolution of the leader seemed imparted to his followers ; and, in a few minutes after, they were portioned into three boats, which, with celerity and in silence, glided towards the meal sloop.

The first was piloted by Sandison.   It contained nearly two-
thirds of the whole party ; and when the other two boats prepared to moor close to the vessel, one on each side, and their
crews, as they had been instructed, remained at their respective
posts, Sandison steered under the stern, and laying hold of the
taffrail, leaped aboard.   He was followed by about twelve of
his companions, and the boat then dropped alongside.   Every
manœuvre had been planned with the utmost deliberation and
care.   One of Sandison's apprentices nailed down the forecastle
hatchway, and thus imprisoned the crew ; the others opened
the hold, unslung the tackling on each side, and immediately
commenced lowering the meal-sacks into their boats ; while
Sandison himself, accompanied by a neighbour, groped his way
down the cabin stairs to secure the master.   Simpson, a large
powerful man, had got out of bed, alarmed by the trampling on
deck, and, with no other covering than his shirt, was cautiously
climbing the stairs, when, coming in sudden contact with the
descending mechanic, he lost footing, and rolled down the steps
he had ascended, drawing the other along with him.   " Murder,
murder, thieves !" he roared out ; and a desperate struggle
ensued on the floor of the cabin.   The place was pitch dark,
and when the other Cromarty man rushed into the fray, he
received, all unwittingly, from his Herculean leader, who had
half wrested himself out of the grasp of Simpson, a blow that
sent him reeling against the vessel's side.   Again the combatants
closed in an iron grapple, and rolled over the floor.   But the
mechanic proved the more powerful ; he rose over his antagonist, and then flinging himself upon him, the basket-hilt of
the broad-sword dashed full against his breast.   " Oh, oh, oh !"
he exclaimed ; " mercy, hae mercy—onything but the sweet
life ;" and coiling himself up like a huge snake, he lay passive
under the grasp of the mechanic, who, kneeling by his side,
drew a pistol, which he had taken the precaution to load with
powder only, and discharged it right above his face ; disclosing

to him for a moment the blackened features that frowned over him, and a whole group of dingy faces that now thronged the cabin stairs. Meanwhile the work proceeded; the sloop gradually lightened as the boats became heavier, and at length a signal from the deck informed Sandison that the object of the expedition was accomplished. Before liberating Simpson, however, the Cromarty men forced him upon his knees, and extorted an oath from him that he should not again return to the north of Scotland for meal.

Before morning, about sixty large sacks, the lading of the three boats, were lodged in a cellar, possessed, says my authority, by Mr. James Rabson, a meal and corn merchant of Cromarty; but James, though fully authorized by all his neighbours to dole out the contents to the inhabitants, and account to Simpson for the money, prudently lodged his key under the door, and set out for the country on some pretext of business. In the meanwhile Simpson applied to the Sheriff of the county, a warrant was granted him, the meal was seized in behalf of the proper owner; and the pacific Mr. Donald Sandison was appointed, on the recommendation of the Sheriff, to stand sentry over it. On the following day, a posse of law-officers from the ancient burgh of Tain, the farmers and farm-servants of Easter-Ross, and Simpson and the sailors, were to come, it was said, to transport his charge from the cellar to the vessel. Sandison, with a half-ludicrous, half-melancholy expression of face, took up his station before the door; and enveloped in his greatcoat, but encumbered with neither pistols nor broadsword, he stalked up and down before it until morning.

About two hours after sunrise, four large boats, crowded with people, were seen approaching the town, and, in a few minutes after, seven-eighths of the whole inhabitants, men, women, and children, armed with stones and bludgeons, were drawn out on the beach to oppose their landing. Such an assemblage! *There* were the parish schoolboys, active little fellows, that could hit to a hair's-breadth; and *there* the town apprentices of all denominations, stripped of their jackets, and with their aprons puffed out before them with well-selected pebbles. *There*, too, were the women of the place, ranged tier beyond tier, from the water's edge to the houses behind, and of all ages and aspects, from the girl that had not yet left school, to the crone that had hobbled from her cottage assisted by her crutch. The lanes were occupied by full-grown men, who, armed with bludgeons, reserved themselves for the final charge, and now crouched behind their wives and sisters to avoid being seen from the boats. A few young lads, choice spirits of the place, had climbed up to the ridges of the low cottages, which at that time presented, in this part of the town, a line parallel to the beach. Some of them were armed with pistols, some with satchels full of stones; and farther up the lanes there was a second party of women, who meditated an attack on Rabson's cellar. Dire was the combination of sound. The boys shouted, the girls shrieked, the apprentices, tapping their fingers against their throats, bleated like sheep in mockery of the farmers, the women yelled out their defiance in one continuous howl, interrupted occasionally by the hoarse exclamations and loud huzzas of the men. The boats advanced by inches. After every few strokes, the rowers would pause over their oars, and wrench themselves half round to reconnoitre the myriads of waving arms and threatening faces which thronged the beach. As they creeped onwards, a few stones flung from slings by some of the boys went whizzing over their heads, " Now pull hard, and at once!" shouted out Simpson; " we have to deal with but women and children, and shall disperse them before they have fired half a broadside." The rowers bent them to their oars, the boats started shorewards like arrows from the string, there arose a shout from the assembled multitude, which the distant hills echoed back to them in low thunder, and a shower of stones from the boys, the apprentices, the women, the men—from the shore, the lanes, the cottage roofs, the chimney tops, came hailing down upon them thick and ceaseless, rattling, pattering, crashing, like the *débris* of a mountain rolled over its precipices by an earthquake. The water was beaten into foam as if lashed by a hurricane. Every individual of the four crews disappeared in an instant; the oars swung loose on the gunwales, or slipped overboard. At length, however, the boats, propelled partly by the wind, partly by the force of the missiles, drifted from the shore; and melancholy was the appearance of the people within, when, after the stones began to fall short, they gathered themselves up, and looked cautiously over the sides. There were broken and contused heads among them beyond all reach of reckoning; and one poor man of Easter-Ross, who had been marked out by a young fellow named Junor, the best slinger in town, had carried two good eyes with him into the conflict, and only one out of it. They rowed slowly to the other side, and the victors could see them, until they landed, unfolding neckcloths and handkerchiefs, and binding up heads and limbs.

The attack on the boats had no sooner commenced, than the female party, who had been stationed in the lanes, proceeded to Rabson's cellar. "We maun hae meal!" said the women to Sandison, who was lounging before the door with his arms folded in his greatcoat, and a little black tobacco-pipe in his mouth. "Puff," replied the mechanic, shooting a huge burst of smoke into the face of the fairest of the speakers. "We maun hae meal!" reiterated the women. "Puff—weel nee-bours—puff—I mauna betray trust, ye ken—puff; an' what else am I stationed here for, but just to keep the meal frae you?—puff, puff." "But we maun hae't, an' we will hae't, an' we sall hae't, whether you will or no!" shrieked out a virago armed with a huge axe, which the mechanic at once recognised as his own, and who dealt, as she spoke, a tremen-dous blow on the door. "Gudesake, Jess!" said the mechanic, losing in his fear for his favourite tool somewhat of his self-possession; " Gudesake, Jess, keep the edge frae the nails!" Stepping back a few paces, he leisurely knocked out the ashes of his pipe against his thumb-nail; and with the remark, that " strong han' (force) was a masterfu' argument; and that one puir working man, who hadna got his night's rest, was no match for a score o' idle queans," he relinquished his post, and took sanctuary in his own dwelling. In less than half an hour after, the whole contents of the cellar had disappeared. There was a hale old woman, a pauper of the place, who did not claim her customary goupens for two whole years thereafter; and a shoemaker named Millar was not seen purchasing an ounce of meal for a much longer time.

Ninety years after the year of the meal mob, and when every one who had either shared in it or remembered it were sleeping in their graves, I was amusing myself, one wet day, in turning over some old papers stored up in the drawers of a moth-eaten scrutoire, which had once belonged to Donald Sandison, when a small parcel of manuscripts, wrapped up with a piece of tape, which had once been red, attracted my notice. The first manuscript I drew out bore date 1742, and was entitled, " Representation, Condescendence, and Interlocutors, in the process of Matthew Simpson against the Cromarty men." It contained a grievous complaint made by the town's-folk to the Right Hon. Lord Balmerino. " Simpson was a person of a rancorous and very litigious spirit," urged the paper; " and it was surely not a little unreasonable in him to expect, as he did in the suit, that the people of a whole country-side, indubitably innocent of every act of violence alleged against them, should be compelled to undertake a weary pilgrimage to Edinburgh to answer to his charges, when, from the circumstances of the case, anything they could have to depone anent the spulzie, would yield exactly the same result, whether deponed at Edinburgh, Cromarty, or Japan." It went on to show that the people were miserably depressed by poverty; and that, if compelled to set out on such a journey, they would have to beg by the way; while their wives and children would be reduced to starvation at home, without even the resource of begging itself, seeing that all their neighbours were as wretchedly poor as themselves.   Next in order in the parcel followed the statements of Mr. Matthew Simpson, addressed also to his Lordship. He had been robbed, he affirmed, by the men of the north three several times; twice by the people, and once by the lawyers; and having lost in this way a great deal of money, he could not well afford to lose more.   It was stated, further, by the master, that Edinburgh could not be farther from Cromarty than Cromarty from Edinburgh; and that it was quite as reasonable, and fully as safe for the weaker party, that the conspirators should have to defend themselves in the metropolis, as that he, the prosecutor, should have to assail them in the village.   Both manuscripts seemed redolent of that old school of Scotch law in which joke was so frequently called in to the assistance of argument, and dry technicalities relieved by dry humour.   A third paper of the parcel bore date 1750, and was entitled, " Discharge from Matthew Simpson to Donald Sandison and others."   The fourth and last was a piece of barbarous rhyme, dignified, however, with the name of poetry, and which, after describing mealmongers as " damned rascals," and " the worst of all men," assured them, with a proper contempt for both the law of the land and the doctrine of purgatory, that there is an executive power vested in the people, which enables them to take summary justice on their oppressors, and that the " devil gets villains as soon as they are dead."

Silken Sawney, the first projector of the spulzie, did not escape in the process, though he contrived a few years after to save his coin by running the country.   He was the only person in Cromarty who, in the year 1745, assumed the white cockade; and no sooner had he appeared with it on the street than he

was apprehended by a party of his neighbours, who were kings-
men, and incarcerated in an alehouse. A guard was mounted
before the door, and, on the morrow, the poor man of silk was
to be sent aboard a sloop of war then lying in the bay ; but as
his neighbours, when they took the precaution of mounting
guard, did not think proper to call to memory that his apartment had a door of its own, which opened into a garden behind,
he deemed it prudent, instead of waiting the result, to pass
through it on a journey to the Highlands, and he never again
returned to Cromarty. The other conspirators suffered in proportion, not to what they had perpetrated, but to what they
possessed. A proprietor named Macculloch was stripped of his
little patrimony, while some of his poorer companions escaped
scot-free. Sandison contrived to pay his portion of the fine,
and made chairs and tables for forty years after. He was
deemed one of the most ingenious mechanics in the north of
Scotland. I have spent whole days in the house of his grandson, half buried in dusty volumes and moth-eaten drawings
which had once been his ; and derived my earliest knowledge
of building from Palladio's First Book of Architecture, in the
antique translation of Godfrey Richards, which, as the margins
testified, he had studied with much care. At a sale of household furniture, which took place in Cromarty about thirty years
ago, the auctioneer, after examining a very handsome though
somewhat old-fashioned table with minute attention, recommended it to the purchasers by assuring them, in a form of
speech at least as old as the days of Erasmus, that it was certainly the workmanship of either the Devil or of Donald
Sandison.

CHAPTER XXII.

" Old sithes they had with the rumples set even,
And then into a tree fast driven ;
And some had hatchets set on a pole—
Mischievous weapons, antic and droll.
    Each where they lifted tax and cess,
And did the lieges sore oppress,
And cocks and hens, and churns and cheese,
Did *kill* and eat when they could seize."
                DUGALD GRAHAM'S *History of the Rebellion.*

WITH the solitary exception mentioned in the previous chapter, the whole people of Cromarty were loyal to the house of Hanover. They were all sound Protestants to the utmost of their ability, and never failed doing justice in a bumper to the " best in Christendom" but when the liquor was bad. It was therefore with no feelings of complacency, that, in the autumn of 1745, they learned that the Pretender, after landing in the western Highlands, had set off with a gathering of Gaelic Roman Catholics to take London ‑from the King. They affirmed, however, that the redcoats were too numerous, and London too strong, to leave the enterprise a chance of success ; and it was not until Cope had been set a-scampering, and the bayonets of England proved insufficient to defend it on the Scottish side, that they began to pity George Rex (poor man), and to talk about the downfall of the Kirk. Their attention, however, was called off from all such minor matters to a circumstance connected with the outbreaking which directly affected themselves. Parties of wild Highlanders, taking advantage of the defenceless state of the Lowlands, and the cause of the

Pretender, went prowling about the country, robbing as the smith fought, " every man to his own hand ;" and stories of their depredations began to pour into the town.   They were doing great skaith, it was said, to victual and drink, spulzieing women of their yarn, and men of their shoes and bonnets ; as for money, there was luckily very little in the country.   Nor was it possible to conciliate them by any adaptation whatever of one's politics to the Jacobite code.   A man of Ferindonald, a genuine friend to the Stuart, had gone out to meet with them, and in the fulness of his heart, after perching himself on a hillock by the wayside, he continued to cry out, " You're wel-come ! you're welcome !" from their first appearance until they had come up to him.   " Welcomes or na welcomes," said a bareheaded, barefooted Highlander, as stooping down he seized him by the ankles ; " welcomes or na welcomes, *thoir dho do brougan.*"   (Give me your shoes.)

Every day brought a new story of the marauders ;—a Navity tacksman, who had listened himself half crazy, and could speak or think of nothing else, was enough of himself to destroy the quiet of the whole parish.   Some buried casks of meal under their barn floors, others chests of plaiding and yarn.   The tacksman interred an immense girnal, containing five bolls of oatmeal, which escaped the rebels only to be devoured by the rats.   So thoroughly had he prepared himself for the worst, that, when week after week went by, and still no Highlanders, he seemed actually disappointed.   One morning, however, in the end of January 1746, he was called out to his cottage door to see something unusual on the hill of Eathie ; a number of fairy-like figures seemed moving along the ridge, and then, as they descended in a dark compact body to the hollow beneath, there were seen to shoot out from them, at uncertain intervals, quick, sudden flashes, like lightnings from a cloud.   " Och, och !" exclaimed the tacksman, who well knew what the appa-rition indicated, " the longest day that e'er came, even came at last." And away he went to reside, until the return of quieter times, in a solitary cave of the hill.

The marauders entered the town about mid-day. They were armed every one after his own fashion, some with dirks and broadswords, some with pistols and fowling-pieces, and not a few with scythes, pikes, and Lochaber-axes. Some carried immense bunches of yarn, some webs of plaiding, some bundles of shirts and stockings. Most of the men of the place, who would readily enough have joined issue with them at the cudgel, but bore no marked affection to broadswords and Lochaber-axes, had conveyed themselves out of the way, leaving their wives to settle with them as they best might. They entered the better-looking houses by half-dozens, turned the furniture topsy-turvy, emptied chests and drawers, did wonderful execution on dried salmon and hung beef, and set ale-barrels abroach. One poor woman, in attempting to rescue a bundle of yarn, had her cheek laid open by a fellow who dashed the muzzle of his pistol into her face; another was thrown down and robbed of her shoes. There lived at this time one Nannie Miller, a matron of the place, who sold ale. She was a large-boned, amazon-looking woman, about six feet in height, of immense strength, and no ordinary share of courage. Two of the Highlanders entered her cottage, and with much good-nature (for they had had a long walk, she said) she set down before them a pint of her best ale and a basket of scones, with some dried fish. They ate and drank, and then rose to spulzie; but they were too few, as it proved, for the enterprise: for when one of them was engaged in ransacking a large meal-barrel, and the other in breaking open a chest, Nannie made a sudden onslaught, bundled the one fellow head-foremost into the barrel, and turning on his companion as he rushed in to the rescue, floored him with a single blow. The day was all her own in a twinkling; the Highlanders fled, one of them half-choked by the meal, the other more than half-throttled by Nannie; but glad, notwithstanding, to get off so well.

x

In the middle of the spulzie a sloop of war hove in sight, and
a boat was seen shooting out to meet her from under the rocks
of the hill. Sail after sail was run out on her yards as soon as
the boat touched her side, and she came careering up the Firth
like an angry giant. The Highlanders gathered in the street,
and, according to old Dunbar,

> Fu' loud in Ershe they begowt to clatter,
> And rouped like revin and ruke.

One of them, who seemed to have drunk freely, was hacking
with his broadsword at the rails of a wooden bridge, and swearing furiously at the ship; and a little girl, who chanced to be
passing with a jug of milk, was so terrified that she fell and
broke the jug. "Poor sing, poor sing!" said the Highlander,
as he raised her and wiped her face with the corner of his plaid,
"hersel' widna hurt a pit o' you." The party, in their retreat,
took the road that passes towards the west, along the edge of
the bay; and no sooner had the sloop cleared the intervening
headland, than she began to fire on them. One of the bullets
struck off a piece from a large granite boulder on the shore
termed the *Pindler*, and in less than half a minute the Highlanders were scattered over the face of the hill. They did not
again return to Cromarty. Though they fared better in their
predatory excursions than most of their countrymen who accompanied the Prince, and transferred to their homes much of the
"plenishing" of the Lowlands, it was observed that in few instances did their gains enrich their descendants. I once wrought
in the same shed with an old mason, a native of the parish of
Urquhart, who, in giving me a history of his early life, told me
that his father had left at his death a considerable sum of
money to himself and three brothers, and that not one of them
was sober for two days together until they had squandered the
whole. "And no wonder," remarked another mason from the
same parish, who was hewing beside him; "your father went

out a-harrying in the Forty-Five, and muckle did he bring back with him, but it was ill gotten, and couldna last."

As spring came on, a new set of stories bagan to pass current among the people of the town. The Pretender had failed, it was said, in his enterprise, and was falling back on the Highlands. But there was something anomalous in the stories; for it was affirmed that he was both running away and gaining all the battles. This they could not understand; and when, early in March, Lord Louden entered the town at the head of sixteen hundred men, in full retreat before the rebels, they began to ask whether it was customary for one flying army to pursue another. His Lordship dealt by them more hardly than even the marauders; for, after transporting his men across the ferry, he broke all their boats. "It's a sair time for puir folk," said an old fisherman when witnessing the destruction of his skiff; "gain King, gain Pretender, waes me, I'm the loser gain wha like."

Amid all the surmises and uncertainties of the town's-people, matters were fast drawing to a crisis with the Highlanders. On the 15th of April a sloop from Lossiemouth entered the Firth, and brought intelligence that Duke William and his army had crossed the Spey, and were on the march for Inverness, then occupied by the rebels. On the following morning nearly all the males of the place, and not a few of the women, had climbed the neighbouring hill to watch the progress of their march. The weather was dull and unpleasant. There was a cold breeze from the east, accompanied by a thick drizzling rain, and the hills of Moray and Inverness were girdled with wreaths of mist. The lower grounds, which lie along the Firth, looked dim and blue through the haze, and the eye vainly commanded the whole tract of country which stretches between Inverness and Nairn. A little after noon, however, the weather began to clear up, and a sailor, who had brought with him the ship-glass, thought he could discover something unusual on the moor of Culloden.

Every eye was turned in that direction.   Suddenly there rose a
little dense cloud of smoke, as if a volcano had burst out on the
moor; then succeeded the booming of cannon and the rattle of
musketry.   "They are at it, God wi' the right!" shouted out
Donald Sandison; "look, Sandy Wright, is the smoke no going
the way o' Inverness?"   "It's but the easterly haar," said
Sandy; "auld as I am, Donald, I could wish to be near enough
to gae ae stroke for the king!"   The smoke continued to rise
in clouds that went rolling towards the west, and the roar of
cannon to rebound among the hills.   At length they could hear
only the smart pattering of musketry, and the tide of battle
seemed evidently sweeping towards Inverness.   The cloud passed
from the moor; and when, at intervals, a fresh burst shot up
through the haze, it seemed to rise from among the fields in
the vicinity of the town.   Anon all was silence; and the people,
after lingering till near nightfall, returned to their homes to tell
that Duke William had beaten the rebels, and to drink healths
to the King.   They spoke always of the Duke's army as "our
folk," and his victory as "our victory."   I have heard an old
woman of the place repeat a rude song, expressive of their
triumph on this occasion, which she had learned from her nurse
when almost an infant.   My memory has retained only one of
the verses, and a horrible verse it is :—

> Lovat's head i' the pat,
> 　Horns and a' thegither,
> We'll mak brose o' that,
> 　An' gie the swine their supper.

In after years they thought less hardly of the cause of the
Stuarts; and I have heard some of their old men relate stories
of the poor people who suffered at this time, with a good deal
of feeling.   There was a Highlander named Robertson, a man
of rare wit and humour, who had been crippled of an arm at
Culloden.   He used, in after years, to come to the place as a
sort of travelling merchant, and always met with much kind-

ness from them. So much attached was he to the Prince that he would willingly have lost the other arm for him too. Another Highlander, who had also been wounded on the moor, was a great favourite with them likewise. On seeing the battle irretrievably lost, and further resistance unavailing, he was stealing warily out of the field, when two English dragoons came galloping up to him to cut him down. He turned round, drew a pistol from his belt, shot the foremost through the body, and then hurled his weapon at the head of the other, who immediately drew rein and rode off. The sword of the dying man wounded him in its descent in the fleshy part of the hand, between the thumb and the forefinger; and he retained the scar while he lived. There was another Highlander who resided near Kessock, who had vowed, immediately after the battle of Preston, that he would neither cut nor comb the hair of his head until Charles Stuart was placed on the throne of his ancestors. And he religiously observed his vow. My grandfather saw him twenty years after the battle. He was then a strange, grotesque-looking thing, not very unlike a huge cabbage set a-walking; for his hair stuck out nearly a foot on each side of his head, and was matted into a kind of felt. But truce with such stories! Fifty years ago they formed an endless series; but they have now nearly all passed away, or only live, if I may so express myself, in those echoes of the departed generations which still faintly reveberate among the quieter recesses of the present. Of all the people who witnessed the smoke of Culloden from the hill of Cromarty I remember only three.

About eighteen years ago, when quite a boy, I was brought by a relation to see a very old man then on his death-bed, who resided in a small cottage among the woods of the hill. My kinsman for the twenty preceding years had lived with him on terms of the closest intimacy, and had been with him, about ten months before, when he met with an accident from a falling tree, by which he received so serious an injury that it proved

the occasion of this his final illness. A thick darkness, however, had settled over all the events of his latter life, and he
remembered neither his acquaintance of twenty years nor the
accident. His daughter named the father of my friend, in the
hope of awakening some early train of thought that might lead
him into the more recent period ; but his knowledge of even
the father had commenced during the forty previous years, and
his name sounded as strangely to him as that of his son. "He
is a great-grandchild," said the woman, "of your old friend
Donald Roy, the Nigg elder." "Of Donald Roy !—a great-
grandchild of Donald Roy !" he exclaimed, holding out his hard
withered hand ; "oh, how glad I am to see him ! How kind
it is of him," he added, "thus to visit a poor bedridden old
man ! I have now lived in the world for more than a hundred
years, and during my long sojourn have known few men I could
compare with Donald Roy."

The old man raised himself in his bed, for his strength had
not yet quite failed him, and began to relate to my friend, in a
full unbroken voice, some of the stories regarding the Nigg elder,
which I have imparted to the reader in a former chapter. His
mind was full of the early past, and he seemed to see its events
all the more clearly from the darkness of the intervening period
—just as the stars may be discerned at noonday at the bottom
of a deep mine, impenetrably gloomy in all its nearer recesses,
when they are invisible from the summit of a hill. He ran over
the incidents of his early life. He told how, in his thirtieth
year, when the country resounded with the clash of arms, he
had quitted his peaceful avocations as a gardener, and joined the
army of the king. He fought at Culloden, and saw the clans
broken before the bayonets of Cumberland. His heart bled, he
said, for his countrymen. They lay bleeding on the moor, or
were scattered over it ; and he saw the long swords of the
horsemen plied incessantly in the pursuit. Still more melancholy were his feelings, when, from a hill of Inverness-shire, he

looked down on a wide extent of country, and saw the smoke of a hundred burning cottages ascending in the calm morning air.—He died a few weeks after our visit, aged a hundred years and ten months. His death took place in winter ;—it was an open, boisterous winter, that bore heavy on the weak and aged ; and in less than a month after, two very old men besides were also gathered to their fathers. And they, too, had had a share in the Forty-five.

The younger was a ship-boy at the time, and the ship in which he sailed was captured with a lading of Government stores, by a party of the rebels. He was named Robertson, and there were several of the Robertsons of Struan among the party. He was soon on excellent terms with them ; and on one occasion, when rallying some of the Struans on their under-taking, he spoke of their leader as the Pretender. " Beware, my boy," said an elderly Highlander, " and do not again repeat that word ; there are men in the ship who, if they but heard you, would perhaps take your life for it ; for remember we are not all Robertsons." The other old man who died at this time, had been an officer, it was said, in the Prince's army ; but he was a person of a distant, reserved cast of character ; and there was little known of his history, except that he had been bred to the profession of medicine, and had been unfortunate through his adherence to the Prince. It was remarked by the town's-people that his spirit and manners were superior to his condition.

Among the old papers in Sandison's scrutoire, I found a curious version of the 137th Psalm, the production of some unfortunate Jacobite of this period. It seems to have been written at Paris shortly after the failure of the enterprise, and when the Prince and his party were in no favour at court ; for the author, a man apparently of keen feelings, applies, with all the sorrowful energy of a wounded spirit, the curses denounced against Edom and Babylon to England and France.

PSALM CXXXVII.

By the sad Seine we sat and wept
    When Scotland we thought on:
Reft of her brave and true, and all
    Her ancient spirit gone.

"Revenge," the sons of Gallia said,
    " Revenge your native land;
Already your insulting foes
    Crowd the Batavian strand."

How shall the sons of freedom e'er
    For foreign conquest fight ?
How wield anew the luckless sword
    That fail'd in Scotland's right ?

If thee, O Scotland ! I forget
    Till fails my latest breath,
May foul dishonour stain my name,
    Be mine a coward's death !

May sad remorse for fancied guilt
    My future days employ,
If all thy sacred rights are not
    Above my chiefest joy !

Remember England's children, Lord,
    Who on Drumossie day,
Deaf to the voice of kindred love,
    " Raze, raze it quite," did say.

And thou, proud Gallia ! faithless friend,
    Whose ruin is not far,
Just Heaven on thy devoted head
    Pour all the woes of war !

When thou thy slaughter'd little ones
    And ravish'd dames shalt see,
Such help, such pity mayst thou have
    As Scotland had from thee.

CHAPTER XXIII.

"*Mop.*—Is it true, think you ?
Aut.—Very true ;—why should I carry lies abroad ?"
WINTER'S TALE.

IN perusing in some of our older Gazetteers the half page devoted to Cromarty, we find that, among the natural curiosities of the place, there is a small cavern termed the Dropping-Cave, famous for its stalactites and its petrifying springs. And though the progress of modern discovery has done much to lower the wonder, by rendering it merely one of thousands of the same class—for even among the cliffs of the hill in which the cavern is perforated, there is scarcely a spring that has not its border of coral-like petrifactions, and its moss and grass and nettle-stalks of marble—the Dropping-Cave may well be regarded as a curiosity still. It is hollowed, a few feet over the beach, in the face of one of the low precipices which skirt the entrance of the bay. From a crag which overhangs the opening there falls a perpetual drizzle, which, settling on the moss and lichens beneath, converts them into stone ; and on entering the long narrow apartment within, there may be seen by the dim light of the entrance a series of springs, which filter through the solid rock above, descending in so continual a shower, that even in the sultriest days of midsummer, when the earth is parched and the grass has become brown and withered, we may hear the eternal drop pattering against the rough stones of the bottom, or tinkling in the recess within, like the string of a harp struck to ascertain its tone. A stone flung into the interior, after rebounding from side to side of the rock, falls with a deep hollow plunge, as if thrown into the sea. Had the Dropping Cave been a cavern of Greece or Sicily, the classical mythology of these countries would have tenanted it with the goddess of rains and vapour.

The walk to the cave is one of the most agreeable in the vicinity of the town, especially in a fine morning of midsummer, an hour or so after the sun has risen out of the Firth. The path to it has been hollowed out of the hill-side by the feet of men and animals, and goes winding over rocks and stones—now in a hollow, now on a height, anon lost in the beach. In one of the recesses which open into the hill, a clump of forest-trees has sprung up, and, lifting their boughs to the edge of the precipice above, cover its rough iron features as if with a veil; while, from the shade below, a fine spring, dedicated in some remote age to "Our Ladye," comes bubbling to the light with as pure and copious a stream as in the days of the priest and the pilgrim. We see the beach covered over with sea-shells and weeds, the cork buoys of the fishermen, and fragments of wrecks. The air is full of fragrance. Only look at yonder white patch in the hollow of the hill; 'tis a little city of flowers, a whole community of one species—the meadow-sweet. The fisherman scents it over the water, as he rows homeward in the cool of the evening, a full half-mile from the shore. And see how the hill rises above us, roughened with heath and fern and foxglove, and crested a-top with a dark wood of fir. See how the beeches which have sprung up on the declivity recline in nearly the angle of the hill, so that their upper branches are only a few feet from the soil; reminding us, in the midst of warmth and beauty, of the rough winds of winter and the blasting influence of the spray. The insect denizens of the heath and the wood are all on wing; see, there is the red bee, and there the blue butterfly, and yonder the burnet-moth with its wings of vermilion, and the large birdlike dragon-fly, and a thousand others besides, all beautiful and all happy. And then the birds;—But why attempt a description? The materials of thought and imagination are scattered profusely around us; the wood the cliffs and

the spring—the flowers the insects and the birds—the shells
the broken fragments of wreck and the distant sail—the sea
the sky and the opposite land—are all tones of the great instrument Nature, which need only to be awakened by the mind
to yield its sweet music.   And now we have reached the cave.

The Dropping-Cave ninety years ago was a place of considerable interest ; but the continuous shower which converted into
stone the plants and mosses on which it fell, and the dark recess
which no one had attempted to penetrate, and of whose extent
imagination had formed a thousand surmises, constituted some
of merely the minor circumstances that had rendered it such.
Superstition had busied herself for ages before in making it a
scene of wonders.   Boatmen, when sailing along the shore in
the night-time, had been startled by the apparition of a faint
blue light, which seemed glimmering from its entrance : on
other occasions than the one referred to in a former chapter,
the mermaid had been seen sitting on a rock a few yards before
it, singing a low melancholy song, and combing her long yellow
hair with her fingers ; and a man who had been engaged in
fishing crabs among the rocks, and was returning late in the
evening by the way of the cave, almost shared the fate of its
moss and lichens, when, on looking up, he saw an old greyheaded man, with a beard that descended to his girdle, sitting
in the opening, and gazing wistfully on the sea.

I find some of these circumstances of terror embodied in verse
by the provincial poet whom I have quoted in an early chapter
as an authority regarding the Cromarty tradition of Wallace;
and now, as then, I will avail myself of his description :—

> ————"When round the lonely shore
> The vex'd waves toil'd with deaf'ning roar,
> And Midnight, from her lazy wain,
> Heard wild winds roar and tides complain,
> And groaning woods and shrieking sprites ;—
> Strange sounds from thence, and fearful lights,
> Had caught the sailor's ear and eye,
> As drove his storm-press'd vessel by.

> More fearful still, Tradition told
> Of that dread cave a story old—
> So very old, ages had pass'd
> Since he who made had told it last.
> 'Twas thus it ran :—Of strange array
> An aged man, whose locks of grey,
> Like hill stream, flow'd his shoulders o'er,
> For three long days on that lone shore
> Sat moveless as the rocks around.
> Moaning in low unearthly sound ;
> But whence he came, or why he stay'd,
> None knew, and none to ask essay'd.
> At length a lad drew near and spoke,
> Craving reply.   The figure shook
> Like mirror'd shape on dimpling brook,
> Or shadow flung on eddying smoke—
> And the boy fled.   The third day pass'd—
> Fierce howl'd at night the angry blast
> Brushing the waves ; wild shrieks of death
> Were heard these bristling cliffs beneath,
> And cries for aid.   The morning light
> Gleam'd on a scene of wild affright.
> Where yawns the cave, the rugged shore
> With many a corse lay cover'd o'er,
> And many a gorgeous fragment show'd
> How fair the bark the storm subdued."

There was a Cromarty mechanic of the last age, named Willie Millar, who used to relate a wonderful adventure which befell him in the cave.   Willie was a man of fertile invention, fond of a good story, and zealous in the improvement of bad ones ; but his zeal was evil spoken of—the reformations he effected in this way being regarded as little better than sinful, and his finest inventions as downright lying.   There was a smithy in the place, which, when he had become old and useless, was his favourite resort.   He would take up his seat on the forge each evening, regularly as the evening came, and relate to a group of delighted but too incredulous youngsters, some new passage in his wonderful autobiography ; which, though it seemed long enough to stretch beyond the flood, received new accessions every night.   So little, indeed, had he in common with the small-minded class who, possessed of only a limited number of narratives and ideas, go over and over these as the hands of a clock pass continually over the same figures, that, with but one exception in favour of the adventure of the cave, he hardly ever told the same story twice.

There was a tradition current in Cromarty, that a town's-man had once passed through the Dropping-Cave, until he heard a pair of tongs rattle over his head on the hearth of a farmhouse of Navity, a district of the parish which lies fully three miles from the opening; and Willie, who was, it seems, as hard of belief in such matters as if he himself had never drawn on the credulity of others, resolved on testing the story by exploring the cave. He sewed sprigs of rowan and wych-elm in the hem of his waistcoat, thrust a Bible into one pocket and a bottle of gin into the other, and providing himself with a torch, and a staff of buckthorn which had been cut at the full of the moon, and dressed without the assistance of iron or steel, he set out for the cave on a morning of midsummer. It was evening ere he returned—his torch burnt out, and his clothes stained with mould and slime, and soaked with water.

After lighting the torch, he said, and taking a firm grasp of the staff, he plunged fearlessly into the gloom before him. The cavern narrowed and lowered as he proceeded; the floor, which was of a white stone resembling marble, was hollowed into cisterns, filled with a water so exceedingly pure that it sparkled to the light like spirits in crystal, and from the roof there depended clusters of richly embossed icicles of white stone, like those which, during a severe frost, hang at the edge of a water-fall. The springs from above trickled along their channelled sides, and then tinkled into the cisterns, like rain from the eaves of a cottage after a thunder-shower. Perhaps he looked too curiously around him when remarking all this; for so it was, that at the ninth and last cistern he missed his footing, and, falling forwards shattered his bottle of gin against the side of the cave. The liquor ran into a little hollow of the marble, and, unwilling to lose what he regarded as very valuable, and what certainly had cost him some trouble and suffering to procure (for he had rowed half way across the Firth for it in terror of the customhouse and a cockling sea), he stooped down and drank till his breath failed him. Never was there better Nantz; and, pausing to recover himself, he stooped and drank, again and again. There were strange appearances when he rose. A circular rainbow had formed round his torch; there was a blue mist gathering in the hollows of the cave; the very roof and sides began to heave and reel, as if the living rock were a Flushing lugger riding on the ground-swell; and there was a low humming noise that came sounding from the interior, like that of bees in a hawthorn thicket on an evening of midsummer. Willie, however, had become much less timorous than at first, and, though he could not well account for the fact, much less disposed to wonder. And so on he went.

He found the cavern widen, and the roof rose so high that the light reached only the snowy icicles which hung meteor-like over his head. The walls were formed of white stone, ridged and furrowed like pieces of drapery, and all before and around him there sparkled myriads of crystals, like dewdrops in a spring morning. The sound of his footsteps was echoed on either hand by a multitude of openings, in which the momentary gleam of his torch was reflected, as he passed, on sheets of water and ribs of rock, and which led, like so many arched corridors, still deeper into the bowels of the hill. Nor, independently of the continuous humming noise, were all the sounds of the cave those of echo. At one time he could hear the wind moaning through the trees of the wood above, and the scream of a hawk as if pouncing on its prey; then there was the deafening blast of a smith's bellows, and the clang of hammers on an anvil; and anon a deep hollow noise resembling the growling of a wild beast. All seemed terribly wild and unnatural; a breeze came moaning along the cave, and shook the marble drapery of the sides, as if it were formed of gauze or linen;

the entire cave seemed turning round like the cylinder of an engine, till the floor stood upright and the adventurer fell heavily against it; and as the torch hissed and sputtered in the water, he could see by its expiring gleam that a full score of dark figures, as undefined as shadows by moonlight, were flitting around him in the blue mist which now came rolling in dense clouds from the interior. In a moment more all was darkness, and he lay insensible amid the chill damps of the cave.

The rest of the adventure wonderfully resembled a dream. On returning to consciousness, he found that the gloom around him had given place to a dim red twilight, which flickered along the sides and roof like the reflection of a distant fire. He rose, and grasping his staff staggered forward. " It is sun- light," thought he, " I shall find an opening among the rocks of Eathie, and return home over the hill." Instead, however, of the expected outlet, he found the passage terminate in a wonderful apartment, so vast in extent, that though an immense fire of pine-trees, whole and unbroken from root to branch, threw up a red wavering sheet of flame many yards in height, he could see in some places neither the walls nor the roof. A cataract, like that of Foyers during the long-continued rains of an open winter, descended in thunder from one of the sides, and presenting its broad undulating front of foam to the red gleam of the fire, again escaped into darkness through a wide broken-edged gulf at the bottom. The floor of the apartment appeared to be thickly strewed with human bones, half-burned and blood-stained, and gnawed as if by cannibals ; and directly in front of the fire there was a low tomblike erection of dark- coloured stone, full twenty yards in length, and roughened with grotesque hieroglyphics, like those of a Runic obelisk. An enormous mace of iron, crusted with rust and blood, reclined against the upper end ; while a bugle of gold hung by a chain of the same metal from a column at the bottom. Willie seized the bugle, and winded a blast till the wide apartment shook with the din ; the waters of the cataract disappeared, as if arrested at their source ; and the ponderous cover of the tomb began to heave and crackle, and pass slowly over the edge, as if assailed by the terrific strength of some newly-awakened giant below.   Willie again winded the bugle ; the cover heaved upwards, disclosing a corner of the chasm beneath ; and a hand covered with blood, and of such fearful magnitude as to resemble only the conceptions of Egyptian sculpture, was slowly stretched from the darkness towards the handle of the mace.   Willie's resolution gave way, and, flinging down the horn, he rushed hurriedly towards the passage.   A yell of blended grief and indignation burst from the tomb, as the immense cover again settled over it; the cataract came dashing from its precipice with a heavier volume than before; and a furious hurricane of mingled wind and spray that rushed howling from the interior, well-nigh dashed the adventurer against the sides of the rock. He succeeded, however, in gaining the passage, sick at heart and nearly petrified with terror; a state of imperfect consciousness succeeded, like that of a feverish dream, in which he retained a sort of half conviction that he was lingering in the damps and darkness of the cave, obstinately and yet unwillingly; and, on fully regaining his recollection, he found himself lying across the ninth cistern, with the fragments of the broken bottle on the one side, and his buckthorn staff on the other.   He could hear from the opening the dash of the advancing waves against the rocks, and on leaping to the beach below, found that his exploratory journey had occupied him a whole day.

The adventure of Willie Millar formed at one time one of the most popular traditions of Cromarty.   It was current among the children not more than eighteen years ago, when the cave was explored a second time, but with a very different result, by a boy of the school in which the writer of these legends had the misfortune of being regarded as the greatest dunce and truant of his time.   The character of Willie forms the best possible commentary on *his* story—the character of the boy may perhaps throw some little light on his. When in his twelfth year, he was by far the most inquisitive little fellow in the place. His curiosity was insatiable. He had broken his toys when a child, that he might see how they were constructed; and a watch which the owner had thoughtlessly placed within his reach, narrowly escaped sharing a similar fate. He dissected frogs and mice in the hope of discovering the seat of life; and when one day found dibbling at the edge of a spring, he said he was trying to penetrate to the source of water. His schoolmaster nicknamed him " The *Senachie*," for the stories with which he beguiled his class-fellows of their tasks were without end or number; the neighbours called him *Philosopher*, for he could point out the star of the pole, with the Great Bear that continually walks round it; and he used to affirm that there might be people in the moon, and that the huge earth is only a planet. Having heard the legend of Willie Millar, he set out one day to explore the cave; and when he returned he had to tell that the legend was a mere legend, and that the cave, though not without its wonders, owed, like the great ones of the earth, much of its celebrity to the fears and the ignorance of mankind.

In climbing into the vestibule of the recess, his eye was attracted by a piece of beautiful lacework, gemmed by the damps of the place, and that stretched over a hollow in one of the sides. It was not, however, a work of magic, but merely the web of a field-spider, that from its acquaintance with lines and angles, seemed to have discovered a royal road to geometry. The petrifying spring next attracted his notice. He saw the mosses hardening into limestone—the stems already congealed, and the upper shoots dying that they might become immortal. And there came into his mind the story of one Niobe, of whom he had read in a school-book, that, like the springs of the cave, wept herself into stone, and the story too of the half-man half-marble prince of the Arabian tale. " Strange," thought the

Y

boy, "that these puny dwarfs of the vegetable kingdom should become rock and abide for ever, when its very giants, the chestnut trees of Etna and the cedars of Lebanon, moulder away in the deep solitude of their forests, and become dust or nothing." Lighting his torch, he proceeded to examine the cavern. A few paces brought him to the first cistern. He found the white table of marble in which it is hollowed raised knee-height over the floor, and the surface fretted into little cavities by the continual dropping, like the surface of a thawing snow-wreath when beaten by a heavy shower. As he strided over the ledge, a drop from above extinguished his torch ;—he groped his way back and rekindled it. He had seen the first cistern described by the adventurer ; and of course all the others, with the immense apartment, the cataract, the tomb, the iron mace, and the golden bugle, lay in the darkness beyond. But, alas ! when he again stepped forward, instead of the eight other hollows he found the floor covered with one continuous pool, over which there rose fast-contracting walls and a descending roof ; and though he pressed onward amid the water that splashed below, and the water that fell from above—for his curiosity was unquenchable, and his clothes of a kind which could not be made worse—it was only to find the rock closing hopelessly before him, after his shoulders had at once pressed against the opposite sides, and the icicles had passed through his hair. There was no possibility of turning round, and so, creeping backwards like a crab, he reached the first cistern, and in a moment after stood in the lighted part of the cave. His feelings on the occasion were less melancholy than those of the traveller, who, when standing beside the two fountains of the Nile, "began in his sorrow to treat the inquiry concerning its source as the effort of a distempered fancy." But next to the pleasure of erecting a system, is the pleasure of pulling one down ; and he felt it might be so even with regard to a piece of traditionary history. Besides, there was a newly-fledged thought which had come fluttering round him for the first time, that more than half consoled him under his disappointment. He remembered that when a child no story used to please him that was not both marvellous and true—that a fact was as nothing to him disunited from the wonderful, nor the wonderful disunited from fact. But the marvels of his childhood had been melting away, one after one—the ghost, and the wraith, and the fairy had all disappeared ; and the wide world seemed to spread out before him a tame and barren region, where truth dwelt in the forms of commonplace, and in these only. He now felt for the first time that it was far otherwise ; and that so craving an instinct, instead of perishing for lack of sustenance, would be fed as abundantly in the future by philosophy and the arts, as it had been in the past by active imaginations and a superstitious credulity.

The path which, immediately after losing itself on the beach where it passes the cave, rises by a kind of natural stair to the top of the precipices, continues to ascend till it reaches a spring of limpid water, which comes gushing out of the side of a bank covered with moss and daisies : and which for more than a century has been known to the town's-people by the name of Fiddler's Well. Its waters are said to be medicinal, and there is a pretty tradition still extant of the circumstance through which their virtues were first discovered, and to which the spring owes its name.

Two young men of the place, who were much attached to each other, were seized at nearly the same time by consumption. In one the progress of the disease was rapid—he died two short months after he was attacked by it ; while the other, though wasted almost to a shadow, had yet strength enough left to follow the corpse of his companion to the grave. The name of the survivor was Fiddler—a name still common among the seafaring men of the town. On the evening of the interment he felt oppressed and unhappy ; his imagination was haunted

by a thousand feverish shapes of open graves with bones mouldering round their edges, and of coffins with the lids displaced;
and after he had fallen asleep, the images, which were still the
same, became more ghastly and horrible. Towards morning,
however, they had all vanished; and he dreamed that he was
walking alone by the sea-shore in a clear and beautiful day of
summer. Suddenly, as he thought, some person stepped up
behind, and whispered in his ear, in the voice of his deceased
companion, "Go on, Willie; I shall meet you at *Stormy.*"
There is a rock in the neighbourhood of Fiddler's Well, so called
from the violence with which the sea beats against it when the
wind blows strongly from the east. On hearing the voice he
turned round, and seeing no one, he went on, as he thought, to
the place named, in the hope of meeting his friend, and sat
down on a bank to wait his coming; but he waited long—
lonely and dejected; and then remembering that he for whom he
waited was dead, he burst into tears. At this moment a large
field-bee came humming from the west, and began to fly round
his head. He raised his hand to brush it away; it widened
its circle, and then came humming into his ear as before. He
raised his hand a second time, but the bee would not be scared
off; it hummed ceaselessly round and round him, until at
length its murmurings seemed to be fashioned into words, ar-
ticulated in the voice of his deceased companion—" Dig, Willie,
and drink!" it said; "Dig, Willie, and drink!" He accordingly set himself to dig, and no sooner had he torn a sod
out of the bank than a spring of clear water gushed from the
hollow; and the bee taking a wider circle, and humming in a
voice of triumph that seemed to emulate the sound of a distant
trumpet, flew away. He looked after it, but as he looked the
images of his dream began to mingle with those of the waking
world; the scenery of the hill seemed obscured by a dark cloud,
in the centre of which there glimmered a faint light; the rocks,
the sea, the long declivity, faded into the cloud; and turning

round he saw only a dark apartment, and the faint beams of
morning shining in at a window. He rose, and after digging
the well, drank of the water and recovered. And its virtues
are still celebrated ; for though the water be only simple water,
it must be drunk in the morning, and as it gushes from the bank ;
and with pure air, exercise, and early rising for its auxiliaries,
it continues to work cures.

CHAPTER XXIV.

" Fechtam memorate blodæam,
Fechtam terribilem."—DRUMMOND'S *Polemo Middinia.*

" Tulzies lang-remember'd an' bluidy,
Terrible tulzies."—*Muckle-Vennel Translation.*

IT is well for human happiness in the ruder ages, that cowardice is rarely or never the characteristic of a people who have either no laws, or laws that cannot protect them ; for, in the more unsettled stages of society, personal courage is a necessary policy, and no one is less safe than he who attempts to escape danger by running away. During the early part of the last century, Cromarty was well-nigh as rude a village of the kingdom as any it contained. The statute-book had found its way into the place at a much remoter period, but its authority had not yet travelled so far ; and so the inhabitants were left to protect themselves by their personal courage and address, in the way their ancestors had done for centuries before. It was partly a consequence of the necessity, and partly from the circumstance that two or three families of the place were deeply imbued for several generations with a warlike spirit, which seemed born with them, that for years, both before and after the Rebellion, the prowess of the people, as exhibited in their quarrels with folk of the neighbouring districts, was celebrated all over the country. True it was, they had quailed before the rebels, but then the best soldiers of the crown had done the same. On one occasion two of them, brothers of the name of Duff—gigantic fellows of six feet and a half—had stood back to back for an entire hour in the throng of a Redcastle market, defending themselves against half the cudgels of Strathglass. On another, at the funeral of a town's-man, who was interred in the burial-ground of Kilmuir, a party of them had fought with the people of the parish, and defeated them in their own territories. On a third, after a battle which lasted for several hours, they had beaten off the men of Rosemarkie and Avoch from a peat-moss in an unappropriated moor; and this latter victory they celebrated in a song, in which it was humorously proposed that, as their antagonists had been overpowered by the *men* of the parish, they should, in their next encounter, try their chance of war with the women. In short, their frays at weddings, funerals, and markets, were multiplied beyond number, until at length the cry of " Hiloa! Help for Cromarty!" had become as formidable as the war-cry of any of the neighbouring clans.

But there are principles which are good or evil according to the direction in which they operate; and of this class is that warlike principle whose operations I am attempting to describe. It was well for the people of Cromarty that, when there was no law powerful enough to protect them, they had courage enough to protect themselves; and particularly well at a period when the neighbouring Highlanders were still united by the ties of clanship into formidable bodies, ready to assert to a man the real or pretended rights of any individual of their number. It was not well, however, that these men of Cromarty should have broken the heads of half the men of Kilmuir, for merely insisting on a prescriptive right of carrying the corpse of a native to the churchyard when it had entered the limits of their own parish, and such was the sole occasion of the quarrel; or that, after appropriating to themselves, much at the expense of justice, the moss of the Maolbuoy Common, they should have deemed it legitimate sport to insult, in bad rhyme, the poor people whom they had deprived of their winter's fuel, and who were starving for want of it. Occasionally, however, they avenged on themselves the wrongs done to their neighbours; for, though no tribe of men could be more firmly united at a market or tryst, where an injury done to any one of them was regarded as an injury done to every one, they were not quite so friendly when in town, where their interests were separate, and not unfrequently at variance. Their necessities abroad had taught them how to fight, and their resentments at home often engaged them in repeating the lesson. Their very enjoyments had caught hold of it, and Martinmas and the New-Year were not more the festivals of good ale than of broken heads. The lesson, sufficiently vexatious at any time, except when conned in its proper school, became peculiarly a misfortune to them upon the change which began to take place in the northern counties about the year 1740, when the law of Edinburgh— as it was termed by a Strathcarron freebooter—arrived at the ancient burgh of Tain, and took up its seat there, much to the terror and annoyance of the neighbouring districts.

Subsequent to this unfortunate event, a lawyer named Macculloch fixed his place of residence among the people of Cromarty, that he might live by their quarrels; and, under the eye of this sagacious personage, the stroke of a cudgel became as potent as that of the wand of a magician. Houses, and gardens, and corn-furrows vanished before it. Law was not yet sold at a determined price. It was administered by men who, having spent the early part of their lives amid feuds and bickerings, were still more characterized by the leanings of the partisan than the impartiality of the judge; and, under these men, the very statute-book itself became a thing of predilections and antipathies; for while in some instances justice, and a great deal more, cost almost nothing, in others it was altogether beyond price. Macculloch, however, who dealt it out by retail, rendered it sufficiently expensive, even when at the cheapest. Fines and imprisonments, and accounts which his poor clients could not read, but which they were compelled to pay, were only the minor consequences of his skill; for on one occasion he contrived that almost half the folk of the town should be cited, either as pannels or witnesses, to the circuit court of Inverness; where, through the wrongheadedness of a jury, and the obstinacy of a judge, a good town's-man and powerful combatant, who would willingly harm no one, but fight with anybody, ran a very considerable risk of being sent to the plantations. The people were distressed beyond measure, and their old antagonists of Kilmuir and Rosemarkie fully avenged.

In course of time, however, they became better acquainted with law; and their knowledge of the lawyer (which, like every other species of knowledge, was progressive), while it procured him in its first stages much employment, prevented him latterly from being employed at all. He was one of the most active of village attorneys. No one was better acquainted with the whole art of recovering a debt, or of entering on the possession of a legacy—of reclaiming property, or of conveying it; but it was ultimately discovered that his own particular interests could not always be identified with those of the people who employed him; and that the same lawsuit might be gained by him and lost by his client. It was one thing, too, for Macculloch to recover a debt, and quite another for the person to whom it had been due. In cases of the latter description he was an adept in the art of promising. Day after day would he fix his term of settlement; though the violation of the promise of yesterday proved only a prelude to the violation of that of to-day, and though both were found to be typical of the promise which was to be passed on the morrow. He had determined, it was obvious, to render his profession as lucrative as possible; but somehow or other—it could only be through an excess of skill —he completely overshot the mark. No one would, at length, believe his promises, or trust to his professions; his great skill began to border in its effects, as these regarded himself, on the opposite extreme; and he was on the eve of being starved out

of the place, when Sir George Mackenzie, the proprietor, made
choice of him as his factor, and intrusted to him the sole
management of all his concerns.

Sir George in his younger days had been, like his grandfather
the Earl, a stirring, active man of business.   He was a stanch
Tory, and on the downfall of Oxford, and the coming in of the
Whigs, he continued to fret away the energies of his character,
in a fruitless, splenetic opposition; until at length, losing heart
in the contest, he became, from being one of the most active,
one of the most indolent men in the country.   He drank hard,
lived grossly, and seemed indifferent to everything.   And never
were there two persons better suited to each other than the
lawyer and Sir George.   The lawyer was always happiest in his
calculations when his books were open to the inspection of no
one but himself; and the laird, though he had a habit of reckoning over the bottle, commonly fell asleep before the amount
was cast up.   But an untoward destiny proved too hard for
Macculloch in even this office.   Apathetical as Sir George was
deemed, there was one of his feelings which had survived the
wreck of all the others;—that one a rooted aversion to the
town of Cromarty, and in particular to that part of the country
adjacent which was his own property.   No one—least of all
himself—could assign any cause for the dislike, but it existed
and grew stronger every day: and the consequences were ruinous
to Macculloch; for in a few years after he had appointed him
to the factorship, he disposed of all his lands to a Mr. William
Urquhart of Meldrum—a transaction which is said to have had
the effect of converting his antipathy into regret.   The factor
set himself to seek out for another master; and in a manner
agreeable to his character.   He professed much satisfaction that
the estate should have passed into the hands of so excellent a
gentleman as Mr. Urquhart; and proposed to some of the
town's-folk that they should eat to his prosperity in a public
dinner, and light up a constellation of bonfires on the heights

which overlook the bay. The proposal took; the dinner was attended by a party of the more respectable inhabitants of the place, and the bonfires by all the children.

A sister of Sir George's, the Lady Margaret, who a few years before had shared in the hopes of her attainted cousin, Lord Cromartie, and had witnessed, with no common sensations of grief, the disastrous termination of the enterprise in which he had been led to engage, was at this time the only tenant of Cromarty Castle. She had resided in the house of Lord George previous to his attainder, but on that event she had come to Cromarty to live with her brother. His low habits of intemperance proved to her a fruitful source of vexation; but how was the feeling deepened when, in about a week after he had set out on a hasty journey, the purpose of which he refused to explain, she received a letter from him, informing her that he had sold all his lands! She saw, in a step so rash and unadvised, the final ruin of her family, and felt with peculiar bitterness that she had no longer a home. Leaning over a window of the castle, she was indulging in the feelings which her circumstances suggested, and looking with an unavailing but natural regret on the fields and hamlets that had so soon become the property of a stranger, when Macculloch and his followers came marching out on the lawn below from the adjoining wood, and began to pile on a little eminence in front of the castle the materials of a bonfire. It seemed, from the effect produced on the poor lady, that, in order entirely to overpower her, it was only necessary she should be shown that the circumstance which was so full of distress to her, was an occasion of rejoicing to others. For a few seconds she seemed stupified by the shouts and exultations of the party below; and then, clasping her hands upon her breast, she burst into tears and hurried to her apartment. As the evening darkened into night, the light of the huge fire without was reflected through a window on the curtains of her bed. She requested her attendant to shut it out;

but the wild shouts of Macculloch's followers, which were echoed until an hour after midnight by the turrets above and the vaults below, could not be excluded. In the morning Lady Margaret was in a high fever, and in a few days after she was dead.

The first to welcome the new laird to his property was Macculloch the factor. Urquhart of Meldrum, or Captain Urquhart, as he was termed, had made his money on sea—some said as a gallant officer in the Spanish service, some as the master of a privateer, or even, it was whispered, as a pirate. He was a rough unpolished man, fond of a rude joke, and disposed to seek his companions among farmers and mechanics, rather than among the people of a higher sphere. But, with all his rudeness, he was shrewd and intelligent, and qualified, by a peculiar tact, to be a judge of men. When Macculloch was shown into his room, he neither returned his bow nor motioned him to a seat, though the lawyer, no way daunted, proceeded to address him in a long train of compliments and congratulations. "Humph!" replied the Captain. "Ah!" thought the lawyer, "you will at least hear reason." He proceeded to state, that as he had been intrusted with the sole management of Sir George's affairs, he was better acquainted than any one else with the resources of the estate and the character of the tenants; and that, should Mr. Urquhart please to continue him in his office, he would convince him he was the fittest person to occupy it to his advantage. "Humph!" replied the Captain; "for how many years, Sir lawyer, have you been factor to Mackenzie?" "For about five," was the reply. "And was he not a good master?" "Yes, sir, rather good, certainly—but his unfortunate habits." "His habits!—he drank grog, did he not? and served it out for himself? So do I. Mark me, Sir factor! You are a —— mean rascal, and shall never finger a penny of mine. You found in Mackenzie a good simple fellow, who employed you when no one else would; but no sooner had he unshipped himself than you hoisted colours for me, —— you, whom, I suppose, you could tie up to the yard-arm for somewhat less than a bred hangman would tie up a thief for ;—ay, that you would ! I have heard of your dinner, sir, and your bonfires, and of the death of Lady Margaret (had you another bonfire for that ?) and now tell you once for all, that I despise you as one of the meanest —— rascals that ever turned tail on a friend in distress. Off, sir—there is the door !" Such was the reward of Macculloch. In a few years after, he had sunk into poverty and contempt ; one instance of many, that rascality, however profitable in the degree, may be carried to a ruinous extreme, and that he who sets out with a determination of cheating every one, may at length prove too cunning for even himself.

The people of the town, not excepting some of those who had shouted round the bonfires and sat down to the dinner, were much gratified by the result of Macculloch's application ; and for some time the laird was so popular that there was no party in opposition to him. An incident soon occurred, however, which had the effect of uniting nine-tenths of the whole parish into a confederacy, so powerful and determined, that it contended with him in a lawsuit for three whole years.

The patronage of the church of Cromarty, on the attainder of Lord George Mackenzie, in whom it had been vested, devolved upon the Crown. It was claimed, however, by Captain Urquhart, and the Crown, unacquainted with the extent of many of the privileges derived to it by the general forfeiture of the late Rebellion, and of this privilege among the others, seemed no way inclined to dispute with him the claim. He therefore nominated to the parish, on the first vacancy, a Mr. Simpson of Meldrum as a proper minister. This Meldrum was a property of Mrs. Urquhart's, and the chief qualification of Mr. Simpson arose from the circumstance of his having been born on it. The Captain was himself a Papist, and had not set a foot within the church of Cromarty since he had come to the estate ; his wife was an Episcopalian, and, more liberal than her husband, she had on one occasion attended it in honour of the wedding of a favourite maid. The people of the town, in the opinion that the presentation could not be in worse hands, and dissatisfied with the presentee, rejected the latter on the ground that Captain Urquhart was not the legitimate patron; and, binding themselves by contract, they subscribed a considerable sum that they might join issue with him in a lawsuit. They were, besides, assisted by the neighbouring parishes; and, after a tedious litigation, the suit was decided in their favour; but not until they had expended upon it, as I have frequently heard affirmed with much exultation, the then enormous sum of five hundred pounds. They received from the Crown their choice of a minister.

Urquhart, whose obstinacy, sufficiently marked at any time, had been roused by the struggle into one of its most determined attitudes, resisted the claims of the people until the last; and, when he could no longer dictate to them as a patron, he set himself to try whether he could not influence them as a landlord. A day was fixed for the parishioners to meet in the church, that they might avail themselves of the gift of the Crown by making choice of a minister; and, before it arrived, the Captain made the round of his estate, visiting his tenants and dependants, and every one whom he had either obliged, or had the power of obliging, with the intention of forming a party to vote for Mr. Simpson. All his influence, however, proved insufficient to accomplish his object. His tenants preserved either a moody silence when he urged them to come into his plans, or replied to his arguments, which savoured sadly of temporal interests, in rude homilies about liberty of conscience and the rights of the people. Urquhart was not naturally a very patient man; he had been trained, too, in a rough school; and, long before he had accomplished the purposed round, he had got into one of his worst moods. His arguments had been converted into threats, and his threats met by sturdy defiances. In the evening of this vexatious day he stood in front of the steadings of

Roderick Ross of the Hill, a plain decent farmer, much beloved by the poor for the readiness with which he imparted to them of his substance, and not a little respected by Urquhart himself for his rough strong sense and sterling honesty. A grey, weather-wasted headstone still marks out his grave ; but of the cottage which he inhabited, of his garden fence, and the large gnarled elms which sprung out of it, of his barns, his cow-houses, and his sheep-folds, there is not a single vestige. They occupied, eighty years ago, the middle of one of the parks which are laid out on the hill of Cromarty where it overlooks the town—the third park in the upper range from the eastern corner. In rainy seasons, the spring which supplied his well comes bursting out from among the furrows. Roderick came from the barn to meet the laird ; and, after the customary greeting, was informed of the cause of his visit. The merits of the case he had discussed at mill and smithy with every farmer on the estate ; and, with his usual bluntness, he now inquired at the laird what interest he, a Papist, could have in the concerns of a Protestant church. " For observe, Captain," said he, " if ye ettle at serving us wi' a minister, sound after your way o' belief, I maun in conscience gie you a' the hinderance I can, as the man must be an unsound Papisher to me ; an' if, what's mair likely, ye only wuss to oblige the callant Simpson wi' a glebe, stipend, an' manse, without meddling wi' ony religion, it's surely my part to oppose ye baith ;—you, for making God's kirk meat an' drink to a hireling ; him, for taking it on sic terms." The Captain, though he used to admire Roderick's natural logic, regarded it with a very different feeling when he found it brandished against himself. " Roderick," said he, and he swore a deadly oath, " you shall either vote for Mr. Simpson or quit your farm at Whitsunday first." " You at least gie me my choice," said the honest farmer, and turning abruptly from him he stalked into the barn.

Roderick left his plough in the furrow on the day fixed for the meeting, and went into the house to prepare for it, by dress-

ing himself in his best clothes. His wife had learned the result
of his conference with the laird, and, in her opinion, the argument of threatened ejection was a more powerful one than any
that could be advanced by the opposite party.   Repeatedly did
she urge it, but to no effect ; Roderick was stubborn as an old
Covenanter. She watched, however, her opportunity ; and when
he went in to dress, which he always did in a small apartment
formed by an outjet of the cottage, she followed him, as if once
more to repeat what she had so often repeated already, but in
reality with a very different intention.   She suffered him to
throw off his clothes, piece by piece, without the slightest attempt to prevent him ; but at the moment when his head and
arms were involved in the intricacies of a stout linen shirt, she
snatched up his holiday bonnet, coat, and waistcoat, together
with the articles of dress he had just relinquished, and rushing
out of the apartment with them, shut and bolted the door behind
her.   To place against it every article of furniture which the
outer room afforded, was the work of the first minute ; and to
advise her liege lord to betake himself to the bed which his
prison contained until the kirk should have *skailed*, was her employment in the second.   Roderick was not to be baulked so.
There was a window in the apartment, which, had the walls
been of stone, would scarcely have afforded passage to an ordi-
nary-sized cat, but luckily they were of turf. Into this opening
he insinuated first his head, next his shoulders, and wriggling
from side to side until the whole wall heaved with the commo-
tion, he wormed himself into liberty ; and then set off for the
church of Cromarty, without bonnet, coat, or waistcoat.   An
angry man was Roderick ; and the anger, which he well knew
would gain him nothing if wreaked on the gudewife, was boiling
up against the Captain and Mr. Simpson. He entered the church,
and in a moment every eye in it was turned on him.   The schoolmaster, a thin serious-looking person, sat in the precentor's desk,
with his writing materials before him, to take down the names

of the voters, hundreds of whom thronged the body of the church. Captain Urquhart, in an attitude between sitting and standing, occupied one of the opposite pews ; about half a dozen of his servants lounged behind him. He was a formidable-looking, dark-complexioned, square-shouldered man, of about fifty ; and over his harsh weather-beaten features, which were in some little degree the reverse of engaging at any time, the occasion of the meeting seemed to have flung a darker expression than was common to them. As Roderick advanced, he started up as if to reconnoitre so terrible an apparition. Roderick's shirt and breeches were stained by the damp mouldy turf of the window, his face had not escaped, and, instead of being marked by its usual expression of quiet good-nature, bore a portentous ferocity of aspect, which seemed to indicate a man not rashly to be meddled with. " In the name of wonder, what brings you here in such plight ?" was the question put to him by an acquaintance in the aisle. " I come here," said Roderick, in a voice sufficiently audible all over the building, " to gie my vote as a free member o' this kirk in the election o' this day ; an' as for the particular plight," lowering his tone into a whisper, " speer about that at the gudewife."—" And whom do you vote for ?" said the schoolmaster, " for the time is up ;—there are two candidates, Simpson and Henderson." " For honest Mr. Henderson," said the farmer ; " an' ill be his luck this day wha votes for ae Roman out o' the fear o' anither, or lets the luve o' warld's gear stan' atween him an' his conscience." The Captain grasped his stick ; Roderick clenched his fist. " Look ye, Captain," he continued, " after flinging awa, for the sake o' the puir kirk, the bonny rigs o' Driemonorie, an' I ken I have done it, ye needna think to daunt me wi' a kent. Come out, Captain, yoursel, or ony twa o' your gang, an' in this quarrel I shall bide the warst. Nay, man, glower as ye list ; I'm no obliged to be feart though ye choose to be angry." The shout of " No Popish patron !—no Popish patron !" which shook the very roof that

z

stretched over the heads of the hundreds who joined in it, served
as a kind of chorus to this fearless defiance. The Captain
suffered his stick to slip through his fingers until the knob
rested on his palm, and then, striding over the pew, he walked
out of church. In less than half an hour after, the popular
candidate was declared duly elected, and at Whitsunday first
Roderick was ejected from his farm. His character, however,
as a man of probity and a skilful farmer, was so well established
throughout the country, that he suffered less on the occasion
than almost any other person would have done. He died many
years after, the tacksman of Peddieston, possessed of ingear and
outgear, and of a very considerable sum of money, with which
he had the temerity to intrust a newfangled kind of money-bor-
rower, termed a bank.

After all they had achieved and suffered on this occasion, the
people of Cromarty were unfortunate in their minister. He
was a person of considerable talent, and an amiable disposition;
and beloved by every class of his parishoners. The young
spoke well of him for his good-nature; the old for the deference which he paid to the opinions of his lay advisers. He
was, besides, deeply read in theology, and acquainted with
the various workings of religion in the various constitutions of
mind. But of all his friends and advisers, there were none
sufficiently acquainted with his character to give him the advice
which he most needed. He was naturally amiable and unassuming, and when he became a convert to Christianity,
scarcely any change took place in his external conduct. He
continued to act from principle in the manner he had previously acted from the natural bent of his disposition. For
the first few years he was much impressed by a sense of the
importance of spiritual concerns, and he became a minister of
the church that he might press their importance upon others;
but there are ebbs and flows of the mind in its moral as certainly as in its intellectual operations; and that flow of zeal

which characterizes the young convert is very often succeeded by a temporary ebb, during which he sinks into comparative indifference. It was thus with Henderson. His first impressions became faint, and he continued to walk the round of his duties, rather from their having become matters of custom to him, and that it was necessary for him to maintain the character of being consistent, than from a due sense of their importance. He continued, too, to instruct his people by delineations of character and expositions of doctrine; but his knowledge of the first was the result of studies which he had ceased to prosecute, and in which he himself had been both the student and the thing studied, and the efficacy of the latter was neutralized by their having become to him less the objects of serious belief than of metaphysical speculation. His peculiar character, too, with all its seeming advantages of natural constitution, was perhaps as much exposed to evil as others of a less amiable stamp. There are passions and dispositions so unequivocally bad, that even indifference itself is roused to oppose them; but when the current of nature and the course of duty seem to run parallel, we suffer ourselves to be borne away by the stream, and are seldom sufficiently watchful to ascertain whether the parallelism be alike exact in every stage of our progress. Henderson's character precluded both suspicion and advice. What were the feelings of his people, when, on summoning the elders of the church, he told them, that, having formed an improper connexion with a girl of the place, he had become a disgrace to the order to which he belonged! He was expelled from his office, and after remaining in town until a neighbouring clergyman had dealt to him the censures of the Church, from the pulpit which he himself had lately occupied, and in presence of a congregation that had once listened to him with pleasure, and now beheld him with tears, he went away, no one knew whither, and was never again seen in Cromarty.

About twenty years after, a young lad, a native of the place, was journeying after nightfall between Elgin and Banff, when he was joined by two persons who were travelling in the same direction, and entered into conversation with them. One of them seemed to be a plain country farmer; the other was evidently a man of education and breeding. The farmer, with a curiosity deemed characteristic of Scotchmen of a certain class, questioned him about the occasion of his journey, and his place of residence. The other seemed less curious; but no sooner had he learned that he was a native of Cromarty, than he became the more inquisitive of the two; and his numberless inquiries regarding the people of the town, showed that at some period he had been intimately acquainted with them. But many of those after whom he inquired had been long dead, or had removed from the place years before. The lad whose curiosity was excited, was mustering up courage to ask him whether he had not at some time or other resided in Cromarty, when the stranger, hastily seizing his hand with the cordiality of an old friend, bade him farewell, and turning off at a cross-road, left him to the company of the farmer. " Who is that gentleman?" was his first question. " The Mr. Henderson," was the reply, " who was at one time minister of Cromarty." The lad learned further, that he supported himself as a country schoolmaster, and was a devout, excellent man, charitable and tender to others, but severe to himself beyond the precedents of Reformed Churches. " I wish," said the farmer, " you had seen him by day;—he has the grey locks and bent frame of old age though he is not yet turned of fifty. There is a hill in a solitary part of the country, near his school, on which he frequently spends the long winter nights in prayer and meditation; and a little below its summit there is a path which runs quite round, and which can be seen a full mile away, that has been hollowed out by his feet."

CHAPTER XXV.

" Unquiet souls
   Risen from the grave, to ease the heavy guilt
   Of deeds in life concealed."—AKENSIDE.

OF all the wilder beliefs of our forefathers, there is none which so truly continues to exist as the belief in the churchyard spectre. Treat it as we may, it has assuredly a fast hold of our nature. We may conceal, but we cannot smother it ;—we may deny it as pointedly as the lackey does his master when the visitor is an unwelcome one, but it is not from that circumstance a whit the less at home. True or false, too, it seems to act no unimportant part in the moral economy of the world. For without a deeper sense of religion to set in its place than most people entertain, men would be greatly the worse for wanting it. There are superstitions which perform, in some measure, the work of the devotional sentiment, when the latter is either undeveloped or misdirected ; and the superstition of the churchyard ghost is unquestionably one of the number.

I am fortunate, so far as the sympathy of place can have any influence on the mind, in the little antique room in which I have set myself to illustrate the belief. Just look round you for one brief minute, and see how the little narrow windows rise into the thatch, and how very profoundly one requires to stoop ere one can enter by the door. The ceiling rises far into the roof. There is a deep recess in the wall occupied by a few pieces of old china, and a set of shelves laden with old books ; and only see how abruptly the hearth-stone rises over the sanded floor, and how well the fashion of yonder old oaken table agrees with that of the old oaken scrutoire in the opposite corner.   Humble as my apartment may seem, it is a place of some little experience in the affairs of both this world and the other.   It has seen three entire generations come into being and pass away, and it now shelters the scion of a fourth.   It has been a frequent scene of christenings, bridals, and lykewakes—of the joys and sorrows, the cares and solicitudes of humble life.   Nor were these all.   There is the identical door at which it is said a great-grand-aunt of the writer saw a sheeted spectre looking in upon her as she lay a-bed ; and there the window at which another and nearer relative was sitting in a stormy winter evening, thinking of her husband far at sea, when, after a dismal gust had howled over the roof, the flapping of a sail and the cry of distress were heard, and she wrung her hands in anguish, convinced that her sailor had perished.   And so indeed it was.   Strange voices have echoed from the adjoining apartment ; the sounds of an unknown foot have been heard traversing its floor ; and I have only to descend the stair ere I stand on the place where a shadowy dissevered hand was once seen beckoning on one of the inmates.   How incalculably numerous must such stories have once been, when the history of one little domicile furnishes so many ?

About sixteen years ago, I accompanied an elderly relative —now, alas ! no more—on a journey through the parishes of Nigg, Fearn, and Tarbat.   He was a shrewd, clear-headed man, of great warmth of heart, who continued to bear a balmy atmosphere of the enthusiasm of early youth about him, despite of the hard-earned experience of fifty-five.   There could not be a more delightful companion even to a boy.   I never knew one half so well acquainted with the traditionary history of the country.   Every hamlet we passed, almost every green mound, had its story ; and there was that happy mixture of point and simplicity in his style of narrative, which almost every one knows how to admire, and scarcely one of a thousand how to imitate. He had, I suspect, a good deal of the sceptic in his composition, and regarded his ghost stories rather as the machinery of a sort of domestic poetry than as pieces of real history; but then, no one could value them more as curious illustrations of human belief, or show less of the coldness of infidelity in his mode of telling them. " Yonder lofty ridge," said he, as we passed along, " is the hill of Nigg, so famous, you know, as a hunting-place of the Fions. Were we on the other side, where it overhangs the sea, I could point out to you the remains of a cottage that has an old ghost story connected with it—a story that dates, I believe, some time in the early days of your grandmother. Two young girls, who had grown up together from the days of their childhood, and were mutually attached, had gone to the lykewake of a female acquaintance, a poor orphan, and found some women employed in dressing the body. There was an indifference and even light-heartedness shown on the occasion that shocked the two friends; and they solemnly agreed before parting, that should one of them outlive the other, the survivor, and no one besides, should lay out the corpse of the departed for the grave. The feeling, however, passed with the occasion out of which it arose, and the mutual promise was forgotten, until several years after, when one of the girls, then the mistress of a solitary farmhouse on the hill of Nigg, was informed one morning, by a chance passenger, that her old companion, who had become the wife of a farmer in the neighbouring parish of Fearn, had died in childbed during the previous night. She called to mind her promise, but it was only to reflect how impossible it was for her to fulfill it. She had her infant to tend, and no one to intrust it to—her maid having left her scarcely an hour before for a neighbouring fair, to which her husband and his ploughman had also gone. She spent an anxious day, and it was with no ordinary solicitude, as she saw the evening gradually darkening, and thought of her promise and her deceased companion, that she went out to a little hillock beside the house, which commanded a view of the moor over which her husband and the servants had to pass on their way from the fair, to ascertain whether any of them were yet returning. At length she could discern through the deepening twilight, a female figure in white coming along the moor, and supposing it to be the maid, and unwilling to appear so anxious for her return, she went into the house. The outer apartment, as was customary at the period, was occupied as a cow-house; some of the animals were in their stalls, and on their beginning to snort and stamp as if disturbed by some one passing, the woman half turned her to the door. What, however, was her astonishment to see, instead of the maid, a tall figure wrapped up from head to foot in a winding-sheet! It passed round to the opposite side of the fire, where there was a chair drawn in for the farmer, and seating itself, raised its thin chalky arms and uncovered its face. The features, as shown by the flame, were those of the deceased woman; and it was with an expression of anger, which added to the horror of the appearance, that the dead and glassy eyes were turned to her old companion, who, shrinking with a terror that seemed to annihilate every feeling and faculty except the anxious solicitude of the mother, strained her child to her bosom, and gazed as if fascinated on the terrible apparition before her. She could see every fold of the sheet; the black hair seemed to droop carelessly over the forehead; the livid, unbreathing lips were drawn apart, as if no friendly hand had closed them after the last agony; and the reflection of the flame seemed to rise and fall within the eyes—varying by its ceaseless flicker the statue-like fixedness of the features. As the fire began to decay, the woman recovered enough of her self-possession to stretch her hand behind her, and draw from time to time out of the child's cradle a handful of straw, which she flung on the embers; but she had lost all reckoning of time, and could only guess at the duration of the visit by finding the straw nearly expended. She was looking forward with a still deepening horror to being left in darkness with the spectre, when voices were heard in the yard without. The apparition glided towards the door; the cattle began to snort and stamp, as on its entrance; and one of them struck at it with its feet in the passing; when it uttered a faint shriek and disappeared. The farmer entered the cottage a moment after, barely in time to see his wife fall over in a swoon on the floor, and to receive the child. Next morning, says the story, the woman attended the lykewake, to fulfil all of her engagement that she yet could; and on examining the body, discovered that, by a strange sympathy, the mark of a cow's hoof was distinctly impressed on its left side."

We passed onwards, and paused for a few seconds where the parish of Nigg borders on that of Fearn, beside an old hawthorn hedge and a few green mounds. "And here," said my companion, "is the scene of another ghost story, that made some noise in its day; but it is now more than a century old, and the details are but imperfectly preserved. You have read, in Johnson's Life of Denham, that Charles II., during his exile in France, succeeded in procuring a contribution of ten thousand pounds from the Scotch that at that time wandered as itinerant traders over Poland. The old hedge beside you, and the few green mounds beyond it, once formed the dwelling-house and garden fence of one of these Polish traders, who had returned in old age to his native country, possessed, as all supposed, of very considerable wealth. He was known to the country folk as the 'Rich Polander.' On his death, however, which took place suddenly, his strong-box was found to contain only a will, bequeathing to his various relations large sums that were vested, no one knew where. Some were of opinion that he had lent money to a considerable amount to one or two neighbouring proprietors; and some had heard him speak of a brother in Poland, with whom he had left the greater part of his capital, and who had been robbed and murdered by banditti, somewhere on the frontier territories, when on his return to Scotland. In the middle of these surmisings, however, the Polander himself returned, as if to settle the point. The field there to the right, in front of the ruins, was at that time laid out as a lawn; there was a gate in the eastern corner, and another in the west; and there ran between them a road that passed the front of the house. And almost every evening the apparition of the Polander, for years after his decease, walked along that road. It came invariably from the east, lingered long in front of the building, and then, gliding towards the west, disappeared in passing through the gateway. But no one had courage enough to meet with it, or address it; and till this day the legacies of the Polander remain unpaid. I was acquainted in my younger days with a very old man, who has assured me that he repeatedly saw the apparition when on its twilight peregrinations along the road; and once as he lay a-bed in the morning in his mother's cottage, long after the sun had risen. There was a broad stream of light falling through an opening in the roof, athwart the grey and mottled darkness of the interior, and the apparition stood partly in the light, partly in the shadow. The richly-embroidered waistcoat, white cravat, and small clothes of crimson velvet, were distinctly visible; but he could see only the faint glitter of the laced hat and of the broad shoe-buckles; and though the thin withered hands were clearly defined, the features were wholly invisible."

We had now entered the parish of Fearn. "And here," continued my companion, as we approached the abbey, "is the scene of two other ghost stories, both, like the last, somewhat meagre in their details, but they may serve to show how, in a rude and lawless age, the cause of manners and of morals must have found no inefficient ally in a deeply-seated belief in the supernatural. A farmer of the parish, who had just buried his wife, had gone on the evening of the funeral to pay his addresses to a young woman who lived in a cottage beside the burying-ground yonder. There was, it would seem, little of delicacy on either side ; and his suit proved so acceptable, that shortly after nightfall he had his new mistress seated on his knee. They were laughing and joking together beside a window that opened to the churchyard, when the mother of the young girl entered the apartment, and, shocked by their levity, reminded him that the corpse of the woman so lately deceased lay in all the entireness and almost all the warmth of life not forty yards from where they sat. 'No, no, mother,' said the man ; 'entire she may be, but she was cold enough in all conscience before we laid her there.' He turned round as he spoke, and saw his deceased wife looking in upon him through the window. And returning home, he took to his bed, and died of a brain fever only a fortnight after. Depend on't, that widowers in this part of the country would be less hasty ever after in courting their second wives.

"The cottage higher up the hill—that one with the roof nearly gone, and the old elm beside it—was occupied about sixty years ago by a farmer of the parish and a harsh-tempered one-eyed woman, his wife. He had a son and daughter, the children of a former marriage, who found the dame a very stepmother. The boy was in but his fifth, the daughter in but her seventh year ; and yet the latter was shrewd enough to remark on one occasion, when beaten by the woman for transferring a little bit of leaven from the baking-trough to her mouth, that her second mother could see better with her one eye than her first mother with her two. The deceased, an industrious housewife, had left behind her large store of blankets and bed-linen ; but the bed of the two children for the summer and autumn after the marriage of their father, was covered by only a few worn-out rags, and when the winter set in, the poor things had to lie in one another's arms for the early part of every night shuddering with cold. For a week together, however, they were found every morning closely wrapt up in some of their mother's best blankets. The stepdame stormed, and threatened, and replaced the blankets in a large store-chest, furnished with lock and hasp ; but it was all in vain—they were found, notwithstanding, each morning on the children's bed regularly as the morning came ; and the poor things, though threatened and beaten, could give no other account of the matter than that they had been very cold when they fell asleep, and warm and comfortable when they awoke. At length, however, the girl was enabled to explain the circumstance in a manner that had the effect of tempering the severity of the stepmother all her life after. Her brother had fallen asleep, she said, but *she* was afraid, and could not sleep ; she was, besides, very cold, and so she lay awake till near the middle of the night, when the door opened, and there entered a lady all dressed in white. The fire was blazing brightly, and she could see as clearly as by day the large chest lying locked in the corner ; but when the lady went to it the hasp flew open, and she took out the blankets and wrapt them carefully round her brother and herself in the bed. The lady then kissed her brother, and was going to kiss her too, when she looked up in her face, and saw it was her first mother. And then she went away without opening the door.

"I remember another ghost story," continued my companion, "the scene of which I shall point out to you when we have entered the parish of Tarbat. There is a little muddy lake in the upper part of the parish which almost dries up in the warmer seasons, and on the further edge of which we shall be able to trace the remains of what was once a farmhouse. Considerably more than a century ago, a young man who travelled the country as a packman suddenly disappeared, no one knew how ; and several years after, in a dry summer, which reduced the lake to less than half its usual size, there was found a human skeleton among the mud and rushes at the bottom.

Long ere the discovery, however, the farmhouse was haunted by a restless, mischievous spectre, wrapped up in a grey plaid. Like most murdered folk of those days, the pedlar walked, restricting his appearance, however, to the interior of the cottage, which at length came to be deserted ; and falling into decay, it lay for the greater part of a half century as a roofless grass-covered ruin. Its old inmates had died off in extreme penury and wretchedness, and both they and the pedlar were nearly forgotten, when a young man, no way related to either, availing himself of the site of the cottage and the portions of its broken walls which still remained, rebuilt it when on the eve of his marriage, and removed to it with his young wife. On the third evening, when all the wedding guests had returned to their respective homes, the young couple were disturbed by strange noises in an adjoining room, and shortly after the door of the apartment fell open, and there entered a figure wrapped up in a grey plaid. 'Who are you?' said the man, leaping out of bed and stretching forth his arms to grapple with the figure. 'The unhappy pedlar,' replied the spectre, stepping backwards, 'who was murdered sixty years ago in this very room, and his body thrown into the loch below. But I shall trouble you no more. The murderer has gone to his place, and in two short hours the permitted time of my wanderings on earth shall be over ; for had I escaped the cruel knife, I would have died in my bed this evening a greyheaded old man.' It disappeared as it spoke ; and from that night was never more seen nor heard by the inmates of the farmhouse." According to Hogg—

> " Certain it is, from that day to this,
> The ghaist of the pedlar was never mair seen."

It seems curious enough that such a story should have been received for many years as true in a district of country in which the people hold, as strict Calvinists, that no man, however sudden or violent his death, can die before his appointed time. It may, however, belong to a somewhat remoter period than that assigned to it—some time in the early half of the last century—and may have originated in the age of the curates, whose theology is understood to have been Arminian. Another of my companion's stories, communicated on this occasion, had its scene laid in a district of country full sixty miles away.

The wife of a Banffshire proprietor, of the minor class, had been about six months dead, when one of her husband's ploughmen, returning on horseback from the smithy in the twilight of an autumn evening, was accosted, on the banks of a small stream, by a stranger lady, tall and slim, and wholly attired in green, with her face wrapped up in the hood of her mantle—who requested to be taken up behind him on the horse, and carried across. There was something in the tones of her voice that seemed to thrill through his very bones, and to insinuate itself in the form of a chill fluid between his skull and the scalp. The request, too, seemed a strange one; for the rivulet was small and low, and could present no serious bar to the progress of the most timid traveller. But the man, unwilling ungallantly to disoblige a lady, turned his horse to the bank, and she sprang up lightly behind him. She was, however, a personage that could be better seen than felt; and came in contact with the ploughman's back, he said, as if she had been an ill-filled sack of wool. And when, on reaching the opposite side of the streamlet, she leaped down as lightly as she had mounted, and he turned fearfully round to catch a second glimpse of her, it was in the conviction that she was a creature considerably less earthly in her texture than himself. She opened with two pale, thin arms, the enveloping hood, exhibiting a face equally pale and thin, which seemed marked, however, by the roguish, half-humorous expression of one who had just succeeded in playing off a good joke. " My dead mistress!" exclaimed the ploughman. " Yes, John, *your mistress*," replied the ghost. " But ride home, my bonny man, for it's

growing late ; you and I will be better acquainted erelong."
John accordingly rode home, and told his story.

Next evening, about the same hour, as two of the laird's
servant-maids were engaged in washing in an out-house, there
came a slight tap to the door. " Come in," said one of the
maids; and the lady entered, dressed, as on the previous night,
in green. She swept past them to the inner part of the wash-
ing-room ; and seating herself on a low bench, from which, ere
her death, she used occasionally to superintend their employment, she began to question them, as if still in the body, about
the progress of their work. The girls, however, were greatly
too frightened to reply. She then visited an old woman who
had nursed the laird, and to whom she used to show, ere her
departure, considerably more kindness than her husband. And
she now seemed as much interested in her welfare as ever.
She inquired whether the laird was kind to her; and, looking
round her little smoky cottage, regretted she should be so indif-
ferently lodged, and that her cupboard, which was rather of the
emptiest at the time, should not be more amply furnished.
For nearly a twelvemonth after, scarce a day passed in which
she was not seen by some of the domestics—never, however,
except on one occasion, after the sun had risen, or before it had
set. The maids could see her in the grey of the morning
flitting like a shadow round their beds, or peering in upon
them at night through the dark window-panes, or at half-open
doors. In the evening she would glide into the kitchen or
some of the out-houses—one of the most familiar and least
dignified of her class that ever held intercourse with mankind
—and inquire of the girls how they had been employed during
the day; often, however, without obtaining an answer, though
from a different cause from that which had at first tied their
tongues. For they had become so regardless of her presence,
viewing her simply as a troublesome mistress who had no longer
any claim to be heeded, that when she entered, and they had

dropped their conversation, under the impression that their visitor was a creature of flesh and blood like themselves, they would again resume it, remarking that the entrant was " only the green lady." Though always cadaverously pale and miserable-looking, she affected a joyous disposition, and was frequently heard to laugh, even when invisible. At one time, when provoked by the studied silence of a servant girl, she flung a pillow at her head, which the girl caught up and returned; at another, she presented her first acquaintance, the ploughman, with what seemed to be a handful of silver coin, which he transferred to his pocket, but which, on hearing her laugh immediately after she had disappeared, he drew out again, and found to be merely a handful of slate-shivers. On yet another occasion, the man, when passing on horseback through a clump of wood, was repeatedly struck from behind the trees by little pellets of turf; and, on riding into the thicket, he found that his assailant was the green lady. To her husband she never appeared; but he frequently heard the tones of her voice echoing from the lower apartments, and the faint peal of her cold unnatural laugh.

One day at noon, a year after her first appearance, the old nurse was surprised to see her enter the cottage, as all her previous visits had been made early in the morning or late in the evening; whereas now, though the day was dark and lowering, and a storm of wind and rain had just broken out, still it *was* day. " Mammie!" she said, " I cannot open the heart of the laird, and I have nothing of my own to give you; but I think I can do something for you now. Go straight to the White House [that of a neighbouring proprietor], and tell the folk there to set out, with all the speed of man and horse, for the black rock at the foot of the crags, or they'll rue it dearly to their dying day. Their bairns, foolish things, have gone out to the rock, and the sea has flowed round them; and if no help reach them soon, they'll be all scattered like seaware on the shore ere the fall of the tide. But if you go and tell your story at the White House, mammie, the bairns will be safe for an hour to come; and there will be something done by their mother to better you, for the news." The woman went as directed, and told her story; and the father of the children set out on horseback in hot haste for the rock—a low, insulated skerry, which, lying on a solitary part of the beach, far below the line of flood, was shut out from the view of the inhabited country by a wall of precipices, and covered every tide by several feet of water. On reaching the edge of the cliffs, he saw the black rock, as the woman had described, surrounded by the sea, and the children clinging to its higher crags. But, though the waves were fast rising, his attempts to ride out through the surf to the poor little things were frustrated by their cries, which so frightened his horse as to render it unmanageable; and so he had to gallop on to the nearest fishing village for a boat. So much time was unavoidably lost, in consequence, that nearly the whole beach was covered by the sea, and the surf had begun to lash the feet of the precipices behind; but, until the boat arrived, not a single wave dashed over the black rock; though immediately after the last of the children had been rescued, an immense wreath of foam rose twice a man's height over its topmost pinnacle.

The old nurse, on her return to the cottage, found the green lady sitting beside the fire. "Mammie," she said, "you have made friends to yourself to-day, who will be kinder to you than your foster-son. I must now leave you: my time is out, and you'll be all left to yourselves; but I'll have no rest, mammie, for many a twelvemonth to come. Ten years ago a travelling pedlar broke into our garden in the fruit season, and I sent out our old ploughman, who is now in Ireland, to drive him away. It was on a Sunday, and everybody else was in church. The men struggled and fought, and the pedlar was killed. But though I at first thought of bringing the case before the laird,

2 A

when I saw the dead man's pack with its silks and its velvets, and this unhappy piece of green satin (shaking her dress), my foolish heart beguiled me, and I bade the ploughman bury the pedlar's body under our ash-tree, in the corner of our garden, and we divided his goods and money between us. You must bid the laird raise his bones, and carry them to the churchyard; and the gold, which you will find in the little *bole* under the tapestry in my room, must be sent to a poor old widow, the pedlar's mother, who lives on the shore of Leith. I must now away to Ireland to the ploughman; and I'll be e'en less welcome to him, mammie, than at the laird's; but the hungry blood cries loud against us both—him and me—and we must suffer together. Take care you look not after me till I have passed the knowe." She glided away as she spoke in a gleam of light; and when the old woman had withdrawn her hand from her eyes, dazzled by the sudden brightness, she saw only a large black greyhound crossing the moor. And the green lady was never afterwards seen in Scotland. But the little hoard of gold pieces, stored in a concealed recess of her former apartment, and the mouldering remains of the pedlar under the ash-tree, gave evidence to the truth of her narrative.

I shall present the reader with one other story under this head—a ghost story of the more frightful class; which, though not at all inexplicable on natural principles, has as many marks of authenticity about it as any of the kind I am acquainted with. For many years the Cromarty Post-office, which, from the peninsular situation of the place, lies considerably out of the line of the mail, was connected with Inverness by a brace of pedestrian postmen, who divided the road between them into two stages; the last, or Cromarty stage, commencing at Fortrose. The post who, about half a century ago, travelled over this terminal stage six times every week was an elderly Highlander of the clan Munro—a staid, grave-featured man, somewhat tinged, it was said, by the constitutional melancholy of his country-folk, and not a little influenced by their peculiar beliefs. He had set out for Fortrose on his way home one evening, when he was overtaken by two acquaintances—the one a miller of Resolis, the other a tacksman of the parish of Cromarty—both considerably in liquor, and loud and angry in dispute. One of the Fortrose fairs had been held that day; and they had quarrelled in driving a bargain. Saunders Munro strove to pacify them, but to little purpose—they bickered idly on with drunken pertinacity; and it was with no little anxiety that, as they reached the Burn of Rosemarkie, where the White-bog and Scarfscraig roads part company, he saw them pause for a moment, as if to determine their route homewards. The miller was a tall athletic Highlander; the tacksman a compact, nervous man, not above the middle size, but resolute and strongly built. He could scarce, however, be deemed a full match for the Highlander; and under some such impression, old Saunders, unluckily as it proved, laid hold of him as he stood hesitating. " You must not go by that White-bog road," he said; " it is the near road for the miller, but not for you; you must come with me by the Scarfs-craig." " No, Saunders," said the tacksman; " I know what you mean; you do not like that I should cross the Maolbuie moor with the miller; but, big as he is, he'll be bigger yet or he daunt me; and I'll just go by the White-bog road to show him that." " Hoot, man," replied Saunders, " I'm no thinking o' that at all; I'm just no very weel to-night, and would be the better for your company; and so ye'll come hame this way with me." " Not a foot," doggedly rejoined the tacksman; and, shaking off the old man, he took the White-bog road with the miller. Saunders stood gazing anxiously after them as they descended the precipitous sides of the burn, until a jutting crag hid them from his sight. And for the rest of the evening, when pursuing his journey homewards, he felt burdened by an overpowering anxiety, which, disproportioned as it seemed to the occasion, he could not shake off.

The tacksman reached his home in less than two hours after he had parted from old Saunders; but two full days elapsed ere any one heard of the miller. In the evening of the second day, two young girls, the miller's sisters, who, after many fruitless inquiries regarding him, had at length come to learn in whose company he had quitted the fair, called at the farmhouse, and found the tacksman sitting moodily beside the fire. He started up, however, as one of them addressed him, and seemed strangely confused on being asked where he had parted from their brother. "I do not remember," he said, "being with your brother at all; and yet, now that I think of it, we must surely have left Rosemarkie together. The truth is, we had both rather too much drink in our heads. But I have some remembrance of passing the Grey Cairn in his company; and—and; —but——I must surely have left him at the Grey Cairn." "It must be ill with my brother," exclaimed one of the girls, "if he be still at the Grey Cairn!" "In truth," replied the tacksman, "I cannot well say where we parted, or whether I did not leave him at Rosemarkie with old Saunders Munro the post."

The evening was by this time merging into night, but the two terrified girls set out for the cairn; and the tacksman, taking down his bonnet, seemed as if he purposed accompanying them. On reaching, however, the outer wall of his yard, he stood for a few seconds as if undecided, and then, turning fairly round, left them to proceed alone. They entered one of the blind pathways that go winding in every direction. through the long heath of the Maolbuie—a bleak, desolate, tumulus-mottled moor—the scene in some remote age of a battle unrecorded by the historian; and its grey cairn, a vast accumulation of lichened stone, is said to cover, as I have already stated in an early chapter, the grave of a Pictish monarch, who, with half his army, perished in the fray. They reached the cairn; but all was silent, save that a chill breeze was moaning through the interstices of the shapeless pile, and sullenly waving the few fir seedlings that skirt its base; and they had turned to leave the spot, when they were startled by the howling of a dog a few hundred yards away. There was a dolorous wildness blent with an ominous familiarity in the sounds, that smote upon their hearts; and they struck out into the moor in the direction whence they proceeded, convinced that they were at length to learn the worst. On coming up to the animal, they found it standing beside the dead body of its master, their brother. The corpse was examined next morning by some of the neighbouring farmers; but nothing could be conclusively determined respecting the manner in which the unfortunate man had met his death. The neckcloth seemed straitened, and the folds somewhat compressed, as if it had been grasped by the hand; but then the throat and neck were scarce at all discoloured, nor were the features more distorted than if the death had been a natural one. The heath and mosses, too, in which the body had half sunk, rose as unbroken on every side of it as if they had never been pressed by the foot. There was no interference of the magistrate in the case, nor examination of parties. The body was conveyed to the churchyard and buried; and a little pile of moor-stones, erected by the herd-boys who tend their cattle on the moor, continued to mark, when I last passed the way, the spot where it had been found.

One evening, a few weeks after the interment, as old Saunders the postman was coming slowly down upon the town of Cromarty through the dark Navity woods, his eye caught a tall figure coming up behind him, and mistaking it in the uncertain light for an acquaintance, a farmer, he paused for a moment by the wayside, and placed his hand almost mechanically on the ready snuff-box. What, however, was his horror and astonishment to find, that what he had mistaken for his acquaintance the farmer was the dead miller of Resolis, attired, as was the wont of the deceased when in holiday trim, in the Highland costume. He could see, scarce less distinctly than when he had parted from him at the Burn of Rosemarkie, the chequers of the tartan and the scarlet of the gay hose garter, and—a circumstance I have never known omitted in any edition of the story—the glimmer of the large brass pin which fastened the kilt at the waist. For an instant Saunders felt as if rooted to the spot ; and then starting forward he hurried homewards, half beside himself with a terror that seemed to obliterate every idea of space and time, but collected enough to remark that the spectre kept close beside him, taking step for step with him as he went, until, at the gate of a burying-ground immediately over the town, it disappeared. On the following evening, when again passing through the Navity woods, nervous with the re-collection of the previous night's adventure, he was startled by a rustling in the bushes ; a shadowy figure came gliding out from among them to the middle of the road, and he found him-self a second time in the presence of the spectre, which accom-panied him, as before, to the gate of the burying-ground. He contrived on the day after to leave Fortrose at so early an hour, that he had reached the outer skirts of the town of Cromarty as the sun was setting ; but on crossing the street to his own house, the spectre started up beside him in the clear twilight, and, regarding him with an expression of grieved anxiety, dis-appeared as he entered the door. An aunt of the writer, who had occasion to call at his house on this evening, found him in bed in a corner of the sitting-room of his domicile, and on in-quiring whether he was ill, was informed by his wife, who sat beside him, the cause of his indisposition.

On his next day's journey, Saunders, instead of following his usual road, struck, on his return, across the fields in the direc-tion of a wooded ravine, which, forming part of the pleasure-grounds of Cromarty House, bears the name of the Ladies' Walk. The evening was cloudless and bright ; and the sun had but just disappeared behind the hill, when he entered the wooded hollow and crossed the little stream which runs along its bottom. But on rising along the opposite acclivity, he found that the apparition of the dead miller, true to him as his shadow, was climbing the hill by his side ; and where the path becomes so narrow—bounded on the one side by a steep descending bank, and on the other by a line of flowering shrubs—that two can hardly walk abreast, it glided onwards through the bushes as lightly as a column of smoke, not a leaf stirring as it passed. On reaching the broken wall which separates the pleasure-grounds from the old parish-churchyard, it stood, and, as Saunders was stepping over the fence, spoke for the first time. " Stop, Saunders," it said, " I must speak to you." " I have neither faith nor strength," replied Saunders, hurrying away, " to speak to the like of you."

The minister of the parish at the time was a gentleman of strong good sense and a liberal tone of mind ; and when the old man waited on him in the course of the evening, and imparted to him his story, he questioned him regarding the state of his nerves and stomach, and gave him an advice which very considerably resembled the prescription of a physician. But though it might be the best possible in the circumstances, it wholly failed to satisfy Saunders ; and so he unburdened his mind on the matter to one of the elders of the parish, a worthy sensible Udoll farmer, a high specimen of the class well known in the north country as "the Men," who, considerably advanced in life, had formed his beliefs at an earlier period than his minister, and was not in the least disposed to treat the case medicinally. He arranged with Saunders a meeting for the following evening at the hill of Eathie, a few miles from his journey's end ; and at Eathie they accordingly met, and passed on through the Navity woods together. But though it was late and long ere they reached town, the details of what befell them by the way they never communicated to any one. Saunders Munro, however, did not again see the apparition, though he travelled for years after at all hours of the day and night. The

elder, when rallied regarding the story by a town's-man whom
I well knew, and who related the circumstance to me, looked
him full in the face, and, with an expression of severe gravity,
" bade him never select that subject for a joke again." "Young
man," he said, " it was no joking business!"

No one, however, evinced so deep an anxiety on the subject
of the miller's ghost, and its supposed interview with the elder,
as the suspected tacksman.   It is known that on one occasion
he placed himself in the elder's way when the latter was returning from a funeral, and solicited a few minutes' private
conversation with him ; but was sternly repelled.   " You can
have but one business with me," the elder said ; " and, if your
conscience be clear from blood, not one itself."   Whatever hand
the tacksman may have had in the miller's death, no one who
knew him, or the circumstances in which he had parted on the
fatal night from old Saunders, could regard him as a murderer ;
though few real murderers ever wore out life in greater apparent
unhappiness than he.   He never after held up his head, but
went about his ordinary labours dejected and spiritless, and in-
vincibly taciturn ; and, some few years subsequent to the event,
he fell into a lingering illness, of which he died.   Were one
making a ghost story, it would be no difficult matter to make
a more satisfactory one.   Never was there a ghost that appeared
to less purpose than that of the miller, or was less fortunate in
securing a publisher for its secret ; but sure I am, never was
there a ghost story more firmly believed in the immediate
scene of it, or narrated with greater truth-like minuteness of
detail, or with less suspicion of at least the *honesty* of the parties on whose testimony it rested.   Nor was it without its
effect in adding strength, within the sphere of its influence, to
the fence set around the sacred tabernacle of the human soul.
Where such stories are credited, the violent spilling of man's life
is never regarded as merely " the diverting of a little red puddle
from its source."

CHAPTER XXVI.

"Oh, many are the poets that are sown
  By nature; men endow'd with highest gifts,
  The vision and the faculty divine,
  Yet wanting the accomplishment of verse."—WORDSWORTH.

DURING even the early part of last century, there were a few of the mechanics of Cromarty conversant in some little degree with books and the pen. They had their libraries of from ten to twenty volumes of sermons and controversial divinity, purchased at auctions or from the booksellers of the south; and I have seen letters and diaries written by them, which would have done no discredit to the mechanics of a more literary age. Donald Sandison's library consisted of nearly a hundred volumes; and his son, whom I remember a very old man, and who had at one time been the friend and companion of the unfortunate Ferguson the poet, had made so good a use of his opportunities of improvement, that in his latter days, when his sight began to fail him, he used to bring with him to church a copy of Beza's Latin New Testament, which happened to be printed in a clearer type than his English one. The people in general, however, were little acquainted with the better literary models. So late as the year 1750, a copy of Milton's Paradise Lost, which had been brought to town by a sailor, was the occasion of much curious criticism among them; some of them alleging that it was heterodox, and ought to be burnt, others deeming it prophetic. One man affirmed it to be a romance, another said it was merely a poem; but a Mr. Thomas Hood, a shopkeeper of the place, set the matter at rest by remarking, that it seemed to him to be a great book, full of mystery like the Revelation of St. John, but certainly no book for the reading of simple unlearned people like him or them. And yet, at even this period, Cromarty had its makers of books and writers of verses ; men of a studious imitative turn—prototypes in some respects of those provincial poets of our own times, who become famous for nearly half an age in almost an entire county. A few brief notices of the more remarkable of my town's-men of this first class may prove not unacceptable to the reader ; for, of all imitators, the poetical imitator is the most eccentric ;— though his verses be imitations, in character he is always an original.

On the southern shore of the Bay of Cromarty, two miles to the west of the town, there stood, about ninety years ago, a meal-mill and the cottage of the miller. The road leading to the country passed in front, between the mill and the beach ; and a ridge of low hills, intersected by deep narrow ravines, and covered with bushes of birch and hazel, rose directly behind. There was a straggling line of alders which marked the course of the stream that turned the mill-wheel ; while two gigantic elms, which rose out of the fence of a little garden, spread their arms over both the mill and the cottage. The view of the neighbouring farm-steadings was shut out by the windings of the coast and the ridge behind ; and to the traveller who passed along the road in front, and saw no other human dwelling nearer him than the little speck-like houses which mottled the opposite shore of the bay, this one seemed to occupy one of the most secluded spots in the parish. Its inmates at this period were John Williamson, the miller, or, as he was more commonly termed, Johnie o' the Shore, and his sister Margaret—two of the best and most eccentric people of their day in the country-side. John was a poet and a Christian, and much valued by all the serious and all the intelligent people of the place ; while his sister, who was remarkable in the little circle of her acquaintance for the acuteness of her judgment in nice points of divinity, was scarcely less esteemed.

The duties of John's profession left him much leisure to write and to pray. During the droughts of summer, his mill-pond would be dried up for months together; and in these seasons he used to retire almost every day to a green hillock in the vicinity of his cottage, which commands an extensive view of the bay and the opposite coast. And there, in a grassy opening among the bushes, would he remain until sunset, with only the Bible and his pen for his companions. He was so much attached to this spot, that he was once heard to say there was no place in which he thought he could so patiently await the resurrection, and he intimated to his friends his wish of being buried in it; but, on his deathbed, he changed his mind, and requested to be laid beside his mother. It is now covered by a fir-wood, and roughened by thickets of furze and juniper, but enough may still be seen to justify his choice. On one side it descends somewhat abruptly into a narrow ravine, through the bottom of which there runs a little tinkling streamlet; on the other, it slopes gently towards the shore. We look on the one hand, and see, through the chance vistas which have been opened in the wood, the country rising above us in long undulations of surface, like waves of the sea after a storm, and variegated with fields, hedge-rows, and clumps of copse-wood. On the other, the wide expanse of the bay lies stretched at our feet, with all its winding shores and blue jutting headlands: we look down on the rower as he passes, and hear the notes of his song and the measured dash of his oars; and when the winds are abroad, we may see them travelling black over the water before they wave the branches that spread over our heads. Many of the poet's happiest moments were passed in the solitude of this retreat; and from the experience derived in it, though one of the most benevolent of men, and at times one of the most sociable, whenever he wished to be happy he sought to be alone.

In going to church every Sabbath, instead of following the public
road, he used invariably to strike across the beach and walk by
the edge of the sea ; and, on reaching the churchyard, he always
retired into some solitary corner, to ponder in silence among the
graves.   To a person of so serious a cast, a life of solitude and
self-examination cannot be a happy, unless it be a blameless one ;
and Johnie o' the Shore was one of the rigidly just.   Like the
Pharisees of old, he tithed mint, and anise, and cumin ; but,
unlike the Pharisees, he did not neglect the weightier matters
of the law.   It is recorded of him, that on descending one
evening from his hillock, he saw his only cow browsing on the
grass-plot of a neighbour, and that, after having her milked as
usual, he despatched his sister with the milk to the owner of
the grass.

Ninety years ago, the press had not found its way into the
north of Scotland, and the people were unacquainted with the
scheme of publishing by subscription.   And so the writings of
Johnie o' the Shore, like those of the ancients before the invention of printing, existed only in manuscript ; and, like them too,
they have suffered from the Goths.   A closely written fragment
of about eighty pages, which once composed part of a bulky
quarto volume, is now all that survives of his works, though at
his death they formed of themselves a little library.   One of
the volumes, written wholly in prose, and which minutely detailed, it is said, all the incidents of his life, with his thoughts
on God and heaven, the world and himself, fell into the hands
of a distant relative who resided somewhere in Easter-Ross.   It
must have been no small curiosity in its way, and for some time
I was flattered by the hope that it still existed and might be
recovered ; but I have come to find that it has shared the fate
of all his other volumes.   The existing fragment is now in my
possession.   It bears date 1743, and is occupied mostly with
hymns, catechisms, and prayers.   His models for the hymns
seem to have been furnished by our Scotch version of the

Psalms ; his catechisms were formed, some on the catechisms of Craig and the Palatine, and some on that of the Assembly Divines ; his prayers remind me of those which are still to be heard in the churches of our northern parishes on "the day of the men." Some of his larger poems are alphabetical acrostics ; —the first line of the first stanza of each beginning with the letter A, and the first line in the last with the letter Z. Most of them, however—and the fact is a singular one, for John and his sister were stanch Presbyterians—are commemorative of the festival-days of the English Church. There are hymns for Passion Friday, for Christ's Incarnation-day, for Circumcision-day, and for Christmas :—a proof that he must have had little in him of that abhorrence of Prelacy which characterized most of the Presbyterians of his time. And he seems, too, to have been of a more tolerating spirit ; and, in the simple benevolence of his heart, to have come perhaps as near the truth on some dark points as men considerably more skilled in dialectics, and more deeply learned. "There are some people," remarks the querist in one of his catechisms, "who say that those who have never heard of Christ cannot be saved ?" "It is surely not our business," is the reply, "to search into the deep things of God, except so far as He is pleased to reveal them ; and, as He has not revealed to us that He condemns all those who have not heard of Christ, it is rash to say so, and uncharitable besides."[1]

One of the most curious poems in the manuscript, is a little piece entitled "An Imagination on the Thunder-claps." It was written before the discoveries of Franklin ; and so the imagination is rather a wild one—not wilder, however, than some of the soberest speculations of the ancients on the same phenomena. The green hillock on this occasion appears to have been

---

[1] Cowper has said quite as much, and rather more, in his "TRUTH."

> "Let heathen worthies, whose exalted mind
> Left sensuality and dross behind,
> Possess for me their undisputed lot,
> And take unenvied the reward they sought."

both his Observatory and his Parnassus ;—he seems to have watched upon it every change of the heavens and earth, from the first rising of the thunder-clouds until they had broken into a deluge, and a blue sky looked down on the red tumbling of streams as they leaped over the ridges, or came rushing from out the ravines.　Though quite serious himself, his uncouth phraseology will hardly fail in eliciting the smile of the reader.

AN IMAGINATION ON THE THUNDER-CLAPS.

Lo ! pillars great of wat'ry clouds
　On firmament appear,
And mounting up with curléd heads,
　Towards the north do steer.
East wind the same doth contradict,
　And round and round they run ;
And earth and sea are dark below,
　And blackness hides the sun.

Like wrestling tides that in the bay
　Do bubble, boil, and foam,
When seas grow angry at the wind,
　And boatmen long for home ;
Ev'n so the black and heavy clouds
　Do fierce together jar—
They meet, and rage, and toss, and whirl,
　And break, and broken are.

Up to the place where fire abides
　These wat'ry clouds have gone
And all the waters which they hold
　Are flung the fire upon.
And the vex'd fire boils in the cloud,
　And lifts a fearful voice,
Like rivers toss'd o'er mighty rocks,
　Or stormy ocean's noise.

It roars, and rolls, and hills do shake,
　And heavens do seem to rend ;
And should the fierce unquenchéd flame
　Through the dark clouds descend,
Like clay 'twould grind the hardest rocks,
　Like dust the strongest brass,
And prostrate pride and strength of man
　Like pride and strength of grass.

And now the broken clouds fall down
　In *groff* rain from on high;
And many streams do rise and roar,
　That heretofore were dry.

And when the red speat will be o'er,
  And wild storm pass'd away,
Rough stones will lie upon the fields,
  And heaps of sand and clay.

But I, though great my sins, am spared,
  These fields to turn and tread :
Which surely had not been the case
  If Jesus had not died.

Quod JOHNIE O' THE SHORE.

Johnie's sister Margaret (after his death she seems to have fallen heir to his title, for she then became Meggie o' the Shore) survived her brother for many years, and died at an extreme old age, about the year 1785. The mill, on its falling into other hands, was thrown down, and rebuilt a full half mile further to the west, but the cottage was spared for Meggie. She had always been characterized by the extreme neatness of her dress and her personal cleanliness, by her taste in arranging the homely furniture of her cottage, and her hospitality : and now, though the death of her brother had rendered her as poor as it is possible for a contented person to become, she was as much marked by her neatness, and as hospitable as ever. On one occasion, a Christian friend who had come to visit her (the late Mr. Forsyth of Cromarty), was so charmed with her conversation, as to prolong his stay from noon until evening, when he rose to go away. She asked him, somewhat hesitatingly, whether he would not first " break bread with her." He accordingly sat down again ; and a half cake of bread and a jug of water (it was all her larder afforded) were set before him. It was the feast of the promise, she said, " Thy bread shall be given thee, and thy water shall be sure." Her circumstances, she added, were not quite so easy as they had been during the lifetime of her brother, but the change was perhaps for the better ; for it had led her to think much oftener than before, when rising from one meal, that God had kindly pledged Himself for the next.

Meggie lived in a credulous age, and she was one of the

credulous herself. Like most of her acquaintance, she heard
at times the voices of spirits in the dash of waves and the roar
of winds, and saw wraiths and dead-lights ; but she was naturally courageous, and had a strong reliance on Providence ; and
so, with all her credulity, she was not afraid to live alone, with,
as she used to say, only God for her neighbour. On a boisterous winter evening, two young girls who were travelling from
the country to the town, were forced by the breaking out of a
fierce snow-storm to take shelter in her cottage. She received
them with her wonted kindness, and entertained them as she
had done her friend. They heard the waves thundering on the
beach, and the wind howling in the woods, but peace and safety
were with them at Meggie's fireside. About midnight there
was a pause in the storm, and they could hear strange sounds,
like the cries of people in distress, mingling with the roar of
the sea. "Raise the window-curtain," said Meggie, "and look
out." The terrified girls raised the curtain. "Do you see
aught ?" she inquired. "There is a bright light," said the
girls, "in the middle of the bay of Udoll. It hangs over the
water at about the height of a ship's mast ; and we can see
something below it like a boat riding at anchor, with the white
sea raging round her." "Now drop the curtain," she replied ;
"I am no stranger, my lassies, to sights and noises like these
—sights and noises of another world ; but I have been taught
that God is nearer to me than any other spirit can be ; and so
have learned not to be afraid." A few nights after, as the story
goes, a Cromarty yawl foundered in the bay of Udoll, and all
on board perished.

Meggie was always a rigid Presbyterian, and jealous of innovations in the Church ; and, as she advanced in years, she
became more rigid and more jealous. She is said to have
regarded with no great reverence the young divines that filled
up in the parishes around her the places of her departed contemporaries ; and who too often substituted, as she alleged, the

learning which they had acquired at college for a knowledge of the human heart and of the Bible. She could ill brook, too, any interference of the State in the concerns of the Kirk :—an Act of Parliament, when read from the pulpit, she deemed little better than blasphemy, and a King's fast a day desecrated above every other. Her zeal in one unlucky instance brought her in contact with the civil law. Her favourite preacher was Mr. Porteous of Kilmuir, a divine of the old and deeply learned cast—eloquent and pious—not unacquainted with the book of nature, and thoroughly conversant with that of God. After hearing him deliver, in the church of Nigg, a powerful and impressive discourse, what was her horror and indignation when she saw him descending from the pulpit to read from the precentor's desk some Proclamation or Act of Council! Had he been less a favourite, or anybody else than Mr. Porteous, she could have shut her ears and sat still; as it was, she sprang from her seat, and twitching the paper out of his hand, flung it to the floor and stamped upon it with her feet. She was apprehended and sent to the jail of Tain; but she found the jail a very comfortable sort of place, and, for the three days during which she was confined to it, she had for her visitors some of the very best people in the country; among the rest, Mr. Porteous himself, who had enough of the old Covenanter in him to feel that she had, perhaps, done only her duty, and that he had very possibly failed in his.

The story of her death is curious and affecting. A friend, in passing her cottage on a journey to the country called in, as usual, to see her. She was as neatly dressed as ever, and the little apartment in which she sat was fastidiously clean; but her countenance was of a deadly paleness, and there was an air of languor about her that seemed the effect of indisposition. "You are unwell, Meggie?" said her friend. "Not quite well, perhaps," she replied, "but I shall be so very soon. You must stay and take breakfast with me." The visitor knew too well

2 B

the value of one of Meggie's breakfasts to refuse, and the simple
fare which her cottage afforded was set before him; but he was
disappointed of the better part of the repast, for she spoke but
little, and seemed unable to eat. "God has been exceedingly
good to me," she remarked, as she rose when he had eaten to
replace in her cupboard the viands which still remained before
him; "with no one to provide for me but Himself, I have not
known what it was to want a meal since the death of my
brother. You return this way in the evening?" said she, addressing her friend. He replied in the affirmative. "Then
promise that you will not pass without coming in to see me;
I am indisposed at present, but I feel—nay, am certain—that
you will find me quite well. Do promise." Her friend promised, and set out on his journey. Twilight had set in before
his return. He raised the latch and entered her apartment,
where all was silent, and the fire dying on the hearth. In a
window which opened to the west, sat Meggie, with her brother's
Bible lying open before her, and her face turned upwards. The
faint light of evening shone full on her features, and their expression seemed to be that of a calm yet joyous devotion. "I
have returned, Meggie," said the man after a pause of a few
minutes. There was no answer. "I have returned, Meggie,"
he reiterated, "and have come to see you, to redeem my promise." Still there was no answer. He went up to her and
found she was dead.

About twenty years after her death, the grave in which she
had been buried was opened to admit the corpse of a distant
relative. A woman of my acquaintance, who was then a little
girl, was at play at the time among the stones of the churchyard; but on seeing an elderly female, a person much of Meggie's
cast of character, go up to the grave, she went up to it too.
She saw the woman looking anxiously at the bones, and there
was one skull in particular which seemed greatly to engage her
attention. It still retained a few locks of silvery hair, and over

the hair there were the remains of a linen cap fastened on by two pins. She stooped down, and drawing out the pins, put them up carefully in a needle case, which she then thrust into her bosom. "Not death itself shall part us!" she muttered, as if addressing herself to the pins; "you shall do for me what you have done for Meggie o' the Shore."

But, in holding this *tête-à-tête* with Meggie, I have suffered myself to lose sight of the poets, and must now return to them. Next in the list to Johnie o' the Shore was David Henderson, a native of Cromarty, born some time in the early part of the last century, and who died in the beginning of the present. He was one of that interesting class, concerning whom Nature and Fortune seem at variance; the one marking them out for a high, the other for a low destiny. They are fitted, by the gifts of mind bestowed upon them by the one, to think and act for themselves and others; and then flung by the other into some obscure lumber-corner of the world, where these gifts prove useless to them at best, and not unfrequently serve only to encumber them. From Nature David received talents of a cast considerably superior to those which she commonly bestows; by Fortune he was placed in one of the obscurest walks of life, and prevented from ever quitting it. He acquired his little education when employed in tending a flock of sheep; the herdboys with whom he associated taught him to read, and he learned to write by imitating the letters of one of the copy-books used in schools upon the smooth flat stones which he found on the sea-shore.

From his earliest years his life was one of constant toil. He was a herd-boy in his seventh, and a ploughman in his sixteenth year. He was then indentured to a mason; and he soon became one of the most skilful workmen in this part of the country, especially in hewing tombstones and engraving epitaphs. There is not a churchyard within ten miles of Cromarty in which there may not be seen some of his inscriptions. His heart was

an affectionate one, and open to love and friendship ; and when
he had served his apprenticeship, and began to be known as a
young man of superior worth and a good clear head, his company came to be much courted by the better sort of people. In
his twenty-fifth year he became attached to a young girl of
Cromarty, named Annie Watson, much celebrated in her day
for her charms personal and mental. She was beautiful to
admiration, rationally yet fervently pious, and possessed of a
mind at once powerful and delicate. It was no wonder that
David should love such a one ; and, as no disparity of condition
formed an obstacle to the union—as she was a woman of sense
and he a man of merit—in all probability she would have made
him happy. But, alas! in the bloom of youth she was taken
from him by that insidious disease, which, while it preys on the
vitals of its victims, renders their appearance more interesting,
as if to make their loss the more regretted. She died of consumption, and David was left behind to mourn over her grave,
and, when his grief had settled into a calm melancholy, to write
a simple ballad-like elegy to her memory. I have heard my
mother say, that it was left by David at the grave of his mistress, where it was afterwards picked up by a person who gave
copies of it to several of his acquaintance; but I do not know
that any of these are now to be found. I have failed in recovering more than a few stanzas of it; and these I took down
as they were repeated to me by my mother, who had committed
them to memory when a child. They may prove interesting,
rude and fragmentary as they are, to such of my readers as love
to contemplate the poetic faculty wrapt up in the dishabille of
an imperfect education. Besides, the writer may be regarded
less as an insulated individual than as representative of a class.
The unknown authors of some of our simpler old ballads, such
as Fdom o' Gordon, Gilmorice, and the Bonny Earl of Moray,
were, it is probable, men of similar acquirements, and a resembling cast of intellect.

ELEGY ON THE DEATH OF A YOUNG WOMAN.

She's slain by death, that spareth none,
  An object worthy love;
And for her sake was many a sigh,—
  No doubt she's now above.
    *   *   *

In dress she lovéd to be neat,
  In handsome trim would go;
She lovéd not to be above
  Her station, nor below.
    *   *   *

But, in brief sentence, to have done
  Of all I have to say—
In midst of all her prospects here,
  She on a deathbed lay.

And when she on a deathbed lay,
  To her were visits made
By good and reverend elders, who
  In her great pleasure had.

For she though in her pleasant youth,
  When time speeds sweetly by,
Esteem'd it, trusting in her God,
  A blessed thing to die.

And she their questions unto them
  Who sought her state to know,
Did answer wisely every one,
  In pleasant words and low.
    *   *   *

Her lykewake it was piously spent
  In social prayer and praise,
Performéd by judicious men,
  Who stricken were in days.

And many a sad and heavy heart
  Was in that mournful place;
And many a weary thought was there
  On her who slept in peace.

And then the town's-folk gather'd all
  To bear her corpse away,
And bitter tears by young and old
  Were shed that mournful day.

And sure, if town's-folk grievéd sore,
  Sore grieve may I and pine;
They much deplored their heavy loss—
  But what was theirs to mine?

For her loved voice, I only hear
   Winds o'er her dust that sigh
For her sweet smile, I only see
   The rank grass waving high.

And I no option have but think
   How I am left alone;
With none on earth to care for me,
   Since she who cared is gone.
      *   *   *
She was the first that ever I
   In beauty's bloom did see
Departing from the stage of time,
   Into eternity.

O may her sex her imitate,
   Example from her take,
And strive t' employ the day of grace,
   And wicked ways forsake!

David survived his mistress for more than forty years. For thirty of these he was an elder of the Church—a man conversant with deathbeds, and a visitor of the fatherless and the widow. Few persons die so regretted as David died, or leave behind them so fair a name; nor will the reader fail to recognise something uncommon in his character when I tell him, that he was steady and prudent though a poet, and of a grave deportment, good-natured, and a Christian, though of a ready wit. He left behind him, treasured up in the memories of his many friends, shrewd, pithy remarks on men and things—specimens of mind, if I may so express myself, which exhibit the quality of the mass from off which they were struck. His wit, too, was equally popular. I have heard some of his *bon-mots* repeated and laughed at more than twenty years after his death; but his writings were so much less fortunate, that there were few of the people with whom I have conversed concerning him, who even knew that he made verses, though none of them were ignorant of his having been a good man.

The last of the Cromarty poets who lived and wrote before the beginning of the present century, was Macculloch of Dun-Loth. He was, for nearly sixty years, a Society schoolmaster in that parish of Sutherlandshire whose name, for some cause or other, is always attached to his own. But I shall attempt introducing him to the reader in the manner in which he has been introduced to myself.

"About twenty-eight years ago," said my informant, "I resided for a few weeks with the late Dr. R—— at the manse of Kiltearn. I was lounging one evening beside the front door, when a singular-looking old man came up to me, and asked for the Doctor. He was such an equivocal-looking sort of person, that it was quite a puzzle to me whether I should show him into the parlour;—he might be little better than a beggar; he might be worth half a million; but whether a rich man or a poor one, no one could look at him and doubt of his being a *particular* man. He was very little, and very much bent, with just such a grotesque cast of countenance as I have seen carved on the head of a walking-stick. His outer man was cased in an old-fashioned suit of raven grey, and he had immense plated buckles in his shoes and in his breeches. I thought of the legend of the Seven Sleepers, and wondered where this fragment of the old world could have lain for the last hundred years. The Doctor relieved me from my perplexity. He had seen him from a window, and, coming out, he welcomed the little old man with his wonted cordiality, and ushered him into the parlour as the poet of Dun-Loth.

"He stayed with us this evening, and never was there a gayer evening spent in the manse. The Doctor had the art of eliciting all that was eccentric in the little man's character, and that was not a little. He plied him with compliments and jokes, and rallied him on his love-adventures and his poetry. The old man seemed swelling like a little toad, only it was with conceit, not venom. He chuckled, every now and then, at the more piquant of the Doctor's good things, with a strange unearthly gaiety that seemed to savour of another world—of another age at least; and then he would jest and compliment in turn. What he said was, to be sure, great nonsense; but then it was the most original nonsense that might be, full of small conceits and quibbles, and so old-fashioned that we all felt it could not be other than the identical nonsense that had flourished in the early days of our great-grandmothers. The young people were all delighted—the little old man seemed delighted too, and laughed as heartily as any of us. Mrs. R——, when a young lady, had been eminently beautiful, and the poet had celebrated her in a song. It was a miserable composition, and some of his neighbours, who wrote nearly as ill as himself, made it the occasion of a furious attack upon him. There were remarks, replies, and rejoinders beyond number; until at length, by mere dint of perseverance, the poet silenced all his opponents, and took to himself the credit of having gained a signal victory. The Doctor brought up the story of the song, and got him to repeat all the replies and rejoinders, which he did with much glee. Next morning he took leave of us, and I never again saw the poet of Dun-Loth."

Macculloch was, as I have stated, a native of the parish of Cromarty, and passed the greater part of a long life as a Society schoolmaster, on a salary of twelve pounds per annum. Out of this pittance he contrived to furnish himself with a library, which, among other works of value, contained the whole of the Encyclopædia Britannica in its second edition. Though full of compliment and gallantry in his younger days, he was for the last forty years of his life, so thoroughly a woman-hater, that he would not suffer one of the sex to enter his cottage, cook his victuals, or wash his linen. His wardrobe consisted of four suits—one of black, one of brown, one of raven grey, and one of tartan; and he wore them week about, without suffering the separate pieces of any one suit to encroach on the week of another. It has been told me that, in his eightieth year, he attended the dispensation of the sacrament in the Highland parish of Lairg, dressed in his tartans—kilt, hose, and bonnet.

I do not well know whether to consider his singularities as those of the rhymer, the most eccentric of all men, or his predilection for rhyming as merely one of his singularities. His compositions were mostly satirical; but his only art of satire was the art of calling names in rhyme; and he seems to have had no positive pleasure in bestowing these, but to have flung them, just as he used to do his taws when in school, at the heads of all who offended him. His death took place about twenty years ago. I subjoin two of the "pasquils" pointed against him in his war with his brother rhymers, and the pieces in which he replied to them. They may show, should they serve no other purpose, what marvellous bad verse could be written in the classical age of Johnson and Goldsmith, and with what justice Dun-Loth piqued himself on having vanquished his opponents.

TO DUN-LOTH.

Dunloth, be wise, take my advice,
  Silence thy muse in time;
For thy thick skull it is too dull
  To furnish prose or rhyme.
But if thy pride will still thee guide
  To sing thy horrid lays;
For any sake, my counsel take,
  And ne'er attempt to praise.
Thy wit's too low, thyself says so,
  In this we both agree;
The Kilmote flower is, I am sure,
  A theme too high for thee.

ANSWER.

To notice much, base trash as such,
  I think it were a crime;
Or yet to stoop, thou nincompoop,
  For thy poor paltry rhyme.
Thy saucy gee shows thee to be
  Like a blind muzzled mole:
Or like a rat chased by the cat
  To a dark muddy hole.
The first time I thy place pass by,
  For thy poetic lesson,
Thou'lt crouch, be sure, behind the door,
  Like a poor yelping messan.

TO ————

How hard is thy lot, fair flower of Kilmote,
  To be sung by a poet so dull;
Thy symmetry fine, is a theme too divine
  For a blockhead with such a thick skull.

ANSWER.

So hard is thy lot, poor scurrilous sot,
  Thy poetry brings thee to shame;
So high to aspire, thou'rt thrust in the mire,
  And laugh'd at by all for the same.

CHAPTER XXVII.

" There are more things in heaven and earth, Horatio,
Than are dreamt of in your philosophy."—HAMLET.

I HAD passed the three first milestones after leaving Forres,
when the clouds began to lour on every side of me, as if earth
and sky were coming together, and the rain to descend in tor-
rents. The great forest of Darnaway looked shaggy and brown
through the haze, as if greeting the heavens with a scowl as
angry as their own ; and a low, long wreath of vapour went
creeping over the higher lands to the left, like a huge snake.
On the right, the *locale* of Shakspere's witch scene, half moor
half bog, with the old ruinous castle of Inshoch standing sentry
over it, seemed ever and anon to lessen its area as the heavily-
laden clouds broke over its farther edge like waves of the sea ;
and the intervening morass—black and dismal at all times—
grew still blacker and more dismal with every fitful thickening
of the haze and the rain. And then, how the furze waved to the
wind, and the few scattered trees groaned and creaked ! The
thunder and the witches were alone wanting.

I passed on, and the storm gradually sank. The evening,
however, was dark and damp, and more melancholy than even
the day, and I was thoroughly wet, and somewhat fatigued to
boot. I could not, however, help turning a little out of my
way to pause for a few minutes amid the ruins of the old farmhouse of Minitarf, just as I had paused in the middle of the
storm to fill my mind with the sublimities of the Harmoor, and

do homage to the genius of Shakspere. But why at Minitarf? Who is not acquainted with the legend of the "Heath near Forres"—who knows anything of the history of the Farm-house? Both stories, however, are characteristic of the very different ages to which they belong ; and the moral of the humbler story is at once the more general in its application, and the more obvious of the two.

Isabel Rose, the gudewife of Minitarf, was a native of Easter-Ross, and having lost both her parents in infancy, she had passed some of the earlier years of her life with a married sister in the town of Cromarty. She had been famed for her beauty, and for being the toast of three parishes ; and of all her lovers, and few could reckon up more, she had been lucky enough to lose her heart to one of the best. The favoured suitor was a handsome young farmer of the province of Moray—a person somewhat less shrewd, perhaps, than many of his countrymen, but inflexibly honest, and perseveringly industrious ; and, as he was a namesake of her own, she became his wife and the mistress of Minitarf, and yet remained Isabel Rose as before. The wife became a mother—the mother of two boys. Years passed by ; the little drama of her life, like one of the dramas of antiquity, had scarce any change of circumstance, and no shifting of scenes ; and her two sons grew up to maturity, as unlike one another in character as if they had not been born to the same parents, nor brought up under the same roof.

John, the elder son, was cautious and sensible, and of great kindliness of disposition. There was nothing bright or striking about him ; but he united to his father's integrity and firmness of purpose much more than his father's shrewdness, and there was a homely massiveness in the character that procured him respect. He was of a mechanical turn ; and making choice of the profession of a house-carpenter—for he was as little ambitious as may be—he removed to Glasgow, where his steadiness and skill recommended him to the various contractors of the place, until in the course of years he became, a good deal to his own surprise, a contractor himself. Sandy, the younger son, was volatile and unsettled, and impatient of labour and restraint, and yet no piece of good fortune could have surprised Sandy. He had somehow come to the conclusion that he was born to be a gentleman, and took rank accordingly, by being as little useful, and dressing as showily as he could. His principles were of a more conventional cast than those of his brother, and his heart less warm ; still, however, there was no positive vice in the character ; and as he was decidedly cleverer than John, and a great deal more genteel, his mother could not help sharing with him in the hope that he was born to be the gentleman of the family—a hope which, of course, was not lessened when she saw him bound apprentice in his seventeenth year to a draper in a neighbouring town.

Sandy's master was what is termed a clever man of business ; one of those smart fellows who want only honesty, and that soundness of judgment which seems its natural accompaniment to make headway in the world. He had already threaded his way through the difficulties of three highly respectable failures ; he had thrice paid his debts at the rate of fifteen shillings per pound, and had thus realized on each occasion a profit of twenty-five per cent. on the whole. And yet, from some inexplicable cause, he was not making more money than traders much less fertile in expedient than himself. His ordinary gains were perhaps the less considerable from the circumstance, that men came to deal with him as completely on their guard as if they had come to fight with him ; and, though a match for any single individual, he was, somehow, no match for every body, even though, after the manner of Captain Bobadil's opponents, they came only one at a time. His scheme, too, of occasionally suspending his payments, had this disadvantage, that the oftener it was resorted to, the risk became greater and the gain less.

The shop of such a person could not be other than a rare school of ingenuity—a place of shifts and expedients—and where, according to the favourite phrase of its master, things were done in a business-like manner ; and Sandy Rose was no very backward pupil.   There are ingenious young men who are a great deal too apt to confound the idea of talent itself with the knavish exercise of it; and who, seeing nothing very knowing in simple honesty, exert their ingenuity in the opposite tract, rather out of a desire of doing clever things than from any very decided bias to knavery.   And Sandy Rose was unfortunately one of the number.   It is undoubtedly an ingenious thing to get possession of a neighbour's money without running the risk of stealing it ; and there can be no question that it requires more of talent to overreach another than to be overreached one's-self.   The three years of Sandy's apprenticeship came to their close, and with the assistance of his father, who in a long course of patient industry had succeeded in saving a few hundred pounds, he opened shop for himself in one of the principal streets of the town.

Sandy's shop, or *warehouse*, as he termed it—for the latter name was deemed the more respectable of the two—was decidedly the most showy in the street.   He dealt largely in fancy goods, and no other kind in the " soft way" show equally well in a window.   True, the risk was greater, for among the ordinary chances of loss he had to reckon on the continual changes of fashion ; but then, from the same cause, the profits were greater too, and Sandy had a decided turn for the more adventurous walks of his profession.   Nothing so respectable as a large stock in trade; the profits of a thousand pounds are necessarily greater than the profits of five hundred.   And so, what between the ready money advanced to him by his father, and the degree of credit which the money procured for him, Sandy succeeded in rendering his stock a large one.   He had omitted only two circumstances in his calculation—the proportion which one's stock should bear to one's capital, and the proportion which it should bear to the trade of the place in which one has settled. When once fairly behind his counter, however, no shopkeeper could be more attentive to his customers, or to the appearance of his shop ; and all allowed that Sandy Rose was a clever man of business. He wrote and figured with such amazing facility, and made such dashes at the end of every word ! He was so indefatigable in his assertions, too, that he made it a rule in every case to sell under prime cost ! He was, besides, so amazingly active—a squirrel in its cage was but a type of Sandy ! He was withal so unexceptionably genteel ! His finest cloths did not look half so well on his shelves as they did on his dapper little person ; and it was clear, from his everyday appearance, that he was one of his own best customers.

Sandy's first half year of business convinced him that a large stock in trade may resemble a showy equipage in more points than one : it may look as respectable in its way, but then it may cost as much. Bills were now falling due almost every week, and after paying away the money saved during the earlier months, the everyday custom of the shop proved too little to meet the everyday demand. Fortunately, however, there were banks in the country—"*more banks than one ;*" and his old master was content to lend him the use of his name, simply on the condition of being accommodated with Sandy's name in turn. Bill, therefore, was met by bill, and the paper of one bank pitted against the paper of another ; and as Sandy was known to have started in trade with a few hundreds, there was no demur for the first twelvemonth or so on the part of the bankers. They then, however, began to demand indorsations, and to hint that the farmer, his father, was a highly respectable man. Sandy expressed his astonishment that any such security should be deemed necessary ; his old master expressed his astonishment too ; nothing could be more unbusiness-like, he said ; but the bankers, who were quite accustomed to the astonishment of all their more doubtful customers, were inflexible notwithstanding, and the old man's name was procured. The indorsation was quite a matter of course, he was told—a thing "neither here nor there," but necessary just for form's sake ; and from that day forward all the accommodation-bills of Sandy and his master bore the name of the simple-minded old man.

I have said that Sandy was one of the most indefatigable of shopkeepers. It was but for the first few months, however, when all was smooth water and easy sailing ; in a few months more, when the tide had begun to set in against him, he became less attentive. Some of his fancy goods were becoming old-fashioned, and in consequence unsaleable, and his stock, large at first, was continuing large still. What between the price of stamps, too, the rate of discount, and the expense of travelling to the several banks in which he did business, he found that the profits of his trade were more than balanced by the expenditure. Sandy's heart, therefore, began to fail him ; and, setting himself to seek amusement elsewhere than behind his counter, he got a smart young lad to take charge of the shop in his absence ; and, as it could not add very materially to the inevitable expense, he provided himself with a horse. He was now every day on the road doing business as his own traveller. He rode twenty miles at a time to secure a five-shilling order, or crave payment of a five-shilling debt. He attended every horse-race and fox-hunt in the country, and paid the king's duty for a half-starved greyhound : Sandy was happy outside his shop, and his lad was thriving within. Matters went on in this train for so long as two years, and the hapless shopkeeper began to perceive that the few hundreds advanced him by his father had totally disappeared in the time, and to wonder what had become of them. Still, however, his stock in trade, though somewhat less showy than at first, was nearly equal in value to one-third his liabilities ; the other two-thirds were debts incurred by his old master; and at worst there lay no other obstacle between him and a highly respectable settlement with his creditors than the unlucky indorsations of his father. He rose, however, one morning to learn that his master had absconded during the night, leaving the shop-key under the door-sill; in a few days after, Sandy had absconded too; and his poor father, who had paid all his debts till now, and had taken a pride in paying them, found that his unfortunate indorsations had involved him in irretrievable ruin. Bankruptcy was a very different matter to the rigidly honest old man from what it was to either Sandy or his master.

For the first few days after the shock, he went wandering about his fields, muttering ceaselessly to himself, and wringing his hands. His whole faculties seemed locked up in a feeling of bewilderment and terror, and every packet of letters which the postman brought him—letters urging the claims of angry creditors, or intimating the dishonour of bills—added to his distress. His son was in hiding no one knew where; and though it was perhaps well that he should have kept out of the way at such a time, poor Isabel could not help feeling that it was unkind. He might surely be able to do something, she thought, to lighten the distress of which he had been so entirely the cause, were it but to tell them what course yet remained for them to pursue. It was in vain that, almost broken-hearted herself, she strove by soothing the old man to restore him to himself: he remained melancholy and abstracted as at first, as if the suddenness of his ruin had deprived him of his faculties. He hardly ever spoke, took scarce any food during the day, and scarce any sleep during the night; and, finally, taking to his bed, he died after a few days' illness—died of a broken heart. On the evening after the interment, his son John Rose, the carpenter, arrived from Glasgow, and found his mother sitting alone in the farmhouse, wholly overwhelmed with grief for the loss of her husband, and the utter ruin which she saw closing around her.

2 c

Their meeting was a sad one; but after the widow's first burst of sorrow was over, her son strove to comfort her, and in part succeeded. She might yet look forward, he said, to better days. He was in rather easy circumstances, employing about half-a-dozen workmen, and at times finding use for more. And though he could not well be absent from them, he would remain with her until he saw how far it was possible to wind up his father's affairs, and she would then go with him, and find what he trusted she should deem a comfortable home in Glasgow. Isabel was soothed by his kindness; but it did not escape the anxious eye of the mother, that her son, at one time so robust and strong, had grown thin, and pale, and hollow-eyed, like a person in the latter stages of consumption, and that, though he seemed anxious to appear otherwise, he was evidently much exhausted by his journey. He rallied, however, on the following day. The sale of his father's effects was coming on in about a week; and as the farmhouse at such a time could be no comfortable home for the widow, he brought her with him across the Firth to her sister's in Cromarty, and then returned to Minitarf.

Her sister's son was a saddler, a sagacious, well-informed man, truthful and honest, and as little imaginative as may be. He was employed at the time at the *Mains* of Invergordon— some six or seven miles from Cromarty—and slept in an apartment of the old castle, since burnt down. No one could be less influenced by superstitious beliefs of the period; and yet when, after scaling the steep circular stair that led to his solitary room, he used to shut the ponderous door and pass his eye along the half-lighted walls, here and there perforated by a narrow arched window, there was usually something in the tone of his feelings which served to remind him that there is a dread of the supernatural too deeply implanted in man's nature to be ever wholly eradicated. On going to bed one evening, and awakening as he supposed after a short slumber, he was much surprised to see the room filled as with a greyish light, in which the walls and the floor could be seen nearly as distinctly as by day. Suddenly the door fell open and there entered a tall young man in black, his hat wrapped up in crape, and with muslin weepers on his sleeves. Another and another entered, attired after the same fashion, until their number might, as he supposed, amount to about fifty. He lay gazing at them in astonishment, conscious of a kind of indistinct wish to ascertain whether he was in reality waking or asleep—a feeling of common enough experience in the dreams of imperfect slumber—when the man who had first come in, gliding up to his bedside, moved his lips as if addressing him, and passing off entered the staircase and disappeared. A second then came up, and heartily shaking him by the hand, also quitted the apartment, followed by all the others in the order in which they had entered, but without shutting the door ; and the last recollection of the sleeper was of an emotion of intense terror, which seemed wholly to overpower him when gazing on the dark opening of the stair beyond. It was broad daylight ere he awoke, and his first glance, as the dream of the previous evening flashed on his mind, was at the door, which sure enough lay open. " I must have missed slipping on the latch," he said, " or some of the servants must have entered during the night ;—but how strange a coincidence !" The particulars of his dream—and it cost him no slight effort to deem it such—employed his thoughts until evening ; when, setting out for his mother's, he found his aunt Isabel, in much grief and dejection, seated beside the fire. He had taken his place beside her, and was striving as he best could to lighten the melancholy which he saw preying on her spirits, when a young man, bespattered with travel, and apparently much fatigued, entered the apartment. Isabel started from her seat, and clasping her hands with a fearful presentiment of some overwhelming calamity, inquired of him what had happened at Minitarf? He stood speechless for a few

seconds as if overcome by some fearful emotion, and then bursting into tears, " Your son John," he said, " died this morning !"
The poor woman fainted away.

" For the two last days of the sale," said the messenger,
" there was a marked alteration in John's manner and appearance.   There was a something so fixed-like in his expression,
and so mournful in his way of looking at things ; and then his
face was deadly pale, and he took scarce any food.   It was
evident that the misfortunes of his family preyed deeply on
his mind.   Yester evening," continued the lad, " he complained
for the first time of being unwell, and retired to bed before
the usual hour.   The two servant-maids rose early in the morning to prepare for leaving the place, and were surprised, on
entering the ' ha',' to find him sitting in the great arm-chair
fronting the fire.   His countenance had changed during the
night ; he looked much older, and very like his father ; and he
was so weak that he could hardly sit up in the chair.   The
girls were alarmed, and would have called for assistance, but
he forbade them.   ' My watch,' he said, ' hangs over my pillow ;
go tell me what o'clock it is.'   It was just twenty minutes past
four.   ' Well,' said he, when they had told him, ' it is the last
hour to me ! there is a crook in my lot ; but it's God's doing,
not man's.'   And, leaning back in the chair, he never spake
more."   The messenger had seen the corpse laid on the bed,
and wrapped up in a winding-sheet, before setting out on his
melancholy journey.   Need I say aught of the feelings of
Isabel ?   The saddler and his mother strove to persuade her to
remain with them till at least after the funeral, but she would
not ; she would go and take one last look of her son, she said
—of her *only* son, for the other was a murderer.   Early, therefore, on the following morning, the saddler hired a small yawl
to bring her across the Firth, and, taking his place in the stern
beside her, the boatmen bent them to their oars, and the hill
of Nigg soon lessened behind them.

After clearing the bay, however, their progress was much impeded by adverse currents; there came on a chill drizzling rain, and the wind, which was evidently rising, began, after veering about oftener than once, to blow right ahead, and to raise a short tumbling sea. Grief of itself is cold and comfortless, and the widow, wrapped up in her cloak, sat shivering in the bottom of the yawl, drenched by the rain and the spray. But she thought only of her son and her husband. The boatmen toiled incessantly till evening; and when night came on, dark and boisterous, they were still two long miles from their landing-place—the effluence of the Nairn. Directly across the mouth of the river there runs a low dangerous bar, and as they approached they could hear the roaring of the breakers above all the hoarse sighings of the wind, and the dash of the lesser waves that were bursting around them. "There," said the saddler, as his eye caught a few faint lights that seemed twinkling along the beach; "there is the town of Nairn right abreast of us; but has not the tide fallen too low for our attempting the bar?" The boatmen replied in the negative, and in a minute after they were among the breakers. For a single instant the skiff seemed riding on the crest of an immense wave, which came rolling from the open sea, and which, as it folded over and burst into foam, dashed her forward like an arrow from the string. She sank, however, as it receded, till her keel grated against the bar beneath. Another huge wave came rolling behind, and, curling its white head like the former, rushed over her stern, filling her at once to the gunwale, and at the same instant propelling her into the deep water within. The saddler sprang from his seat, and raising his aunt to the hinder thwart, and charging her to hold fast, he shouted to the boatmen to turn the boat's head to the shore. In a few minutes after, they had landed.

Poor Isabel, well-nigh insensible—for grief and terror, added to cold and fatigue, had prostrated all her energies, bodily and mental—was carried to the town and lodged in the house of an acquaintance. When morning came she was unable to leave her bed, and so the saddler had to set out for Minitarf alone, which he reached about noon; and on being recognised as a cousin of the deceased, he was ushered into the room where the body lay. He seated himself on the edge of the bed, and raising the coffin-lid, gazed for a few seconds on the face of the dead; on hearing a footstep approaching the door, he replaced the cover. There entered a genteel-looking young man dressed for the funeral; but not the apparition of an inhabitant of the other world would have started the saddler more. He recognised in the stranger the young man of his dream. Another person entered, and him he also recognised as the man who had shaken hands with him; and who now, on being introduced to him as a relative of the deceased tacksman of Minitarf, sure enough, grasped him warmly by the hand. As the room filled around him with the neighbouring farmers attired in their soberest and best, he felt as if he still dreamed, for these were the very men whom he had seen in the old castle; and it was almost mechanically, when the coffin was carried out and laid on the bier, that, as the nearest relative of the dead he took his place as chief mourner. As the funeral proceeded, however, he collected his scattered thoughts. "Have I indeed had experience," said he to himself, "of one of those mysterious intimations of coming evil, the bare possibility of which few thinking men, in these latter times, seem disposed to credit on testimony alone? And little wonder, truly, that they should be so sceptical; for, for what purpose could such a warning have been given? It has enabled me to ward off no impending disaster; —nay, it has told its story so darkly and doubtfully, that the event alone has enabled me to interpret it. Could a purpose so idle have employed an agent of the invisible world? And yet," thought he again, as the train of his cogitations found way into the deeper recesses of his mind, "an end *has* been accomplished by it, and a not unimportant end either. The evil has befallen as certainly and heavily as if there had been no previous warning; but, is *my mind* in every respect the same? Something *has* been accomplished. And surely He who in His providence cares for all my bodily wants, without sinking, in the littleness of the object cared for, aught of the *greatness* of His character, might, without lessening in aught His character for *wisdom*, have taken this way of making me see, more distinctly than in all my life before, that there is indeed an invisible world, and that all the future is known to Him." There was seriousness in the thought, and never did he feel more strongly that the present scene of things is not the last, than when bending over the open grave he saw the corpse lowered down and heard the earth falling hollow on the coffin-lid.

But why dwell longer on the details of a story so mournful! The saddler, on his return to Nairn, found the widow in the delirium of a fever, from which she never recovered. Her younger son was seen in the West Indies ten years after, a miserable slave-driver, with a broken constitution and an unquiet mind. And there he died—no one caring where or how. I am not fond of melancholy stories; but "to purge the heart by pity and terror" is the true end of tragedy—an end which the gorgeous creations of the poets are not better suited to accomplish than the domestic tragedies which we see every day enacting around us. It is well, too, to note how immensely the folly and knavery of mankind add to the amount of human suffering; and how, according to the wise saying of the Preacher, "One sinner destroyeth much good."

CHAPTER XXVIII.

"Alack-a-day! It was the school-house indeed; but, to be sure, sir, the squire has pulled
it down, because it stood in the way of his prospects."—MACKENZIE.

THE old school of Cromarty was situated in a retired little
corner, behind the houses where the parish burying-ground bordered on the woods of the old castle. It was a low, mean-
looking building, with its narrow latticed windows, which were
half buried in the thatch, opening on the one side to the uncouth
monuments of the churchyard, and on the other, through a
straggling line of willows which fringed the little stream in
front, to the ancient timeworn fortalice perched on the top of
the hill. Mean, however, as it seemed—and certainly no public
edifice could owe less to the architect—it formed one of Knox's
strongholds of the Reformation, and was erected by the united
labours of the parishioners, agreeably to the scheme laid down
in the First Book of Discipline, long previous to the Education
Act of 1646. It had become an old building ere the Restoration, and fell into such disrepair during the reign of Episcopacy,
that for a time it no longer sheltered the scholars. I find it
enacted in the summer of 1682, by the Kirk-Session—for,
curious as it may seem, even the curates in the north of Scotland had their kirk-sessions and their staffs of elders—that
" the hail inhabitants of the burgh, especially masons and such
as have horse, do repayre and bigg the samin in the wonted
place, and that the folk upland do provide them with feal and
diffiot." And, in the true spirit of the reign of Charles II., a
penalty of four pounds Scots enforced the enactment.

The scheme of education drawn up by our first Reformers was stamped by the liberality of men who had learned from experience that tyranny and superstition derive their chief support from ignorance. Almost all the knowledge which books could supply at the time was locked up in the learned languages; and so it was necessary that these languages even the common people should acquire. It was appointed, therefore, " that young men who purposed to travail in some handicraft for the good of the commonwealth, should first devote ane certaine time to Grammar and the Latin tongue, and ane certaine time to the other tongues and the study of philosophy." Even long after the enactment, when we had got authors of our own in every department of literature, and a man could have become learned, if knowledge be learning, simply as an English reader, an acquaintance with Latin formed no unimportant part of a common Scotch education. Our fathers pursued the course which circumstances had rendered imperative in the days of their great-grandfathers, merely because their great-grandfathers had pursued it, and because people find it easier to persist in hereditary practices than to think for themselves. And so the few years which were spent in school by the poorer pupils of ordinary capacity, were absurdly frittered away in acquiring a little bad Latin and a very little worse Greek. So strange did the half-learning of our common people (derived in this way) appear to our southern neighbours, that there are writers of the last century who, in describing a Scotch footman or mechanic, rarely omit making his knowledge of the classics an essential part of his character. The barber in Roderick Random quotes Horace in the original ; and Foote, in one of his farces, introduces a Scotch valet, who, when some one inquires of him whether he be a Latinist, indignantly exclaims, " Hoot awa, man ! a Scotchman and no understand Latin!"

The school of Cromarty produced, like most of the other

schools of the kingdom, its Latinists who caught fish and made shoes; and it is not much more than thirty years since the race became finally extinct. I have heard stories of an old house-painter of the place, who, having survived most of his school-fellows and contemporaries, used to regret, among his other vanished pleasures, the pleasure he could once derive from an inexhaustible fund of Latin quotation, which the ignorance of a younger generation had rendered of little more value to him than the paper-money of an insolvent bank; and I have already referred to an old cabinetmaker whom I remember, who was in the practice, when his sight began to fail him, of carrying his Latin New Testament with him to church, as it chanced to be printed in a clearer type than any of his English ones. It is said, too, of a learned fisherman of the reign of Queen Anne, that when employed one day among his tackle, he was accosted in Latin by the proprietor of Cromarty, who, accompanied by two gentlemen from England, was sauntering along the shore, and that, to the surprise of the strangers, he replied with no little fluency in the same language.

The old castle rose, I have said, direct in front of the old school, about three hundred yards away; and, tall itself, and elevated by the green hill on which it stood, it formed, with all its timeworn turrets, and all its mouldering bartisans, a formidable spectre of the past. Little thought the proud hereditary sheriffs of the stern old tower, that the humble building at the foot of the hill was a masked battery raised against their authority, which was to burst open their dungeon door and to beat down their gallows. But a very formidable battery it proved. There is a class of nature's aristocracy that has but to arise from among the people, in order that the people may become influential and free; and the lowly old school did its part in separating from the general mass its due proportion of these, as mercury separates gold from the pulverized rock in which it is contained. If, in passing along the streets, we see a handsomer domicile than the low tenements around it, we may safely conclude that the builder spent his boyhood in the old school; that if he went out to some of the colonies, he carried with him as his stock in trade a knowledge of figures and the pen, and returned with both that and a few thousands on which to employ it; or if his inclination led him to sea, that he became, through his superior intelligence, the commander of a vessel; if to London, that he rose into wealth as a merchant; or if he remained at home, that he gained a competency as a shopkeeper, general trader, or master mechanic. I am not making too much of my subject when I affirm, that the little thatched hovel at the foot of the Castlehill gave merchants to the Exchange, ministers to the Church, physicians to the Faculty, professors to Colleges, and members to Parliament.

One of the pupils reared within its walls—the son of old Clerk Davidson, a humble subordinate of the hereditary sheriff —became a wealthy London merchant, and, after establishing in the city a respectable firm, which still exists, represented his native county in Parliament. Another of its boys, the late Mr. William Forsyth, to whom I have already had occasion to refer, revived the sinking trade of the town; and, though the son of a man who had once worked as a mechanic, he took his well-merited place among the aristocracy of the district, not less from the high tone of his character, and the liberality of his views and sentiments, than from the extent of his resources. Yet another of its boys, a Mr. James Ross, entered life as a common sailor, and, after rising by his professional skill to a command in the navy, published a work on the management of nautical affairs, which attracted a good deal of notice at the time among the class to which it was specially addressed. The late Dr. James Robertson, librarian of the University of Edinburgh, and its Professor of the Oriental Tongues, was a native of Cromarty, of humble parentage, and experienced his first stirrings of scholastic ambition in the old school.   He was the author of a Hebrew grammar, to which the self-taught linguist, Dr. Alexander Murray, owed, as he tells us in his interesting Autobiography, his first introduction to Hebrew; and we learn from Boswell, in his Journal of a Tour to the Hebrides, that Dr. Johnson, when in Edinburgh, " was much pleased with the College library, and with the conversation of Dr. James Robertson, the librarian."   Provost Hossack of Inverness, whom the author of the " Jacobite Memoirs " terms, in relating his spirited remonstrance with the Duke of Cumberland in behalf of the conquered rebels, " a man of humanity, and the Sir Robert Walpole of Inverness, under the direction of President Forbes," was also a Cromarty man, the child of seafaring parents, and received the education through which he rose, in its school.   And his namesake and contemporary, Dr. Hossack of Greenwich, one of the first physicians of his time, was likewise a native of Cromarty—not of the town, however, but of the landward part of the parish; and owed his first knowledge of letters to the charity of the schoolmaster.   There is, unfortunately, not much of the Doctor's story known; but to the little which survives there attaches a considerable amount of interest.

He had lost both his parents when an infant; all his other nearer kindred were also dead: and so he was dependent in his earlier years for a precarious subsistence on the charity of a few distant relatives, not a great deal richer than himself. Among the rest there was a poor widow, a namesake of his own, who earned a scanty subsistence by her wheel, but who had heart enough to impart a portion of her little to the destitute scholar.   The boy was studious and thoughtful, and surpassed in his tasks most of his schoolfellows; and after passing with singular rapidity through the course pursued at school, he succeeded in putting himself to college.   The struggle was arduous and protracted; sometimes he wrought as a common labourer, sometimes he taught an adventure school; he deemed no honest employment too mean or too laborious that forwarded his scheme; and thus he at length passed through the University course. His town's-people then lost sight of him for nearly twenty years. It was understood, meanwhile, that some nameless friend in the south had settled a small annuity on poor old Widow Hossack; and that a Cromarty sailor, who had been attacked by a dangerous illness when at London, had owed his life to the gratuitous attentions of a famous physician of the place, who had recognised him as a town's-man. No one, however, thought of the poor scholar; and it was not until his carriage drove up one day through the main street of the town, and stopped at the door of his school-fellow, William Forsyth, that he was identified with "the great Doctor" who had attended the seaman, and the benefactor of the poor widow. On entering the cottage of the latter, he found her preparing gruel for supper, and was asked, with the anxiety of a gratitude that would fain have rendered him some return, " O Sir! will ye no tak' *brochan?*" He is said to have been a truly excellent and benevolent man—the Abercromby of a former age; and the ingenious and pious Moses Browne (a clergyman who, to the disgrace of the English Church, was suffered to languish through life in a curacy of fifty pounds per annum) thus addresses him in one of his larger poems, written immediately after the recovery of the author from a long and dangerous illness :—

> " The God I trust with timeliest kind relief
> Sent the beloved physician to my aid,
> (Generous, humanest, affable of soul,
> Thee, dearest Hossack ;—Oh ! long known, long loved,
> Long proved ; in oft found tenderest watching cares,
> The Christian friend, the man of feeling heart ;)
> And in his skilful, heaven-directed hand,
> Put his best pleasing, *only* fee, my cure."
>
> SUNDAY THOUGHTS, Part iv.

The reputation of the old school necessarily varied with the character and acquirements of its several teachers. About a century ago, it was one of the most celebrated in this part of the kingdom, and was attended by the children of country gentlemen for sixty miles round. The teacher, a Mr. David Macculloch, was a native of the parish; and so highly were his services appreciated by the people, especially by such among them as kept lodgers, that they used to allege he was the means of circulating more money among them than all their shopkeepers and tradesfolk put together. He was a licentiate of the Church, and was lost to the place by receiving an appointment to a semi-Highland parish somewhere in Perthshire; when his fame as a teacher was transferred for half an age to the parish schoolmaster of Fortrose, a Mr. Smith. It was under this man, who is said to have done for the burghers of Chanonry and Rosemarkie all that Mr. Macculloch had done for the householders of Cromarty, that Sir James Mackintosh, so well known in after years as a statesman and philosopher, received the rudiments of his education. Next in course the burgh of Nairn became famous for the skill of its parish teacher, a Mr. Strath; and there still survive a few of his pupils to testify to his merits and to express their gratitude. Since his death, however, the fame of educational ability has failed to be associated in any very marked degree with our northern parochial schools—in part a consequence, it is probable, of that change in the tactics of tuition which, by demanding a division of labour in the educational as in other departments, at once lessens the difficulty and increases the efficiency of teaching. It is at least obvious that few succeed well in what is very difficult; and that every improvement in any art must add either to the value of what the art produces, or, what seems to have happened in this case, to the facility of production.

The successor of Mr. Macculloch in the old school—a Mr. Russel—though not equally celebrated as a teacher, was in other respects a more remarkable man. About twelve years after his appointment, he relinquished his pedagogical charge for a chapel in Kilmarnock, and there he came in contact and quarrelled with our great national poet, who, bold and unyielding as he was, seems to have regarded the stern pedagogue of the north as no weak or puny antagonist; at least, against none of his other clerical opponents did he open so powerful a battery. We find him figuring in the " Holy Fair," in the " Ordination," in the " Kirk's Alarm," and in the " Twa Herds," one of whom was the " wordy Russel." Some degree of interest must necessarily attach to the memory of a man who seems destined never to be wholly forgotten ; and as I have known and often conversed with several of his pupils, and remember even some of his mature contemporaries, I must communicate to the reader a few of their more characteristic recollections of the man of whom they were accustomed to speak and think as Russel the " hard schoolmaster."

It is now somewhat more than eighty years since John Russel, a native of Moray, and one of the Church's probationers, was appointed to the parish school of Cromarty. He was a large, robust, dark-complexioned man, imperturbably grave, and with a singularly stern expression stamped on his dusky forehead, that boded the urchins of the place little good. And in a few months he had acquired for himself the character of being by far the most rigid disciplinarian in the country. He was, I believe, a good, conscientious man, but unfortunate in a temper at once violent and harsh, and in sometimes mistaking its dictates for those of duty. At any rate, whatever the nature of the mistake, never was there a schoolmaster more thoroughly feared and detested by his pupils ; and with dread and hatred did many of them continue to regard him long after they had become men and women. His memory was a dark morning cloud resting on their saddened boyhood, that cast its shadows into after life. I have heard of a lady who was so overcome by sudden terror on unexpectedly seeing him, many years after she had quitted school, in one of the pulpits of the south, that she fainted away in the pew ; and of another of his scholars named M'Glashan—a robust, daring young man of six feet—who, when returning to Cromarty from some of the colonies, solaced himself by the way with thoughts of the hearty drubbing with which he was to clear off all his old scores with the dominie.   Ere his return, however, Mr. Russel had quitted the parish ; nor, even if it had chanced otherwise, might the young fellow have gained much in an encounter with one of the boldest and most powerful men in the country.

But Polyphemus himself, giant as he was, and a demigod to boot, could not always be cruel with impunity.   The schoolmaster had his vulnerable point ; he was a believer in ghosts ; at all events he feared them very heartily, whether he believed in them or no ; and some of his boys, much as they dreaded him, contrived on one occasion to avenge themselves upon him through his fears.   In the long summer evenings he was in the habit of prosecuting his studies to a late hour in the schoolroom ; from which, in returning to his lodgings, he had to pass through the churchyard.   And when striding homewards one night, laden with books and papers, so affrighted was he by a horrible apparition, all over white, which started up beside him from beneath one of the tombstones, that, casting his burden to the winds, and starting off like wildfire, he never once looked behind him until he had gained his landlady's fireside.   It is said that he never after prosecuted his evening studies in the school.   The late minister of Knockbain, Mr. Roderick M'Kenzie, for many years father of the Presbytery of Chanonry, used to tell with much glee that he knew a very great deal about the urchin who, in behalf of the outraged youthhood of the place, wore the white sheet on this interesting occasion.   " I was quite as much afraid of ghosts," he used to say, " as Mr. Russel himself ; but three of my companions lay fast ensconced, to keep me in heart and countenance, under a neighbouring gravestone."

There was among Russel's pupils a poor boy named Skinner, who, as was customary in Scottish schools of the period, blew the horn for gathering the scholars, and kept the catalogue and the key, and who, in return for his services, was educated by the master, and received some little gratuity from the boys besides. To the south of the Grampians he would have been termed the Janitor of the school; whereas in the north, in those days, the name attached to him, in virtue of his office, was the humbler one of " The Pauper." Unluckily, on one occasion, the key dropped out of his pocket; and, when school time came, the irascible dominie had to burst open the door with his foot. He raged at the boy with a fury so insane, and beat him with such relentless severity, that in the extremity of the case, the other boys rose up shrieking around him as if they were witnessing the perpetration of a murder; and the tyrant, brought suddenly to himself by so strange an exhibition, flung away the rod and sat down. And such, it is said, was the impression made on the mind of poor " Pauper Skinner," that though he quitted the school shortly after, and plied the profession of a fisherman until he died an old man, he was never from that day seen disengaged for a moment, without mechanically thrusting his hand into the key-pocket. If excited too, by any unexpected occurrence, whatever its nature, he was sure to grope hastily, in his agitation, for the missing key. One other anecdote illustrative of Mr. Russel's temper. He was passing along the main street of the town, in a day of wind and rain from the sea, with his head half-buried in his breast, when he came violently in contact with a thatcher's ladder, which had been left sloping from the roof of one of the houses. A much less matter would have sufficed to awaken the wrath of Mr. Russel: he laid hold of the ladder, and, dashing it on the pavement, broke with his powerful foot, ere he quitted it, every one of the " rounds."

For at least the last six years of his residence in Cromarty

2 D

he was not a little popular as a preacher. His manner was strong and energetic, and the natural severity of his temper seems to have been more than genius to him when expatiating, which he did often, on the miseries of the wicked in a future state. The reader will scarce fail to remember the picture of the preacher dashed off by Burns in the Holy Fair; or to see that the poet's arrows, however wickedly shot, came from no bow drawn at venture:—

> " Black Russel is nae spairin';
> His piercing words, like Highland swords,
>   Divide the joints an' marrow
> His talk o' hell, where devils dwell,
>   Our verra sauls does harrow
>   Wi' fright that day.
>
> " A vast unbottom'd, boundless pit,
>   Fill'd fou o' lowin' brunstane,
> Whase ragin' flame and scorchin' heat
>   Wad melt the hardest whunstane.
> The half-asleep start up wi' fear,
>   And think they hear it roarin',
> When presently it does appear
>   'Twas but some neebor snorin'
>   Asleep that day."

I have seen one of Russel's sermons in print; it is a controversial one, written in a bold rough style, and by no means inferior as a piece of argument; but he was evidently a person rather to be listened to than read. He was quite as stern in Church matters, it is said, as in those of the school; but men are less tractable than boys; and his severity proved more effectual in making his pupils diligent than in reforming the town's-people. He converted a few rather careless boys into not very inferior scholars; but though he set himself so much against the practice of Sabbath-evening walking, that he used to take his stand every Sunday, after the church had dismissed, full in the middle of the road which leads from the town to the woods and rocks of the Southern Sutor, and sometimes turned back the walkers by the shoulders after he had first shaken them by the breast, the practice of Sabbath-evening walking became even more common than before. Instead of addressing himself to the moral sense of the people, he succeeded in but arousing their combative propensities; and these, once awakened, took part against a good cause, simply because it had been unwisely and unjustifiably defended.

I have an uncle in Cromarty, now an elderly man, who, when residing in Glasgow in the year 1792, walked about ten miles into the country to attend a sacramental *occasion*, at which he was told Mr. Russel was to officiate, and which proved to be such a one as Burns has described in his " Holy Fair." There were excellent sermons to be heard from the *tent*, and very tempting drink to be had in an ale-house scarcely a hundred yards away; and between the tent and the ale-house were the people divided, according to their tastes and characters. A young man preached in the early part of the day—his discourse was a long one; and, ere it had come to a close, the mirth of the neighbouring topers, which became louder the more deeply they drank, had begun to annoy the congregation. Mr. Russel was standing beside the tent. At every fresh burst of sound he would raise himself on tiptoe, look first, with a portentous expression of countenance, towards the ale-house, and then at the clergyman; who at length, concluding his part of the service, yielded to him his place. He laid aside the book, and, without psalm or prayer, or any of the usual preliminaries, launched at once into a powerful extempore address, directed, over the heads of the people, at the ale-house. I have been assured by my relative that he never before or since heard any thing half so energetic. His ears absolutely tingled, as the preacher thundered out, in a voice almost superhuman, his solemn and terrible denunciations. Every sound of revelry ceased in a moment; and the Bacchanals, half-drunk, as most of them had ere now become, were so thoroughly frightened as to be fain to steal out through a back window, and slink away along bypaths through the fields. Mr. Russel was ultimately appointed one of the ministers of Stirling. A Cromarty man, a soldier in a Highland regiment, when stationed in Stirling Castle, had got involved one day in some street quarrel, and was swearing furiously, when a tall old man in black came and pulled him out of the crowd. "Wretched creature that ye are!" said the old man; "come along with me." He drew him into a quiet corner, and began to expostulate with him on his profanity, in a style to which the soldier, an intelligent though by no means steady man, and the child of religious parents, could not but listen. Mr. Russel—for it was no other than he—seemed pleased with the attention he paid him; and on learning whence he had come, and the name of his parents, exclaimed with much feeling, "Wae's me! that your father's son should be a blackguard soldier on the streets of Stirling! But come awa." He brought him home with him, and added to the serious advice he had given him an excellent dinner. The temper of the preacher softened a good deal as he became old; and he was much a favourite with the more serious part of his congregation. He was, with all his defects, an honest, pious man; and had he lived in the days of Renwick or Cargill, or, a century earlier, in the days of Knox or Wishart, he might have been a useful one. But he was unlucky in the age in which he lived, in his temper, and in coming in contact with as hard-headed people as himself.

The parish schools of Scotland had their annual saturnalian feast, of what may be well deemed an extraordinary character, if we consider their close connexion with the National Church, and that their teachers were in so many instances licensed clergymen waiting for preferment. On Fasten's-eve, just when all Rome was rejoicing in the license of the Carnival, the school-master, after closing the service of the day with prayer, would call on the boys to divide and choose for themselves "*Head-stocks*," *i.e.*, leaders, for the yearly cock-fight of the ensuing Shrove-Tuesday. A sudden rush would immediately take place among the pigmy population of the school to two opposing desks, which, piled up with urchin a-top of urchin half-way to the rafters, would straightway assume the appearance of two treacled staves, covered with black-bottle flies in a shopkeeper's yard, on a day of midsummer. The grave question of leadership soon settled, in consequence of previous out-of-door arrangement, the master, producing the catalogue, would next proceed to call the boys in alphabetical order; and each boy to intimate, in reply, under what "head-stock" he purposed fighting his cocks, and how many cocks he intended bringing into the pit. The master, meanwhile, went on recording both items in a book —in especial the number of the cocks—as, according to the registered figure, which always exceeded the array actually brought into the fight, he received, as a fixed perquisite of his office, a fee of twopence per head. The school then broke up; and for the two ensuing days, which were given as holidays for the purpose of preparation, the parish used to be darkened by wandering scholars going about from farmhouse to farmhouse in quest of cocks. Most boys brought at least one cock to the pit; and "head-stocks"—selected usually for the wealth of their parents, and with an eye to the entertainment with which the festival was expected to close—would sometimes bring up as many as ten or twelve. The cock-fight ball, given by the victorious "head-stock" on the eve of his victory, was always regarded as the crowning item in the festival.

On the morning of Shrove Tuesday, the floor of the school, previously cleared of all the forms, and laid out into a chalked circle, representative of the cockpit, became a scene of desperate battle. The master always presided on these occasions as umpire; while his boys clustered in a ring, immediately under his eye, a little beyond the chalked line. The cocks of the lads who ranged under the one "head-stock" were laid down one after one on the left, those of the other, as a bird dropped exhausted or ran away, upon the right; and thus the fight went on from morning till far in the evening ; when the "head-stock" whose last bird remained in possession of the field, and whose cocks had routed the greatest number in the aggregate, was declared victor, and formally invested with a tinsel cap, in a ceremony termed the "crowning." The birds, however, were permitted to share in the honour of their masters—and in many schools there was a small silver bell, the property of the institution, attached to the neck of the poor cock who had beaten the largest number of opponents ; but very rarely did he long survive the honour. I remember seeing one gallant bird, who had vanquished six cocks in succession, stand in the middle of the pit, one of his eyes picked out, and his comb and bells all in a clot of blood, and then, in about half a minute after his last antagonist had fled, fall dead upon the floor. It is really wonderful how ingenious boys can be made, in even the more occult mysteries of the cockpit, when their training has been good. Some hopeful scholars had learned to provide themselves with medicated grains for drugging, as the opportunity offered, the birds of an opponent ; and it was no unusual thing for a lad who carried his cock under his arm in the crowd, to find the creature rendered unfit for the combat by the skilful application of the pin of an antagonist, who, having stolen stealthily upon him from behind, succeeded in serving the poor animal as the minions of Mortimer served the hapless Edward II. Game-birds who, in inconsistency with their previous character, refused to fight, were often found, on examination, to have pins thrust up more than two inches into their bowels. The birds who, without any such apology, preferred running away to fighting, were converted into *droits*, under the ill-omened name of *fugies*, and forfeited to the master of the school. And these were rendered by him the subject of yet another licensed amusement of the period. The *fugies* were fastened to a stake in the play-ground, and destroyed, one after one, in the noble game of cock-throwing, by such of the pupils or of the town's-people as could indulge in the amusement at the rate of a halfpenny the throw. The master not only pocketed all the halfpennies, but he also carried home with him all the carcases. It is perhaps not very strange that good men, of naturally severe temper, like Mr. Russel, should have said grace over their *cock-a-leekie* thus procured, without once suspecting that there was anything wrong in the practice ; but that schoolmasters like M'Culloch, who was a person of humanity, should have done so, serves strikingly to show how blinding and tyrannical must be that influence which custom exercises over even the best of men ; and that not only does religion exert a beneficial effect on civilisation, but that civilisation may, in turn, react with humanizing influence on the religious. The very origin of the festival is said to have been ecclesiastical. It was instituted, we find it intimated in the *Clavis Calendaria*, " in allusion to the indignities offered to our Saviour by the Jews before the crucifixion ;" but how it should have survived the Reformation, and been permitted not only to shelter, like the Gibeonites of old, in the house of the enemy, but have also become an object of the direct patronage of many of our best men of the evangelical school, seems a problem of somewhat difficult solution. It is just possible, however, that the Reformers, who were well enough acquainted with human nature to be aware of the necessity of relaxation, might have seen nothing very barbarous in the practice ; seeing that the tone of men's feelings in such matters depends more on the degree of refinement which has been attained to by the age or country in which they live, than on the severity of their general morals or the purity of their creed. I may add, that the practice of cock-throwing was abolished in the old school of Cromarty by Mr. Russel's immediate successor—the late Rev. Mr. Macadam of Nigg ; but the annual cock-fight survived until put down, a few years ago,[1] by the present incumbent of the parish.

[1] It was abolished by the late Rev. Mr. Stewart, in the second year of his incumbency (1826.)

There was one other Cromarty man of the last century who became eminent in his own walk and day, and to whom I must therefore refer; but I know not that he owed much, if anything, to the old school of the burgh.

In the *Scots Magazine* for May 1789, there is a report by Captain Philip d'Auvergne, of the Narcissus frigate, on the practical utility of Kenneth M‘Culloch's sea compasses. The captain, after an eighteen months' trial of their merits, compared with those of all the other kinds in use at the time, describes them as immensely superior, and earnestly recommends to the Admiralty their general introduction into the navy. In passing, on one occasion, through the race of Alderney in the winter of 1787, there broke out a frightful storm, and so violent was the opposition of the wind and tide, that while his vessel was sailing at the rate of eleven miles on the surface, she was making scarce any headway by the land. The sea rose tremendously—at once short, high, and irregular; and the motions of the vessel were so fearfully abrupt and violent, that scarce a seaman aboard could stand on deck. At a time so critical, when none of the compasses supplied from his Majesty's stores *would stand*, but vacillated more than three points on each side, "it commanded," says the captain, "the admiration of the whole crew, winning the confidence of even the most timorous —to see how quickly and readily M‘Culloch's steering compass recovered the vacillations communicated to it by the motion of the ship and the shocks of the sea, and how truly in every brief interval of rest the needle pointed to the Pole." It is further added, that on the Captain's recommendation these compasses were tried on board the Andromeda, commanded at the time by Prince William Henry, our present king, and so satisfied was the Prince of the utility of the invention, that he too became a strenuous advocate for their general introduction, and testified his regard for the ingenious inventor, by appointing him his compass-maker. M‘Culloch, however, did not long survive the honour, dying a few years after, and I have been unable to trace with any degree of certainty the further history of his improved compasses. But though only imperfectly informed regarding his various inventions, and they are said to have been many, and singularly practical, I am tolerably well acquainted with the story of his early life; and as it furnishes a striking illustration of that instinct of genius, if I may so express myself, which leads the possessor to exactly the place in which his services may be of most value to the community, by rendering him useless and unhappy in every other, I think I cannot do better than communicate it to the reader.

There stood, about forty years ago, on the northern side of the parish of Cromarty, an old farm-house—one of those low, long, dark-looking erections of turf and stone which still survive in the remoter districts of Scotland, as if to show how little man may sometimes improve, in even a civilized country, on the first rude shelter which his necessities owed to his ingenuity. A worn-out barrel, fixed slantwise in the ridge, served as a chimney for the better apartment (the spare room of the domicile), which was also furnished with a glazed window; but in the others the smoke was suffered to escape, and the light to enter, as chance or accident might direct. The eaves, overhung by stonecrop and studded by bunches of the houseleek, drooped heavily above the small blind openings and low door; and a row of ancient elms, which rose from out the fence of a neglected garden, spread their gnarled and ponderous arms over the roof. Such was the farmhouse of Woodside, in which Kenneth M'Culloch, the son of the farmer, was born some time in the early half of the last century. The family from which he sprang—a race of honest, plodding tacksmen—had held the place from the proprietor of Cromarty for considerably more than a hundred years before, and it was deemed quite a matter of course that Kenneth, the eldest son, should succeed his father in the farm. Never was there a time, in at least this part of

the country, in which agriculture stood more in need of the
services of original and inventive minds.   There was not a
wheeled cart in the parish, nor a plough constructed on the
modern principle.   There was no changing of seed to suit the
varieties of soil, no green cropping, no rotatory system of production; it almost seemed as if the main object of the farmer
was to raise the least possible amount of grain at the greatest
possible expense of labour.   The farm of Woodside was primitive enough in its usages and modes of tillage to have formed a
study to the antiquary.   Towards autumn, when the fields vary
most in colour, it resembled a rudely executed chart of some
large island, so irregular were the patches which composed it,
and so broken on every side by a surrounding sea of brown
sterile moor, that went here and there winding into the interior
in long river-like strips, or expanded within into firths and
lakes.   In one corner there stood a heap of stones, in another
a thicket of furze—here a piece of bog—there a broken bank
of clay.   The implements, too, with which the fields were
tilled, were quite as uncouth in their appearance as the fields
themselves.   There was the single-stilted plough, that did little
more than scratch the surface; the wooden-toothed harrow, that
did hardly so much; the cumbrous sledge—no inconsiderable
load of itself, for carrying home the corn in harvest; and the
basket-woven conical cart, with its rollers of wood, for bearing
out the manure in spring.   With these, too, there was the
usual misproportion to the extent and produce of the farm, of
lean inefficient cattle—four half-starved animals performing,
with incredible labour, the work of one.   And yet, now that a
singularly inventive mind had come into existence on this very
farm, and though its attentions had been directed, as far as external influences could direct them, on the various employments
of the farmer, the interests of husbandry were to be in no degree improved by the circumstance.   Nature, in the midst of
her wisdom, seems to cherish a dash of the eccentric.   The

ingenuity of the farmer's son was to be employed, not in facilitating the labours of the farmer, but in inventing binnacle
lamps, which would yield an undiminished light amid the agita-
tions of a tempest, and in constructing mariners' compasses on a
new principle. There are instances of similar character furnished by the experience of almost every one. In passing, some
years since, over a dreary moor in the interior of the country,
my curiosity was excited by a miniature mast, furnished, like
that of a ship, with shrouds and yards, and bearing a-top a gaudy
pinnet, which rose beside a little Highland cottage. And on
inquiring regarding it at the door, I was informed that it was
the work of the cottager's son, a lad who, though he had
scarcely ever seen the sea, had taken a strange fancy to the life
of a sailor, and had left his father only a few weeks before, to
serve aboard a man-of-war.

Kenneth's first employment was the tending of a flock of
sheep, the property of his father; and wretchedly did he acquit
himself of the charge. The farm was bounded on the eastern
side by a deep bosky ravine, through the bottom of which a
scanty runnel rather trickled than flowed; and when it was
discovered on any occasion that Kenneth's flock had been left to
take care of themselves, and of his father's corn to boot—and
such occasions were wofully frequent—Kenneth himself was
almost invariably to be found in the ravine. There would he
sit for hours among the bushes, engaged with his knife in carving uncouth faces on the heads of walking-sticks, or in constructing little water-mills, or in making Liliputian pumps of the
dried stalks of the larger hemlock, and in raising the waters of
the runnel to basins dug in the sides of the hollow. Sometimes
he quitted his charge altogether, and set out for a meal-mill
about a quarter of a mile from the farm, where he would linger
for half a day at a time watching the motion of the wheels.
His father complained that he could make nothing of him—
" The boy," he said, " seemed to have nearly as much sense as

other boys of his years, and yet for any one useful purpose he was nothing better than an idiot." His mother, as is common with mothers, and who was naturally an easy kind-hearted sort of woman, had better hopes of him. Kenneth, she affirmed, was only a little peculiar, and would turn out well after all. He was growing up, however, without improving in the slightest, and when he became tall enough for the plough, he made a dead stand. He would go and be a tradesman, he said—a mason, or smith, or house-carpenter—anything his friends chose to make him; but a farmer he would not be. His father, after a fruitless struggle to overcome his obstinacy, carried him with him to a friend in Cromarty, our old acquaintance, Donald Sandison, and after candidly confessing that he was of no manner of use at home, and would, he was afraid, be of little use any-where, bound him by indenture to the mechanic for four years.

Kenneth's new master, as I have already had occasion to state, was one of the best workmen in his profession in the north of Scotland. His scrutoires and wardrobes were in repute up to the close of the last century, and in the ancient art of wains-cot-carving he had no equal in the country. He was an intelli-gent man too, as well as a superior mechanic; but with all his general intelligence, and all his skill, he failed to discover the latent capabilities of his apprentice. Kenneth was dull and absent, and had no heart to his work; and though he seemed to understand the principles on which his master's various tools were used and the articles of his trade constructed, as well as any workman in the shop, there were none among them who used the tools so awkwardly, or constructed the articles so ill. An old botching carpenter who wrought in a little shop at the other end of the town, was known to the boys of the place by the humorous appellation of "Spull (*i. e.* spoil)-the-wood," and a lean-sided, ill-conditioned, dangerous boat which he had built, as "the Wilful Murder." Kenneth came to be regarded as a sort of second "Spull-the-Wood," as a fashioner of rickety tables,

ill-fitted drawers, and chairs that, when sat upon, creaked like
badly-tuned organs; and the boys, who were beginning to regard him as fair game, sometimes took the liberty of asking
him whether he, too, was not going to build a Wilful Murder?
Such, in short, were his deficiencies as a mechanic, that in the
third year of his apprenticeship his master advised his father
to take him home with him and set him to the plough—an
advice, however, on which the farmer, warned by his previous
experience, sturdily refused to act.

It was remarked that Kenneth acquired more of his profession in the last year of his apprenticeship than in all the others.
His skill as a workman came to rank but little below the
average ability of his shopmates; and he seemed to enjoy more,
and had become less bashful and awkward. His master on one
occasion brought him aboard a vessel in the harbour, to repair
some injury which her bulwarks had sustained in a storm; and
Kenneth, for the first time in his life, was introduced to the
mariner's compass. The master in after days, when his apprentice had become a great man, used to relate the circumstance
with much complacency, and compare him, as he bent over the
instrument in wonder and admiration, to a negro of the Kanga
tribe worshipping the elephant's tooth. On the close of his
apprenticeship he left this part of the country for London, accompanied by his master's eldest son, a lad of a rather careless
disposition, but, like his father, a first-rate workman.

Kenneth soon began to experience the straits and hardships
of the inferior mechanic. His companion found little difficulty
in procuring employment, and none at all in retaining it when
once procured. Kenneth, on the contrary, was tossed about
from shop to shop, and from one establishment to another; and
for a full twelvemonth, during the half of which he was wholly
unemployed, he did not work for more than a fortnight together
with any one master. It would have fared worse with him
than it did, had it not been for his companion, Willie Sandison,

who generously shared his earnings with him every time he stood in need of his assistance. In about a year after they had gone to London, however, Willie, an honest and warmhearted but thoughtless lad, was inveigled into a disreputable marriage, and lost in consequence his wonted ability to assist his companion. I have seen one of Kenneth's letters to his old master, written about this time, in which he bewails Willie's mishap, and dwells gloomily on his own prospects. How these first began to brighten I am unable to say, for there occurs about this period a wide gap in his story, which all my inquiries regarding him have not enabled me to fill; but in a second letter to his master, now before me, which bears date 1772, just ten years after the other, there are the evidences of a surprising improvement in his circumstances and condition.

He writes in high spirits. Just before sitting down to his desk he had heard from his old friend Willie, who had gone out to one of the colonies, where he was thriving in spite of his wife. He had heard, too, by the same post from his mother, who had been so kind to him during his luckless boyhood; and the old woman was well. He had, besides, been enabled to remove from his former lodgings to a fine airy house in Duke's Court, opposite St. Martin's Church, for which he had engaged, he said, to pay a rent of forty-two pounds per annum, a very considerable sum nearly sixty years ago. Further, he had entered into an advantageous contract with Catherine of Russia, for furnishing all the philosophical instruments of a new college then erecting in Petersburgh—a contract which promised to secure about two years' profitable employment to himself and seven workmen. In the ten years which intervened between the dates of his two letters, Kenneth M'Culloch had become one of the most skilful and inventive mechanicians of London. He rose gradually into affluence and celebrity, and for a considerable period before his death his gains were estimated at about a thousand a year. His story, however, illustrates rather

the wisdom of nature than that of Kenneth M‘Culloch.   We
think all the more highly of Franklin for being so excellent a
printer, and of Burns for excelling all his companions in the
labours of the fields; nor did the skill or vigour with which
they pursued their ordinary employments hinder the one from
taking his place among the first philosophers and first statesmen of the age, nor prevent the other from achieving his wide-
spread celebrity as the most original and popular of modern
poets.   Be it remembered, however, that there is a narrow and
limited cast of genius, unlike that of either Burns or Franklin,
which, though of incalculable value in its own sphere, is of no
use whatever in any other; and to precipitate it on its proper
object by the pressure of external circumstances, and the general
inaptitude of its possessor for other pursuits, seems to be part
of the wise economy of Providence.   Had Kenneth M‘Culloch
betaken himself to the plough, like his father and grandfather,
he would have been, like them, the tacksman of Woodside, and
nothing more; had he found his proper vocation in cabinet-
making, he would have made tables and chairs for life, like his
ingenious master, Donald Sandison.

CHAPTER XXIX.

"To a mysteriously consorted pair
This place is consecrate, to Death and Life,
And to the best affections that proceed
From their conjunction."—WORDSWORTH.

WERE I to see a person determined on becoming a hermit, through a disgust of the tame aspect of manners and low tone of feeling which seem characteristic of what is termed civilized society, I should be inclined to advise that, instead of retiring into a desert, he should take up his place of residence in a country churchyard.

Perhaps no personage of real life can be more properly regarded as a hermit of the churchyard than the itinerant sculptor, who wanders from one country burying-ground to another, recording on his tablets of stone the tears of the living and the worth of the dead. If possessed of an ordinary portion of feeling and imagination, he can scarce fail of regarding his profession as a school of benevolence and poetry. For my own part, I have seldom thrown aside the hammer and trowel of the stone-mason for the chisel of the itinerant sculptor, without receiving some fresh confirmation of the opinion. How often have I suffered my mallet to rest on the unfinished epitaph, when listening to some friend of the buried expatiating, with all the eloquence of grief, on the mysterious warning—and the sad deathbed—on the worth that had departed—and the sorrow that remained behind! How often, forgetting that I was merely an auditor, have I so identified myself with the mourner as to feel my heart swell, and my eyes becoming moist! Even the very aspect of a solitary churchyard seems conducive to habits of thought and feeling. I have risen from my employment to mark the shadow of tombstone and burial-mound creeping over the sward at my feet, and have been rendered serious by the reflection, that as those gnomons of the dead marked out no line of hours, though the hours passed as the shadows moved, so, in that eternity in which even the dead exist, there is a nameless tide of continuity, but no division of time. I have become sad, when, looking on the green mounds around me, I have regarded them as waves of triumph which time and death have rolled over the wreck of man ; and the feeling has deepened, when, looking down with the eye of imagination through this motionless sea of graves, I have marked the sad remains of both the long-departed and the recent dead thickly strewed over the bottom. I have grieved above the half-soiled shroud of her for whom the tears of bereavement had not yet been dried up, and sighed over the mouldering bones of him whose very name had long since perished from the earth.

Not long ago I wrought for about a week in the burying-ground of Kirk-Michael, a ruinous chapel in the eastern extremity of the parish of Resolis, distant about six miles from the town of Cromarty. It is a pleasant solitary spot, lying on the sweep of a gentle declivity. The sea flows to within a few yards of the lower wall; but the beach is so level, and so little exposed to the winds, that even in the time of tempest there is heard within its precincts only a faint rippling murmur, scarcely loud enough to awaken the echoes of the ruin. Ocean seems to muffle his waves in approaching this field of the dead. A row of elms springs out of the fence, and half encircles the building in the centre. Standing beside the mouldering walls, the foreground of the scene appears thickly sprinkled over with graves and tablets ; and we see the green moss creeping round the rude sculptures of a primitive age, imparting lightness and beauty to that on which the chisel had bestowed a very opposite character. The flake-like leaves and gnarled trunks of the elms fill up what a painter would term the midground of the picture ; and seen from between the boughs, the Bay of Cromarty, shut in by the Sutors so as to present the appearance of a huge lake, and the town beyond half enveloped in blue smoke —the windows sparkling through the cloud like spangles on a belt of azure—occupy the distance.

The western gable of the ruin is still entire, though the very foundations of part of the walls can no longer be traced on the sward, and it is topped by a belfry of hewn stone, in which the *dead bell* is still suspended. From the spires and balls with which the cornice is surmounted, the moss and lichens which bristle over the mouldings, and the stalks of ragweed which shoot out here and there from between the joints, the belfry, though designed in a barbarous style of architecture, is rich in the true picturesque. It furnished me, when the wind blew from the east, with an agreeable music, not, indeed, either gay or very varied, but of a character which suited well with that of the place. I wrought directly under it, and frequently paused in my labours to hearken the blast moaning amid its spires, and whistling through its apertures ; and I have occasionally been startled by the mingling deathlike tones produced by the hammer, when forced by the wind against the sides of the bell. I was one day listening to this music, when, by one of those freaks which fling the light of recollection upon the dark recesses of the past, much in the manner that I have seen a child throwing the gleam of a mirror from the sunshine into the shade, there were brought before me the circumstances of a dream, deemed prophetic of the death of him whose epitaph I was then inscribing. It was one of those auguries of contingency which, according to Bacon, men mark when they hit, and never mark when they miss.

In the latter part of 1822 a young lad, a mason's apprentice, was employed with his master in working within the policies of Pointzfield—a gentleman's seat about a mile from the burying-ground. He wished much to visit the tombs and chapel, but could find no opportunity; for the day had so shortened that his employments engaged him from the first peep of light in the morning until half an hour after sunset. And perhaps the wish was the occasion of the dream. He had no sooner fallen asleep, after the fatigues of the day, than he found himself approaching the chapel in one of the finest of midsummer evenings. The whole western heavens were suffused with the blush of sunset—the hills, the woods, the fields, the sea, all the limbs and members of the great frame of nature, seemed enveloped in a mantle of beauty. He reached the burying-ground, and deemed it the loveliest spot he had ever seen. The tombs were finished after the most exquisite designs, chastely Grecian, or ornately Gothic; and myriads of flowering shrubs winded around the urns, and shaded the tablets in every disposition of beauty. The building seemed entire, but it was so encrusted with moss and lichens as to present an appearance of extreme antiquity; and on the western gable there was fixed a huge gnomon of bronze, fantastically carved like that of an antique dial, and green with the rust of ages. Suddenly a low breeze began to moan through the shrubs and bushes, the heavens became overcast, and the dreamer, turning towards the building, beheld with a sensation of fear the gnomon revolving slowly as on an axis, until the point rested significantly on the sward. He fled the place in deep horror, the night suddenly fell, and when floundering on in darkness and terror, through a morass that stretches beyond the southern wall of the chapel, he awoke, and lo! it was a dream. Only five weeks elapsed from this evening, until he followed to the burying-ground the corpse of a relative, and saw that the open grave occupied the identical spot on which the point of the gnomon had rested.

During the course of the week which I spent in the burying-ground, I became acquainted with several interesting traditions connected with its mute inhabitants. There are some of these which show how very unlike the beliefs entertained in the ages which have departed, are to those deemed rational in the present; while there are others which render it evident that though men at different eras think and believe differently, human nature always remains the same. The following partakes in part of the character of both.

There lived, about a century ago, in the upper part of the parish of Cromarty, an elderly female of that disposition of mind which Bacon describes as one of the very errors of human nature. Her faculties of enjoyment and suffering seemed connected by some invisible tie to the fortunes of her neighbours; but the tie, unlike that of sympathy, which binds pleasure to pleasure, and sorrow to sorrow, united by a strange perversity the opposite feelings; for she was happy when the people around her were unfortunate, and miserable when they prospered. So decided a misanthropy was met by a kindred feeling in those acquainted with her; nor was she regarded with only that abhorrence which attaches to the evil wish, and the malignant intention, but also with the contempt due to that impotency of malice which can only wish and intend.

Her sphere of mischief, however, though limited by her circumstances, was occupied to its utmost boundary; and she frequently made up for her want of power by an ingenuity, derived from what seemed an almost instinctive knowledge of the weaknesses of human nature. It was difficult to tell how she effected her schemes, but certain it was that in her neighbourhood lovers became estranged, and families divided. Late in the autumn of her last year, she formed one of a band of reapers employed in cutting down the crops of a Cromarty farmer. Her partner on the ridge was a poor widow, who had recently lost her husband, and who, though wasted by grief and sickness, was now toiling for her three fatherless children.

Every person on the field pitied her but one ; and the malice of even that one, perverted as her dispositions were, would probably have been disarmed by the helplessness of its object, had it not chanced, that about five years before, when the poor woman and her deceased husband were on the eve of their marriage, she had attempted to break off the match, by casting some foul aspersions on her character. Those whom the wicked injure, says the adage, they never forgive ; and with a demoniac abuse of her knowledge of the dispositions of the people with whom she wrought, she strained beyond her strength to get ahead of them, knowing that a competition would necessarily take place, in which, she trusted, the widow would have either to relinquish her employment as above her strength, or so exhaust herself in the contest as to relapse into sickness. The expected struggle ensued, but, to the surprise of every one, the widow kept up her place in the foremost rank until evening, when she appeared less fatigued than almost any individual of the party. The wretch who had occasioned the contest, and who had fallen behind all the others, seemed dreadfully agitated for the two last hours it continued ; and she was heard by the persons who bound up the sheaves, muttering, during the whole time, words apparently of fearful meaning, which were, however, drowned amid the rustling of the corn, and the hurry and confusion of the competition. Next morning she alone of all the reapers was absent ; and she was found by the widow, who seemed the only one solicitous to know what had become of her, and who first entered her hovel to inquire after her, tossing in the delirium of fever. The poor woman, though shocked and terrified by her ravings and her agony, tended her till within half an hour of midnight, when she expired.

At that late hour a solitary traveller was passing the road which winds along the southern shore of the bay. The moon, in her last quarter, had just risen over the hill on his right, and, half-veiled by three strips of cloud, rather resembled a heap of ignited charcoal seen through the bars of a grate, than the orb which only a few nights before had enabled the reaper to prosecute his employments until near morning. The blocks of granite scattered over the neighbouring beach, and bleached and polished by the waves, were relieved by the moonshine, and resembled flocks of sheep ruminating on a meadow ; but not a single ray rested on the sea beyond, or the path or fields before ; —the beam slided ineffectually along the level ;—it was light looking at darkness. On a sudden, the traveller became conscious of that strange mysterious emotion which, according to the creed of the demonologist, indicates the presence or near approach of an evil spirit. He felt his whole frame as if creeping together, and his hair bristling on his head ; and, filled with a strange horror, he heard, through the dead stillness of the night, a faint uncertain noise, like that of a sudden breeze rustling through a wood at the close of autumn. He blessed himself, and stood still. A tall figure, indistinct in the darkness, came gliding along the road from the east, and inquired of him in a voice hollow and agitated, as it floated past, whether it could not reach Kirk-Michael before midnight ? " No living person could," answered the traveller ; and the appearance, groaning at the reply, was out of sight in a moment. The sounds still continued, as if a multitude of leaves were falling from the boughs of a forest, and striking with a pattering sound on the heaps congregated beneath, when another figure came up, taller, but even less distinct, than the former. It bore the appearance of a man on horseback. " Shall I reach Kirk-Michael before midnight ?" was the query again put to the terrified traveller ; but before he could reply, the appearance had vanished in the distance, and a shriek of torment and despair, which seemed re-echoed by the very firmament, roused him into a more intense feeling of horror. The moon shone out with supernatural brightness ; the noise, which had ceased for a moment, returned, but the sounds were different—for they now seemed to be those of faint laughter, and low indistinct murmurings in the tone of ridicule; and the gigantic rider of a pale horse, with what appeared to be a female shape bent double before him, and accompanied by two dogs, one of which tugged at the head and the other at the feet of the figure, was seen approaching from the west. As this terrible apparition passed the traveller, the moon shone full on the face of the woman bent across the horse, and he distinctly perceived, though the features seemed convulsed with agony, that they were those of the female who, unknown to him, had expired a few minutes before. None of the other stories are of so terrible a character.

Attached to the eastern gable of the ruin, there is a tomb which encloses several monuments; among the rest a plain slab of marble bearing an epitaph, the composition of which would reflect honour on the pen even of Pope. Like most of the other tablets of the burying-ground, it has its history. Somewhat more than fifty years ago, the proprietor of Newhall, an estate in the neighbourhood, was a young man of very superior powers of mind, and both a gentleman and a scholar. When on a visit at the house of his uncle, the proprietor of Invergordon, he was suddenly taken ill, and died a few hours after, leaving behind him a sister, who entertained for him the warmest affection, and the whole of his tenants, who were much attached to him, to regret his loss. He was buried in the family vault of his uncle, who did not long survive him; and whose estate, including the vault, was sold soon after by the next of kin—a circumstance which aggravated, in no slight degree, the grief of his sister. There was one gloomy idea that continually occupied her mind—the idea that even the dust of her brother had, like the earth and stones of his cemetery, become the property of a stranger. Sleeping or waking, the interior of the vault was continually before her. I have seen it. It is a damp melancholy apartment of stone, so dimly lighted that the eye cannot ascertain its extent, with the sides hollowed into recesses, partly occupied by the dead, and with a few rusty iron lamps suspended from the ceiling, that resemble in the darkness a family of vampire bats clinging to the roof of a cavern. A green hillock, covered with moss and daisies, would have supplied the imagination of the mourner with a more pleasing image, and have associated better with the character of the dead.

His sister was the wife of a gentleman who was at that time the proprietor of Braelanguil. One evening, about half a year after the sale of her uncle's property, she was prevailed upon by her husband to quit her apartment, to which she had been confined for months before, and to walk with him in a neighbouring wood. She spoke of the virtues and talents of the deceased, the only theme from which she could derive any pleasure; and she found that evening in her companion a more deep and tender sympathy than usual. The walk was insensibly prolonged, and she was only awakened from her reverie of tenderness and sorrow, by finding herself among the graves of Kirk-Michael. The door of her husband's burying-ground lay open. On entering it, she perceived that a fresh grave had been added to the number of those which had previously occupied the space, and that one of the niches in the wall was filled up by a new slab of marble. It was the grave and monument of her brother. The body had been removed from the vault, and re-interred in this place by her consort; and it would perhaps be difficult to decide whether the more delicate satisfaction was derived by the sister or the husband from the walk of the evening. The epitaph is as follows :—

> What science crown'd him, or what genius blest,
> No flattering pencil bids this stone attest;
> Yet may it witness with a purer pride,
> How many virtues sunk when Gordon died.
> Clear truth and nature, noble rays of mind,
> Open as day, that beam'd on all mankind;
> Warm to oblige, too gentle to offend,
> He never made a foe nor lost a friend.
> Nor yet from fortune's height, or learning's shade,
> It boasts the tribute to his memory paid;

But that around, in grateful sorrow steep'd,
The humble tenants of the cottage wept;
Those simple hearts that shrink from grandeur's blaze,
Those artless tongues that know not how to praise,
Feel and record the worth that hallow here
A friend's remembrance, and a sister's tear.[1]

Half-way between the chapel and the northern wall of the burying-ground, there is a square altar-like monument of hewn ashlar, enclosing in one of its sides a tablet of grey freestone. It was erected about sixty years ago by a baronet of Fowlis to the memory of his aunt, Mrs. Gordon of Ardoch, a woman whose singular excellence of character is recorded by the pen of Doddridge. She was the only sister of three brothers—men who ranked among the best and bravest of their age, and all of whom died in the service of their country—two in the field of battle, the third when pursuing a flying enemy.

The eldest son of the family was Sir Robert Munro, twenty-seventh baronet of Fowlis, a man whose achievements, as recorded by the sober pen of Doddridge, seem fitted to associate

[1] These fine couplets were written, I have since learned, by Henry Mackenzie, " The Man of Feeling," an attached friend of the deceased. Mackenzie has also dedicated to his memory one of his most characteristic *Mirrors*—the ninetieth. After making a few well-turned remarks on the unhappiness of living too long, " I have been led to these reflections," we find him saying, " by a loss I lately sustained in the sudden and un-looked-for death of a friend, to whom, from my earliest youth, I have been attached by every tie of the most tender affection. Such was the confidence that subsisted between us, that in his bosom I was wont to repose every thought of my mind, and every weakness of my heart. In framing him, nature seemed to have thrown together a variety of opposite qualities, which, happily tempering each other, formed one of the most engaging characters I have ever known ;—an elevation of mind, a manly firmness, a Castilian sense of honour, accompanied with a bewitching sweetness, proceeding from the most delicate attention to the feelings of others. In his manners, simple and unassuming ; in the company of strangers, modest to a degree of bashfulness ; yet possessing a fund of knowledge and an extent of ability, which might have adorned the most exalted station. But it was in the small circle of his friends that he appeared to the highest advantage ; there the native benignity of his soul diffused, as it were, a kindly influence on all around him, while his conversation never failed at once to amuse and instruct.

" Not many months ago, I paid him a visit at his seat in a remote part of the kingdom. I found him engaged in embellishing a place, of which I had often heard him talk with rapture, and the beauties of which I found his partiality had not exaggerated. He showed me all the improvements he had made, and pointed out those he had meant to make. He told me all his schemes and all his projects. And while I live I must ever retain a warm remembrance of the pleasure I then enjoyed in his society.

" The day I meant to set out on my return he was seized with a slight indisposition,

rather with ideas derived from the high conceptions of poetry and romance, than with those which we usually acquire from our experience of real life. He was a person of calm wisdom, determined courage, and unassuming piety. On quitting the university, which he did when very young, he passed into Flanders, where he served for several years under Marlborough, and became intimate with the celebrated James Gardiner, then a cornet of dragoons. And the intimacy ripened into a friendship which did not terminate until death; perhaps not even then. On the peace of 1712 he returned to Scotland; and the Rebellion broke out three years after. At the head of his clan, the Munros, in union with the good Earl of Sutherland, he so harassed a body of three thousand Highlanders, who, under the Earl of Seaforth, were on the march to join the insurgents at Perth, that the junction was retarded for nearly two months—a delay which is said to have decided the fate of the Stuarts in Scotland. In the following year he was appointed one of the commissioners of inquiry into the forfeited estates of the attainted; and he exerted himself in this office in erecting parishes in the

which he seemed to think somewhat serious; and indeed, if he had a weakness, it consisted in rather too great anxiety with regard to his health. I remained with him till he thought himself almost perfectly recovered; and, in order to avoid the unpleasant ceremony of taking leave, I resolved to steal away early in the morning, before any of the family should be astir. About daybreak I got up and let myself out. At the door I found an old and favourite dog of my friend's, who immediately came and fawned upon me. He walked with me through the park. At the gate he stopped and looked up wistfully in my face; and though I do not well know how to account for it, I felt at that moment, when I parted with the faithful animal, a degree of tenderness, joined with a melancholy so pleasing, that I had no inclination to check it. In that frame of mind I walked on (for I had ordered my horses to wait me at the first stage) till I reached the summit of a hill, which I knew commanded the last view I should have of the habitation of my friend. I turned to look back on the delightful scene. As I looked, the idea of the owner came full into my mind; and while I contemplated his many virtues, and numberless amiable qualities, the suggestion arose, if he should be cut off, what an irreparable loss it would be to his family, to his friends, and to society. In vain I endeavoured to combat this melancholy foreboding by reflecting on the uncommon vigour of his constitution, and the fair prospect it afforded of his enjoying many days. The impression still recurred, and it was some considerable time before I had strength of mind sufficient to conquer it.

"I had not been long at home, when I received accounts of his being attacked by a violent distemper; and, in a few days after, I learned it had put an end to his life."

remote Highlands, which derived their stipends from the confiscated lands. In this manner, says his biographer, new presbyteries were formed in counties where the discipline and worship of Protestant Churches had before no footing. It is added, that by his influence with Government he did eminent service to the wives and children of the proscribed. He was for thirty years a member of Parliament, and distinguished himself as a liberal consistent Whig—the friend both of the people and of the king. In the year 1740, when the country was on the eve of what he deemed a just war, though he had arrived at an age at which the soldier commonly begins to think of retiring from the fatigues of the military life, he quitted the business of the senate for the dangers of the field, and passed a second time into Flanders. He now held the rank of lieutenant-colonel, and such was his influence over the soldiers under him, and such their admiration of his character, that his spirit and high sense of honour seemed to pervade the whole regiment. When a guard was granted to the people of Flanders for the protection of their property, they prayed that it should be composed of Sir Robert's Highlanders; and the Elector-Palatine, through his envoy at the English court, tendered to George I. his thanks for this excellent regiment; for the sake of whose lieutenant-colonel, it was added, he would for the future always esteem a Scotchman.

The life of Sir Robert resembled a well-wrought drama, whose scenes become doubly interesting as it hastens to a close. In the battle of Fontenoy he was among the first in the field, and having obtained leave that his Highlanders should fight after the manner of their country, he surprised the whole army by a display of extraordinary yet admirable tactics, directed against the enemy with the most invincible courage. He dislodged from a battery, which he was ordered to attack, a force superior to his own, and found a strong body of the enemy, who were stationed beyond it, preparing to open on him a sweeping fire. Commanding his regiment to prostrate itself to avoid the shot, he raised it when the French were in the act of reloading, and, sword in hand, rushed at its head upon them with so irresistible a charge as forced them precipitately through their lines. Then retreating, according to the tactics of his country, he again brought his men to the charge as before, and with similar effect. And this manœuvre of alternate flight and attack was frequently repeated during the day. When after the battle had become general, the English began to give ground before the superior force of the enemy, Sir Robert's regiment formed the rearguard in the retreat. A strong body of French horse came galloping up behind ; but, when within a few yards of the Highlanders, the latter turned suddenly round, and received them with a fire so well directed and effectual, that nearly one-half of them were dismounted. The rest, wheeling about, rode off, and did not again return to the attack. It was observed, that during the course of this day, when the Highlanders had thrown themselves on the ground immediately as the enemy had levelled their pieces for firing, there was one person of the regiment who, instead of prostrating himself with the others, stood erect, exposed to the volley. That one was Sir Robert Munro. The circumstances of his death, which took place about eight months after, at the battle of Falkirk, were adapted to display still more his indomitable heroism of character. He had recently been promoted to the command of a regiment, which, unlike his brave Highlanders at Fontenoy, deserted him in the moment of attack, and left him enclosed by the enemy. Defending himself with his half-pike against six of their number, two of whom he killed, he was not overpowered, though alone, until a seventh coming up shot him dead with a musket. His younger brother, who accompanied the regiment, and who had been borne along by the current of the retreat, returned in time only to witness his fate and to share it.

It has been told me by a friend, who, about forty years ago, resided for some time in the vicinity of Fowlis, that he could have collected, at that period, anecdotes of Sir Robert from among his tenantry sufficient to have formed a volume. They were all of one character :—tints of varied but unequivocal beauty, which animated into the colour and semblance of life the faint outline of heroism traced by Doddridge. There was an old man who used to sit by my friend for hours together narrating the exploits of his chief. He was a tall, upright, greyhaired Highlander, of a warm heart and keen unbending spirit, who had fought at Dettingen, Fontenoy, Culloden, and Quebec. One day, when describing the closing scene in the life of his almost idolized leader, after pouring out his curse on the dastards who had deserted him, he started from his seat, and grasping his staff as he burst into tears, exclaimed in a voice almost smothered by emotion, " Ochon, ochon, had his ain folk been there ! !"

The following anecdote of Sir Robert, which I owe to tradition, sets his character in a very amiable light. On his return from Flanders in 1712, he was introduced to a Miss Jean Seymour, a beautiful English lady. The young soldier was smitten by her appearance, and had the happiness of perceiving that he had succeeded in at least attracting her notice. So happy an introduction was followed up into an intimacy, and at length, what had been only a casual impression on either side, ripened into a mutual passion of no ordinary warmth and delicacy. On Sir Robert's quitting England for the north, he arranged with his mistress the plan of a regular correspondence, and wrote to her immediately on his arrival at Fowlis. After waiting for a reply with all the impatience of the lover, he sent off a second letter complaining of her neglect, which had no better success than the first, and shortly after a third, which shared the fate of the two others. The inference seemed too obvious to be missed ; and he strove to forget Miss Seymour. He hunted, fished, visited his several friends, involved himself in a multiplicity of concerns, but all to no purpose; she still continued the engrossing object of his affections, and, after a few months' stay in the Highlands, he again returned to England, a very unhappy man. When waiting on a friend in London, he was ushered precipitately into the midst of a fashionable party, and found himself in the presence of his mistress. She seemed much startled by the rencounter; the blood mounted to her cheeks; but, suppressing her emotion, she turned to the lady who sat next her, and began to converse on some common topic of the day. Sir Robert retired, and beckoning on his friend, entreated him to procure for him an interview with Miss Seymour, which was effected, and an explanation ensued. The lady had not received a single letter; and forming at length, from the seeming neglect of her lover, an opinion of him similar to that from which she herself was suffering in his esteem, she attempted to banish him from her affections;—an attempt, however, in which she was scarcely more successful than Sir Robert. They were gratified to find that they had not been mistaken in their first impressions of each other, and parted more attached, and more convinced that the attachment was mutual, than ever. And in less than two months after Miss Seymour had become Lady Munro.

Sir Robert succeeded in tracing all his letters to one point, a kind of post-office on the confines of Inverness-shire. There was a proprietor in the neighbourhood, who was deeply engaged in the interests of the Stuarts, and decidedly hostile to Sir Robert, the scion of a family which had distinguished itself from the first dawn of the Reformation in the cause of civil and religious liberty. There was, therefore, little difficulty in assigning an author to the contrivance; but Sir Robert was satisfied in barely tracing it to a discovery; for, squaring his principles of honour rather by the morals of the New Testament than by the dogmas of that code which regards death as the only expiation of insult or injury, he was no duellist. An opportunity, however, soon occurred of his avenging himself in a manner agreeable to his character and principles. On the breaking out of the Rebellion of 1715, the person who had so wantonly sported with his happiness joined the Earl of Mar, and, after the failure of the enterprise, was among the number of the proscribed. Sir Robert's influence with the Government, and the peculiar office to which he was appointed, gave him considerable power over the confiscated properties, and this power he exerted to its utmost in behalf of the wife and children of the man by whom he had been injured. "Tell your husband," said he to the lady, "that I have now repaid him for the interest he took in my correspondence with Miss Seymour."

Sir Robert's second brother (the other, as has been related, died with him at Falkirk) was killed, about seven months after the battle, in the Highlands of Lochaber. His only sister survived him for nearly twenty years, " a striking example (I use the language of Doddridge) of profound submission and fortitude, mingled with the most tender sensibility of temper." She was the wife of a Mr. Gordon of Ardoch (now Pointzfield), whom she survived for several years ; and her later days were spent in Cromarty, where there are still a few elderly people who remember her, and speak of her many virtues and gentle manners with a feeling bordering on enthusiasm. There was a poor half-witted girl who lived in her neighbourhood, known among the town's-people by the name of *Babble* Hanah. The word in italics is a Scottish phrase applied to persons of an idiotical cast of mind, and yet though poor Hanah had no claim to dispute the propriety of its application in her own case, her faint glimmering of reason proved quite sufficient to light her on the best possible track of life. She had learned from revelation of the immortality of the soul, and the two states of the future ; and experience had taught even her, what indeed it would teach every one, did every one but attend to its lessons, that there is a radical depravity in the nature of man, and a continual succession of evil in the course of life. She had learned, too, that she was one of the least wise of a class of creatures exceedingly foolish at best, and that to escape from evil needed much wisdom. She was, therefore earnest in her prayers to the Great Spirit who was so kind to her—and to even those feeble animals who, though they enjoy no boon of after life, have a wisdom to provide for the winter, and to dig their houses in the rocks—that in this world He would direct her walk agreeably to His own will, and render her wiser in the world to come. Socrates could have taught all this to Xenophon and Plato, but God only could have taught it to Hanah. The people of the place, with dispositions like those of the great bulk of people in every place, were much more disposed to laugh at the poor thing for what she wanted, than to form right estimates of the value of what she had. Not so Lady Ardoch ;—Hanah was one of her friends. The lady's house was a place where, in the language of Scripture, "prayer was wont to be made ;" and no one was a more regular attendant on the meetings held for this purpose, than her friend the half-witted girl. The poor thing always sat at her feet, and was termed by her, her own Hanah. Years, however, began to weigh down the frame of the good lady ; and after passing through all the gradations of bodily decay, with a mind which seemed to brighten and grow stronger as it neared to eternity, she at length slept with her fathers. Hanah betrayed no emotion of grief ; she spoke to no one of the friend whom she had lost ; but she moped and pined away, and became indifferent to everything ; and a few months after, when on her deathbed, she told a friend of the deceased who had come to visit her, that she was going to the country of Lady Ardoch.

CHAPTER XXX.

" Rise, honest muse, and sing the man of Ross."—POPE.

IN the letter in which Junius accuses the Duke of Grafton
of having sold a patent place in the collection of customs to one
Mr. Hine, he informs the reader that the person employed by
his Grace in negotiating the business, was " George Ross, the
Scotch Agent and worthy confidant of Lord Mansfield. And
no sale by the candle," he adds, " was ever conducted with
greater formality." Now, slight as this notice is, there is something in it sufficiently tangible for the imagination to lay hold
of. If the reader thinks of the Scotch Agent at all, he probably thinks of him as one of those convenient creatures so necessary to the practical statesman, whose merit does not consist
more in their being ingenious in a great degree, than in their
being honest in a small one. So mixed a thing is poor human
nature, however, that though the statement of Junius has never
been fairly controverted, no possible estimate of character could
be more unjust. The Scotch Agent, whatever may have been
the nature of his services to the Duke of Grafton, was in reality
a high-minded, and, what is more, a truly patriotic man ; so
good a person indeed, that, in a period of political heats and
animosities like the present, his story, fairly told, may teach us
a lesson of charity and moderation. I wish I could transport
the reader to where his portrait hangs, side by side with that
of his friend the Lord Chief-Justice, in the drawing-room of
Cromarty House. The air of dignified benevolence impressed
on the features of the handsome old man, with his grey hair

2 F

curling round his temples, would secure a fair hearing for
him from even the sturdiest of the class who hate their neighbours for the good of their country. Besides, the very presence
of the noble-looking lawyer, so much more like the Murray
eulogized by Pope and Lyttelton, than the Mansfield denounced
by Junius, would of itself serve as a sort of guarantee for the
honour of his friend.

George Ross was the son of a petty proprietor of Easter-Ross,
and succeeded, on the death of his father, to the few barren acres
on which, for a century or two before, the family had been ingenious enough to live. But he possessed besides what was
more valuable than twenty such patrimonies—an untiring energy
of disposition, based on a substratum of sound good sense ; and,
what was scarcely less important than either, ambition enough
to turn his capacity of employment to the best account. Ross-
shire, a century ago, was no place for such a man ; and as the
only road to preferment at this period was the road that led
south, George Ross left, when very young, his mother's cottage
for England, where he spent nearly fifty years amongst statesmen
and courtiers, and in the enjoyment of the friendship of such
men as President Forbes and Lord Mansfield. At length he
returned, when an old greyheaded man, to rank amongst the
greatest capitalists and proprietors of the county ; and purchased,
with other lesser properties in the neighbourhood, the whole
estate of Cromarty. Perhaps he had come to rest him ere he
died ; but there seems to be no such thing as changing one's
natural bent when confirmed by the habits of half a lifetime ;
and the energies of the Scotch Agent, now that they had gained
him fortune and influence, were as little disposed to fall asleep
as they had been forty years before. As it was no longer necessary, however, that they should be employed on his own
account, he gave them full scope in behalf of his poorer neighbours. The country around him lay dead. There were no
manufactories, no knowledge of agriculture, no consciousness

that matters were ill, and consequently no desire of making them better ; and the herculean task imposed upon himself by the Scotch Agent, now considerably turned of sixty, was to animate and revolutionize the whole. And such was his statesmanlike sagacity in developing the hitherto undiscovered resources of the country, joined to a high-minded zeal that could sow liberally, in the hope of a late harvest for others to reap, that he fully succeeded.

He first established in the town an extensive manufactory of hempen cloth, which has ever since employed about two hundred persons within its walls, and fully twice that number without. He next built an ale brewery, which, at the time of its erection, was by far the largest in the north of Scotland. He then furnished the town at a great expense with an excellent harbour, and set on foot a trade in pork, which, for the last thirty years, has been carried on by the people of the place to an extent of from about fifteen to twenty thousand pounds annually. He set himself, too, to initiate his tenantry in the art of rearing wheat ; and finding them wofully unwilling to become wiser on the subject, he tried the force of example, by taking an extensive farm under his own management, and conducting it on the most approved principles of modern agriculture. He established a nail and spade manufactory ; brought women from England to instruct the young girls in the art of working lace ; provided houses for the poor ; presented the town with a neat substantial building, the upper part of which still serves for a council-room and court-house, and the lower as a prison; and built for the accommodation of the poor Highlanders, who came thronging into the town to work on his lands and his manufactories, a handsome Gaelic chapel. He built for his own residence an elegant house of hewn stone; surrounded it with pleasure-grounds designed in the best style of the art; planted many hundred acres of the less improvable parts of his property, and laid open the hitherto scarcely accessible beauties of the hill of

Cromarty, by crossing and recrossing it with well-nigh as many walks as there are veins in the human body. He was proud of his exquisite landscapes, and of his own skill in heightening their beauty, and fully determined, he said, if he but lived long enough, to make Cromarty worth an Englishman's while coming all the way from London to see.

When Oscar fell asleep, says the old Irish bard, it was impossible to awaken him before his time except by cutting off one of his fingers, or flinging a rock at his head; and wo to the poor man who disturbed him! The Agent found it every whit as difficult to awaken a sleeping country, and in some respects almost as unsafe. I am afraid human nature is nearly the same thing in the people that it is in their rulers, and that both are alike disposed to prefer the man who flatters them to the man who merely does them good. George Ross was by no means the most popular of proprietors—he disturbed old prejudices, and unfixed old habits. The farmers thought it hard that they should have to break up their irregular maplike patches of land, divided from each other by little strips and corners not yet reclaimed from the waste, into awkward-looking rectangular fields; and that they durst no longer fasten their horses to the plough by the tail—a piece of natural harness evidently formed for the purpose. The town's-people deemed the hempen manufactory unwholesome, and found that the English lacewomen, who to a certainty were tea-drinkers, and even not very hostile, it was said, to gin, were in a fair way of teaching their pupils something more than the mere weaving of lace. What could be more heathenish, too, than the little temple covered with cockle-shells which the laird had just reared on a solitary corner of the hill; but the temple they soon sent spinning over the cliff into the sea, a downward journey of a hundred yards. And then his odious pork trade! There was no prevailing on the people to rear pigs for him, and so he had to build a range of offices in an out-of-the-way nook of his lands, which he stocked with hordes of these animals, that he might rear them for himself. The herds increased in size and number, and, voracious beyond calculation, almost occasioned a famine. Even the great wealth of the speculatist proved insufficient to supply them with the necessary food, and the very keepers were in danger of being eaten alive. The poor animals seemed departing from their very nature, for they became long, and lank, and bony as the griffins of heraldry, until they looked more like race-horses than pigs ; and as they descended with every ebb in huge droves to browse on the sea-weed, or delve for shell-fish among the pebbles, there was no lack of music befitting their condition, when the large rock-crab revenged with his nippers on their lips the injuries inflicted on him by their teeth. Now, all this formed a fine subject for joking to people who indulged in a half-Jewish dislike of the pig, and who could not guess that the pork trade was one day to pay the rents of half the widows' cottages in the country. But no one could lie more open than George Ross to that species of ridicule which the men who see further than their neighbours, and look more to the advantage of others than to their own, cannot fail to encounter. He was a worker in the dark, and that at no slight expense; for though all his many projects were ultimately found to be benefits conferred on his country, not one of them proved remunerative to himself. But he seems to have known mankind too well to have expected a very great deal from their gratitude; though, on one occasion at least, his patience gave way.

The town in the course of years had so entirely marched to the west, that, as I have already had occasion to remark, the town's cross came at length to be fairly left behind, with a hawthorn hedge on the one side and a garden fence on the other; and when the Agent had completed the house which was to serve as council-room and prison to the place, the cross was taken down from its stand of more than two centuries, and placed in front of the new building. That people might

the better remember the circumstance, there was a showy procession got up; healths were drunk beside the cross in the
Agent's best wine, and not a little of his crystal broken against
it; and the evening terminated in a ball. It so happened,
however, through some cross chance, that, though all the
gentility of the place were to be invited, three young men, who
deemed themselves as genteel as the best of their neighbours,
were passed over—the foreman of the hemp manufactory had
received no invitation, nor the clever superintendent of the
nail-work, nor yet the spruce clerk of the brewery; and as they
were all men of spirit, it so happened that, during the very
next night, the cross was taken down from its new pedestal,
broken into three pieces, and carried still further to the west,
to an open space where four lanes met; and there it was found
in the morning—the pieces piled over each other, and surrounded by a profusion of broken ale-bottles. The Agent was
amazingly angry—angrier, indeed, than even those who best
knew him had deemed him capable of becoming; and in the
course of the day the town's crier went through the streets
proclaiming a reward of ten pounds in hand, and a free room
in Mr. Ross's new buildings for life, to any one who would
give such information as might lead to the conviction of the
offenders.

In one of his walks a few days after, the Agent met with a
poor miserable-looking Highland woman, who had been picking
a few withered sticks out of one of his hedges, and whose
hands and clothes seemed torn by the thorns. "Poor old
creature!" he said, as she dropped her courtesy in passing; "you
must go to my manager and tell him I have ordered you a
barrel of coals. And stay—you are hungry; call at my house
in passing, and the servants will find you something to take
home with you." The poor woman blessed him, and looked
up hesitatingly in his face. She had never betrayed any one,
she said; but his honour was so good a gentleman—so very

good a gentleman; and so she thought she had best tell him all she knew about the breaking of the cross. She lived in a little garret over the room of Jamie Banks the nailer; and having slept scarcely any all the night in which the cross was taken down, for the weather was bitterly cold, and her bed-clothes very thin, she could hear weighty footsteps traversing the streets till near morning, when the house-door opened and in came Jamie with a tottering unequal step, and disturbed the whole family by stumbling over a stool into his wife's washing tub. Besides, she had next day overheard his wife rating him for staying out to so *untimeous* an hour, and his remark in reply, that she would do well to keep quiet unless she wished to see him hanged. This was the sort of clue the affair required, and in following it up, the unlucky nailer was apprehended and examined; but it was found, that, through a singular lapse of memory, he had forgotten every circumstance connected with the night in question, except that he had been in the very best company, and one of the happiest men in the world.

Jamie Banks was decidedly the most eccentric man of his day in at least one parish; full of small wit and little conceits, and famous for a faculty of invention fertile enough to have served a poet. On one occasion when the gill of whisky had risen to three-halfpence in Cromarty, and could still be bought for a penny in Avoch, he had prevailed on a party of his acquaintance to accompany him to the latter place, that they might drink themselves rich on the strength of the old proverb; and as they actually effected a saving of two shillings in spending six, it was clear, he said, that had not their money failed them, they would have made fortunes apiece. Alas for the littleness of that great passion, the love of fame! I have observed that the tradespeople among whom one meets with most instances of eccentricity, are those whose shops, being places of general resort, furnish them with space enough on which to achieve a humble notoriety, by rendering themselves unlike everybody else. To secure to Jamie Banks due leisure for recollection, he was committed to jail.

He was sitting one evening beside the prison fire with one of his neighbours and the jailer, and had risen to exclude the chill night air by drawing a curtain over the open barred window of the apartment, when a man suddenly started from behind the wall outside, and discharged a large stone with tremendous force at his head. The missile almost brushed his ear as it sung past, and, rebounding from the opposite wall, rolled along the floor. " That maun be Rob Williamson!" exclaimed Jamie, " wanting to keep *me* quiet; out, neebour Jonathan, an' after him." Neebour Jonathan, an active young fellow, sprang to the door, caught the sounds of retreating footsteps as he turned the gate, and dashing after like a greyhound, succeeded in laying hold of the coat-skirts of Rob Williamson, as he strained onwards through the gate of the hemp manufactory. He was immediately secured and lodged in another apartment of the prison ; and in the morning Jamie Banks was found to have recovered his memory.

He had finished working, he said, on the evening after the ball, and was just putting on his coat preparatory to leaving the shop, when the superintendent called him into his writing-room, where he found three persons sitting at a table half covered with bottles. Rob Williamson, the weaver, was one of them; the other two were the clerk of the brewery and the foreman of the hemp manufactory; and they were all arguing together on some point of divinity. The superintendent cleared a seat for him beside himself, and filled his glass thrice in succession, by way of making up for the time he had lost—nothing could be more untrue than that the superintendent was proud! They then all began to speak about morals and Mr. Ross; the clerk was certain that, what with his harbour and his piggery, and his heathen temples and his lacewomen, he would not leave a rag of morality in the place; and Rob was quite as sure he was no friend to the gospel. He a builder of Gaelic kirks, forsooth! had he not, yesterday, put up a Popish Dagon of a cross, and made the silly mason bodies worship it for the sake o' a dram? And then, how common ale-drinking had become in the place since he had built his brewery—in his young days they drank naething but gin;—and what would their grandfathers have said to a *whigmaleerie* of a ball! " I sipped and listened," continued Jamie, " and thought the time couldna have been better spent at an elder's meeting in the kirk; and as the night wore later, the conversation became more edifying still, until at length all the bottles were emptied, when we sallied out in a body, to imitate the old reformers by breaking the cross. ' We may suffer, Jamie, for what we have done,' said Rob to me, as we parted for the night; ' but remember it was duty, Jamie—it was duty. We have been testifying wi' our hands, an' when the hour o' trial comes, we mauna be slow in testifying wi' our tongues too.' He wasna slack, the deceitfu' bodie!" concluded Jamie, " in trying to stop mine." And thus closed the evidence. The Agent was no vindictive man; he dismissed his two superintendents and the clerk, to find for themselves a more indulgent master; but the services of Jamie Banks he still retained, and the first employment which he found for him after his release, was the fashioning of four iron bars for the repair of the cross.

The Agent, in the closing scene of his life, was destined to experience the unhappiness of blighted hope. He had an only son, a weak and very obstinate young man, who, without intellect enough to appreciate his well-calculated schemes, and yet conceit enough to sit in judgment on them, was ever showing his spirit by opposing a sort of selfish nonsense, that aped the semblance of common sense, to the expansive and benevolent philosophy of his father. But the old man bore patiently with his conceit and folly. Like the great bulk of the class who attain to wealth and influence through their own exertions, he was anxiously ambitious to live in his posterity, and be the founder of a family; and he knew it was quite as much according to the nature òf things, that a fool might be the father, as that he should be the son, of a wise man. He secured, therefore, his lands to his posterity by the law of entail; did all that education and example could do for the young man; and succeeded in getting him married to a sweet amiable Englishwoman, the daughter of a bishop. But, alas! his precautions and the hopes in which he indulged, proved equally vain. The young man, only a few months after his marriage, was piqued when at table by some remark of his father regarding his mode of carving—some slight allusion, it is said, to the maxim, that little men cannot afford to neglect little matters; and rising with much apparent coolness from beside his wife, he stepped into an adjoining room, and there blew out his brains with a pistol. The stain of his blood may still be seen in two large brownish-coloured blotches on the floor.

It was impossible that so sad an event should have occurred in this part of the country fifty years ago, without exciting as marked an interest in the supernatural world as in our own. For weeks before, strange unearthly sounds had been heard after nightfall from among the woods of the hill. The forester, when returning homewards in the stillness of evening, had felt the blood curdling round his heart, as low moans, and faint mutterings, and long hollow echoes, came sounding along the pathways, which then winded through the thick wood like vaulted passages through an Egyptian cemetery; and boys of the town who had lingered among the thickets of the lower slopes until after sunset, engaged in digging sweet-knots or pig-nuts, were set a-scampering by harsh sudden screams and loud whistlings, continued in one unvaried note for minutes together. On the evening that preceded the commission of the rash act, a party of school-boys had set out for the hill to select from among the young firs some of the straightest and most slender, for fishing-rods ; and aware that the forester might have serious objections to any such appropriation of his master's property, they lingered among the rocks below till the evening had set in ; when they stole up the hill-side, and applied themselves to the work of choosing and cutting down, in a beautiful little avenue which leads from the edge of the precipices into the recesses of the wood.   All at once there arose, as if from the rock-edge, a combination of the most fearful sounds they had ever heard ;—it seemed as if every bull in the country had congregated in one little spot, and were bellowing together in horrid concert.   The little fellows looked at one another, and then, as if moved by some general impulse—for they were too panic-struck to speak—they darted off together like a shoal of minnows startled from some river-side by a shadow on the bank. The terrible sounds waxed louder and louder, like the sounds of the dread horn which appalled Wallace at midnight in the deserted fortress, after the death of Faudon ; and, long ere they had reached the town, the weaker members of the party began to fall behind.   One little fellow, on finding himself left alone, began to scream in utter terror, scarce less loudly than the mysterious bellower in the wood ; but he was waited for by a bold, hardy boy—a grandchild and name-son of old Sandy Wright the boatman—who had not even relinquished his rod, and who afterwards did his country no dishonour when, in like fashion, he grasped his pike at the landing in Egypt.   To him I owe the story.   He used to say, it was not until he had reached with his companions the old chapel of St. Regulus, a full mile from the avenue, that the sounds entirely ceased.   They were probably occasioned by some wandering bittern, of that species whose cry is said by naturalists to resemble the interrupted bellowings of a bull, but so much louder that it may be distinctly heard at a mile's distance.

George Ross survived his son for several years, and he continued, though a sadder and graver man, to busy himself with all his various speculations as before. It was observed, however, that he seemed to care less than formerly for whatever was exclusively his own—for his fine house and his beautiful lands —and that he chiefly employed himself in maturing his several projects for the good of his country-folk. Time at length began to set his seal on his labours, by discovering their value ; though not until death had first affixed *his* to the character of the wise and benevolent projector. He died full of years and honour, mourned by the poor, and regretted by every one ; and even those who had opposed his innovations with the warmest zeal, were content to remember him, with all the others, as " the good laird."

CHAPTER XXXI.

" Friends, *No-man* kills me; *No-man* in the hour
Of sleep oppresses me with fraudful power.

If no man hurt thee, but the hand divine
Inflict disease, it fits thee to resign."—ODYSSEY.

SOME of the wildest and finest pieces of scenery in the neighbourhood of Cromarty, must be sought for in an upper corner
of the parish, where it abuts on the one hand on the parish of
Rosemarkie, and on the other on the Moray Firth. We may
saunter in this direction over a lonely shore, overhung by picturesque crags of yellow sandstone, and roughened by so fantastic
an arrangement of strata, that one might almost imagine the
riblike bands, which project from the beach, portions of the
skeleton of some huge antediluvian monster. No place can be
more solitary, but no solitude more cheerful. The natural
rampart, that rises more than a hundred yards over the shore,
as if to shut us out from the world, sweeps towards the uplands
in long grassy slopes and green mossy knolls ;—or juts out into
abrupt and weathered crags, crusted with lichens and festooned
with ivy ;—or recedes into bosky hollows, roughened by the
sloethorn, the wild-rose, and the juniper. On the one hand,
there is a profusion of the loveliest light and shadow—the
softest colours and the most pleasing forms ; on the other, the
wide extent of the Moray Firth stretches out to the dim horizon,
with all its veinlike currents and its undulating lines of coast ;
while before us we see far in the distance the blue vista of the
great Caledonian valley, with its double wall of jagged and

serrated hills; and directly in the opening the grey diminished spires of Inverness. We saunter onwards towards the west, over the pebbles and the shells, till where a mossy streamlet comes brattling from the hill; and see, on turning a sudden angle, the bank cleft to its base, as if to yield the waters a passage. 'Tis the entrance to a deeply-secluded dell, of exquisite though savage beauty; one of those hidden recesses of nature, in which she gratefully reserves the choicest of her sweets for the more zealous of her admirers; and mingles for them in her kindliest mood all that expands and delights the heart in the contemplation of the wild and beautiful, with all that gratifies it in the enjoyment of a happy novelty, in which pleasure comes so unlooked for, that neither hope nor imagination has had time to strip it of a single charm.

We enter this singular recess along the bed of the stream, and find ourselves shut out, when we have advanced only a few paces, from well-nigh the entire face of nature and the whole works of man. A line of mural precipices rises on either hand —here advancing in gigantic columns, like those of an Egyptian portico—there receding into deep solitary recesses tapestried with ivy, and darkened by birch and hazel. The cliffs vary their outline at every step, as if assuming in succession all the various combinations of form which constitute the wild and the picturesque; and the pale yellow hue of the stone seems, when brightened by the sun, the very tint a painter would choose to heighten the effect of his shades, or to contrast most delicately with the luxuriant profusion of bushes and flowers that waves over every shelve and cranny. A colony of swallows have built from time immemorial in the hollows of one of the loftiest precipices; the fox and the badger harbour in the clefts of the steeper and more inaccessible banks. As we proceed, the dell becomes wilder and more deeply wooded, the stream frets and toils at our feet—here leaping over an opposing ridge, there struggling in a pool, yonder escaping to the light from under some broken fragment of cliff—there is a richer profusion of flowers; a thicker mantling of ivy and honeysuckle;—and, after passing a semicircular inflection of the bank, which, waving from summit to base with birch and hawthorn, seems suited to remind one of some vast amphitheatre on the morning of a triumph, we find the passage shut up by a perpendicular wall of rock about thirty feet in height, over which the stream precipitates itself in a slender column of foam into a dark mossy basin. The long arms of an intermingled clump of birches and hazels stretch half-way across, trebling with their shade the apparent depth of the pool, and heightening in an equal ratio the whole flicker of the cascade, and the effect of the little bright patches of foam which, flung from the rock, incessantly revolve on the eddy.

There is a natural connexion, it is said, between wild scenes and wild legends; and some of the traditions connected with this romantic and solitary dell illustrate the remark. Till a comparatively late period, it was known at many a winter fireside as a favourite haunt of the fairies, the most poetical of all our old tribes of spectres, and at one time one of the most popular. I have conversed with an old woman, one of the perished volumes of my library, who, when a very little girl, had seen myriads of them dancing as the sun was setting on the further edge of the dell; and with a still older man, who had the temerity to offer one of them a pinch of snuff at the foot of the cascade. Nearly a mile from where the ravine opens to the sea, it assumes a gentler and more pastoral character; the sides, no longer precipitous, descend towards the stream in green sloping banks; and a beaten path, which runs between Cromarty and Rosemarkie, winds down the one side and ascends the other. More than sixty years ago, one Donald Calder, a shopkeeper of Cromarty, was journeying by this path shortly after nightfall. The moon, at full, had just risen, but there was a silvery mist sleeping on the lower grounds that obscured the light, and the dell in all its extent was so overcharged by the vapour, that it seemed an immense overflooded river winding through the landscape. Donald had reached its further edge, and could hear the rush of the stream from the deep obscurity of the abyss below, when there rose from the opposite side a strain of the most delightful music he had ever heard. He stood and listened: the words of a song of such simple beauty, that they seemed, without effort on his part, to stamp themselves on his memory, came wafted on the music, and the chorus, in which a thousand tiny voices seemed to join, was a familiar address to himself. " He! Donald Calder! ho! Donald Calder!" There are none of my Navity acquaintance, thought Donald, who sing like that; "Wha can it be?" He descended into the cloud; but in passing the little stream the music ceased; and on reaching the spot on which the singers had seemed stationed, he saw only a bare bank sinking into a solitary moor, unvaried by either bush or hollow, or the slightest cover in which the musician could have lain concealed. He had hardly time, however, to estimate the marvels of the case when the music again struck up, but on the opposite side of the dell, and apparently from the very knoll on which he had so lately listened to it; the conviction that it could not be other than supernatural overpowered him, and he hurried homewards under the influence of a terror so extreme, that, unfortunately for our knowledge of fairy literature, it had the effect of obliterating from his memory every part of the song except the chorus. The sun rose as he reached Cromarty; and he found that, instead of having lingered at the edge of the dell for only a few minutes —and the time had seemed no longer—he had spent beside it the greater part of the night.

Above the lower cascade the lofty precipitous banks of the dell recede into a long elliptical hollow, which terminates at the upper extremity in a perpendicular precipice, half cleft to its base by a narrow chasm, out of which the little stream comes

bounding in one adventurous leap to the bottom.   A few birch
and hazel bushes have anchored in the crannies of the rock, and
darkened by their shade an immense rounded block of granite
many tons in weight, which lies in front of the cascade.   Immediately beside the huge mass, on a level grassy spot, which
occupies the space between the receding bank and the stream,
there stood about a century ago a meal-mill.   It was a small
and very rude erection, with an old-fashioned horizontal water-
wheel, such as may still be met with in some places of the
remote Highlands ; and so inconsiderable was the power of the
machinery, that a burly farmer of the parish, whose bonnet a
waggish neighbour had thrown between the stones, succeeded
in arresting the whole with his shoulder until he had rescued
his Kilmarnock.   But the mill of Eathie was a celebrated mill
notwithstanding.   No one resided near it, nor were there many
men in the country who would venture to approach it an hour
after sunset ; and there were nights when, though deserted by
the miller, its wheels would be heard revolving as busily as
ever they had done by day, and when one who had courage
enough to reconnoitre it from the edge of the dell, might see
little twinkling lights crossing and recrossing the windows in
irregular but hasty succession, as if a busy multitude were employed within.   On one occasion the miller, who had remained
in it rather later than usual, was surprised to hear outside the
neighing and champing of horses and the rattling of carts, and
on going to the door he saw a long train of basket-woven
vehicles laden with sacks, and drawn by shaggy little ponies of
every diversity of form and colour.   The attendants were slim
unearthly-looking creatures, about three feet in height, attired
in grey, with red caps ; and the whole seemed to have come out
of a square opening in the opposite precipice.   Strange to relate,
the nearer figures seemed to be as much frightened at seeing
the miller as the miller was at seeing them ; but, on one of
them uttering a shrill scream, the carts moved backwards into

2 G

the opening, which shut over them like the curtain of a theatre as the last disappeared.

There lived in the adjoining parish of Rosemarkie, when the fame of the mill was at its highest, a wild unsettled fellow, named M'Kechan. Had he been born among the aristocracy of the country, he might have passed for nothing worse than a young man of spirit; and after sowing his wild oats among gentlemen of the turf and of the fancy, he would naturally have settled down into the shrewd political landlord, who, if no builder of churches himself, would be willing enough to exert the privilege of giving clergymen, exclusively of his own choosing, to such churches as had been built already. As a poor man, however, and the son of a poor man, Tam M'Kechan seemed to bid pretty fair for the gallows; nor could he plead ignorance that such was the general opinion. He had been told so when a herd-boy; for it was no unusual matter for his master, a farmer of the parish, to find him stealing pease in the corner of one field, when the whole of his charge were ravaging the crops of another. He had been told so too when a sailor, ere he had broken his indentures and run away, when once caught among the casks and packages in the hold, ascertaining where the Geneva and the sweetmeats were stowed. And now that he was a drover and a horse-jockey, people, though they no longer told him so, for Tam had become dangerous, seemed as certain of the fact as ever. With all his roguery, however, when not much in liquor he was by no means a very disagreeable companion; few could match him at a song or the bagpipe, and though rather noisy in his cups, and somewhat quarrelsome, his company was a good deal courted by the bolder spirits of the parish, and among the rest by the miller. Tam had heard of the piebald horses and their ghostly attendants; but without more knowledge than fell to the share of his neighbours, he was a much greater sceptic, and after rallying the miller on his ingenuity and the prettiness of his fancy, he volunteered to spend a night at the mill, with no other companion than his pipes.

Preparatory to the trial the miller invited one of his neighbours, the young farmer of Eathie, that they might pass the early part of the evening with Tam ; but when, after an hour's hard drinking, they rose to leave the cottage, the farmer, a kind-hearted lad, who was besides warmly attached to the jockey's only sister, would fain have dissuaded him from the undertaking. "I've been thinking, Tam," he said, "that flyte wi' the miller as ye may, ye would better let the *good people* alone ;—or stay, sin' ye are sae bent on playing the fule, I'll e'en play it wi' you ;—rax me my plaid ; we'll trim up the fire in the killogie thegether ; an' you will keep me in music." "Na, Jock Hossack," said Tam, "I maun keep my good music for the *good people*, it's rather late to flinch now ; but come to the burn-edge wi' me the night, an' to the mill as early in the morning as ye may ; an' hark ye, tak a double caulker wi' you." He wrapt himself up closely in his plaid, took the pipes under his arm, and, accompanied by Jock and the miller, set out for the dell, into which, however, he insisted on descending alone. Before leaving the bank, his companions could see that he had succeeded in lighting up a fire in the mill, which gleamed through every bore and opening, and could hear the shrill notes of a pibroch mingling with the dash of the cascade.

The sun had risen high enough to look aslant into the dell, when Jock and the miller descended to the mill, and found the door lying wide open. All was silent within ; the fire had sunk into a heap of white ashes, though there was a bundle of fagots untouched beside it, and the stool on which Tam had been seated lay overturned in front. But there were no traces of Tam, except that the miller picked up, beside the stool, a little flat-edged instrument, used by the unfortunate jockey in concealing the age of his horses by effacing the marks on their teeth, and that Jock Hossack found one of the drones of his pipes among the extinguished embers.   Weeks passed away and there was still nothing heard of Tam ; and as every one seemed to think it would be in vain to seek for him anywhere but in the place where he had been lost, Jock Hossack, whose marriage was vexatiously delayed in consequence of his strange disappearance, came to the resolution of unravelling the mystery, if possible, by passing a night in the mill.

For the first few hours he found the evening wear heavily away ; the only sounds that reached him were the loud monotonous dashing of the cascade, and the duller rush of the stream as it swept past the mill-wheel.   He piled up fuel on the fire till the flames rose half-way to the ceiling, and every beam and rafter stood out from the smoke as clearly as by day ; and then yawning, as he thought how companionable a thing a good fire is, he longed for something to amuse him.   A sudden cry rose from the further gable, accompanied by a flutter of wings, and one of the miller's ducks, a fine plump bird came swooping down among the live embers.   "Poor bird!" said Jock, "from the fox to the fire ; I had almost forgotten that I wanted my supper."   He dashed the duck against the floor—plucked and embowelled it—and then, suspending the carcass by a string before the fire, began to twirl it round and round to the heat. The strong odoriferous fume had begun to fill the apartment, and the drippings to hiss and sputter among the embers, when a burst of music rose so suddenly from the green without, that Jock, who had been so engaged with the thoughts of his supper as almost to have forgotten the fairies, started half a yard from his seat.   "That maun be Tam's pipes," he said ; and giving a twirl to the duck he rose to a window.   The moon, only a few days in her wane, was looking aslant into the dell, lighting the huge melancholy cliffs with their birches and hazels, and the white flickering descent of the cascade.   The little level green on the margin of the stream lay more in the shade ; but Jock could see that it was crowded with figures marvellously diminutive in stature, and that nearly one-half of them were engaged in dancing. It was enough for him, however, that the music was none of Tam's making; and, leaving the little creatures to gambol undisturbed, he returned to the fire.

He had hardly resumed his seat when a low tap was heard at the door, and shortly after a second and a third. Jock sedulously turned his duck to the heat, and sat still. He had no wish for visitors, and determined on admitting none. The door, however, though firmly bolted, fell open of itself, and there entered one of the strangest-looking creatures he had ever seen. The figure was that of a man, but it was little more than three feet in height; and though the face was as sallow and wrinkled as that of a person of eighty, the eye had the roguish sparkle and the limbs all the juvenile activity of fourteen. " What's your name, man?" said the little thing, coming up to Jock, and peering into his face till its wild elfish features were within a few inches of his. "What's your name?" "*Mysel' an' Mysel'*," —*i. e.*, myself—said Jock, with a policy similar to that resorted to by Ulysses in the cave of the giant. " Ah, *Mysel' an' Mysel' !*" rejoined the creature; " *Mysel' an' Mysel' !* and what's that you have got there, *Mysel' an' Mysel' ?*" touching the duck as it spoke with the tip of its finger, and then transferring part of the scalding gravy to the cheek of Jock. Rather an unwarrantable liberty, thought the poor fellow, for so slight an acquaintance; the creature reiterated the question, and dabbed Jock's other cheek with a larger and still more scalding application of the gravy. " What is it?" he exclaimed, losing in his anger all thought of consequences, and dashing the bird, with the full swing of his arm, against the face of his visitor, " It's that!" The little creature, blinded and miserably burnt, screamed out in pain and terror till the roof rung again; the music ceased in a moment, and Jock Hossack had barely time to cover the fire with a fresh heap of fuel, which for a few seconds reduced the apartment to total darkness, when the

crowd without came swarming like wasps to every door and
window of the mill. "Who did it, Sanachy—who did it?"
was the query of a thousand voices at once. "Oh, 'twas
Mysel' an' Mysel'," said the creature; "'twas *Mysel' an'
Mysel'*." "And if it was yoursel' and yoursel', who, poor
Sanachy," replied his companions, "can help that?" They
still, however, clustered round the mill; the flames began to rise
in long pointed columns through the smoke, and Jock Hossack
had just given himself up for lost, when a cock crew outside the
building, and after a sudden breeze had moaned for a few
seconds among the cliffs and the bushes, and then sunk in the
lower recesses of the dell, he found himself alone. He was
married shortly after to the sister of the lost jockey, and never
again saw the *good people*, or, what he regretted nearly as little,
his unfortunate brother-in-law. There were some, however,
who affirmed, that the latter had returned from fairyland seven
years after his mysterious disappearance, and supported the
assertion by the fact, that there was one Thomas M'Kechan
who suffered at Perth for sheep-stealing a few months after the
expiry of the seventh year.

One other tradition of the burn of Eathie, and I have done.
But I need run no risk of marring it in the telling. More
fortunate than most of its contemporaries, it has been preserved by the muse of one of those forgotten poets of our
country, who, thinking more of their subjects than of themselves, "saved others' names and left their own unsung." And
I have but to avail myself of his ballad.

FAUSE JAMIE.

PART FIRST.

" Whar hae ye been, my dochter deir,
I' the cauld an' the plashy weet ?
There's snaw i' the faulds o' your silken hair,
An' bluid on your bonny feet.

There's grief and fright, my dochter deir,
　I' the wand'rin' blink o' your ee;
An' ye've stayed arout i' the sleet an' the cauld
　The livelang nicht frae me."

" O mither deir ! mak' ye my bed,
　For my heart it's flichtin' sair;
An' oh ! gin I've vexed ye, mither deir,
　I'll never vex ye mair.

I've stayed arout the lang mirk nicht,
　I' the sleet an' the plashy rain;
But, mither deir, mak' ye my bed,
　An' I'll ne'er gang out again.

An' oh, put by that maiden snood,
　Whar nane may evir see;
For Jamie's ta'en a richer joe,
　An' left but shame to me."

An' she has made her dochter's bed,
　An' her auld heart it was wae;
For as the lang mirk hours gaed by,
　Her lassie wore away.

The dead wirk i' her bonny hause
　Was wirkin' a' that day an' nicht;
An' or the morning she was gane,
　Wi' the babe that nevir saw the licht.

The mither grat by her dochter's bed,
　An' she has cursed curses three:
That he wha wrocht her deidly ill
　Ane happy man mocht never be.

———

FAUSE JAMIE.

PART SECOND.

There was licht i' the widow's lonesome shiel,
　An' licht i' the farmer's ha';
For the widow was sewin' her dochter's shroud,
　An' the bride's folk dancin' a'.

But aye or the tither reel was danced out,
　The wae bridegroom begoud to tire;
An' a spale on the candil turn'd to the bride,
　An' a coffin loup'd frae the fire.

An' whan to the kirkin' the twasome went
   Sae trig, i' the burrow's toune below,
Their first feet as they left the kirk
   Was the burial o' Jamie's joe.

Jamie he labourt air an' late,
   An' mickle carit for pleugh an' kye;
But laigher aye he sank i' the warl'
   As the weary years gaed by.

His puir gudewife was dowie an' wae;
   His threesome bairns a grief to see—
The tane it was deaf, the tither blin',
   The third a lamiter like to be.

The burns were rinnin' big wi' spate,
   Lentron win's blew gurly and snell;
Whan Jamie cam to Cromartie town
   Wi' a cart o' bear to sell.

" O why do ye daidle so late i' the toune,
   Jamie, it's time ye were boune to ride ?"
" It's because that I dinna like to gang,
   An' I kenna how to bide.

Pic-mirk nicht it's settin' in,
   The wife at hame sits dowie and wae;
An', Elder, I maunna bide i' the toune,
   An' I kenna how to gae.

It saw'd on my rigs or the drouchts cam on,
   It milk'd i' my byre or my kye did dee;
It follows me aye wharevir I gang,
   An' I see *it* now though ye canna see."

" Gin it follows ye aye wharevir ye gang,
   There's anither Jamie that follows ye too;
An' gin that ye nevir wrangit the dead,
   The dead will nevir be mastir o' you."

Jamie he gripit the elder's han',
   An' syne he slackit the branks to ride,
An' doun he gied to the Eathie burn;
   But he nevir cam up on the ither side.

There's a maisterless colly at Jamie's door,
   Eerie it manes to the wife arin,
There's a gled an' a craw on the Eathie crag,
   And a broken corp at the fit o' the linn.

CHAPTER XXXII.

" —— He heard amazed, on every side
His church insulted, and her priests belied,
The laws reviled, the ruling powers abused,
The land derided, and her foes excused,
He heard and ponder'd. What to men so vile
Should be his language ? For his threatening style
They were too many. If his speech were meek,
They would despise such vain attempts to speak :
—— These were reformers of each different sort."—CRABBE.

IN former times people knocked one another on the head for the sake of their masters—fellows whom they had made too great to care at all about them ; in the present age they have become so much wiser, that they quarrel on their own behalf alone. An entire people might be regarded in the past as an immense engine, with perhaps a single mind for its moving power ; we may now compare every petty district to a magazine, stored like the warehouse of a watchmaker with little detached machines, each one furnished with a moving power of its own. But though politics and party spirit change almost every ten years, human nature is always the same ;—aspects vary, and circumstances alter, but the active principle, through all its windings and amid all its disguises, is ever consistent with itself.

The people of Cromarty who lived ninety years ago were quite as unskilled in politics as their neighbours, and thought as little for themselves. They were but the wheels and pinions of an immense engine ; and regarding their governors as men sent into the world to rule—themselves, as men born to obey

—they troubled their heads no more about the matter. Even the two Rebellions had failed of converting them into politicians; for, viewing these in only their connexion with religion, they exulted in the successes of Hanover as those of Protestantism, and identified the cause of the Stuarts with Popery and persecution. Their Whiggism was a Whiggism of the future world only; and the liberty of preparing themselves for heaven was the only liberty they deemed worth fighting for.

Principles such as these, and the dominancy of the Protestant interest, rendered the people of Cromarty, for two whole reigns, as quiet subjects as any in the kingdom. In latter times, too, there was a circumstance which thoroughly attached them to the Government, by shutting out from among them the Radicalism of modern times for well-nigh a whole age. The Scotch, early in the reign of George III., had risen high at court;— Earl Bute had become Premier, and Mansfield Lord Chief-Justice; and the English, who would as lief have witnessed the return of William and his Normans, grumbled exceedingly. The Premier managed his business like most other premiers;— the Chief-Justice conducted *his* rather better than most other chief-justices; but both gentlemen, says Smollett, " had the misfortune of being born natives of North Britain; and this circumstance was, in the opinion of the people, more than sufficient to counterbalance all the good qualities which human nature could possess." Junius, and Wilkes, and Churchhill, and hundreds more, who, with as much ill-nature, but less wit, were forgotten as soon as the public ceased to be satisfied with ill-nature alone, opened in full cry against the King, the Ministry, and the Scotch. The hollo reached Cromarty, and the town's-folk were told, with all the rest of their countrymen, that they were proud, and poor, and dirty, and not very honest, and that they had sold their King; all this, too, as if they hadn't known the whole of it before. Now it so happened, naturally enough I suppose, that they could bear to be dirty, but not to be told of it, and poor, but not to be twitted with their poverty, and that they could be quite as angry as either Junius or Church- hill, though they could not write letters like the one, nor make verses like the other. And angry they were—desperately angry at Whiggism and the English, and devotedly attached to the King, poor man, who was suffering so much for his attachment to the Scotch. Nothing could come amiss to them from so thorough a friend of their country ; and when, on any occasion, they could not wholly defend his measures, they contented them- selves with calling him an honest man.

On came the ill-fated, ill-advised American War, and found the people of Cromarty as loyal as ever. Washington, they said, was a rascal ; Franklin, an ill-bred mechanic ; and the people of the United States, rebels to a man. There was a ballad, the composition of some provincial poet of this period, which narrated, in very rude verse, the tragical death of two brothers, natives of Ross-shire, who were killed unwittingly by their father, a soldier of the Republic; and this simple ballad did more for the cause of the King among the people of Cromarty, than all the arguments in Locke could have done for that of the Americans ; there was not an old woman in either town or parish who did not thoroughly understand it. The unfortunate father, Donald Munro, had emigrated to America, says the ballad, many years before ; leaving his two infant sons with his brother, a farmer of Ross-shire. The children had shot up into active young men, when the war broke out ; and, unable to pay for their passage, had enlisted into a regiment destined for the colonies, in the hope of meeting with their father. They landed in America ; and finding themselves one evening, after a long and harassing march, within a few miles of the place where he resided, they set out together to pay him a visit ; but in pass- ing through a wood on their way, they were shot at from among the trees, and with so fatal an aim that the one was killed, and the other mortally wounded. A stout elderly man, armed with

a double-barrelled rifle, came pressing towards them through
the bushes, as a fowler would to the game he had just knocked
down. It was their father, Donald Munro ; and the ballad
concludes with the ravings of his horror and despair on ascertaining the nature of his connexion with his victims, blent with
the wild expressions of his grief and remorse for having joined
in so unnatural a rebellion.

Even in this age, however, as if to show that there can be
nothing completely perfect that has human nature in it, Cromarty had its one Whig ;—a person who affirmed that Franklin
was a philosopher, and Washington a good man, and that the
Americans were very much in the right. Could anything be
more preposterous ? The town's-folk lacked patience to reason
with a fellow so amazingly absurd. He was a slater, and his
name was John Holm ;—a name which became so proverbial
in the place for folly, that, when any one talked very great
nonsense, it was said of him that he talked like John Holm.
The very children, who had carried the phrase with them to the
play-ground and the school, used to cut short the fudge of a
comrade, or, at times, even some unpopular remark of the
master, with a " Ho ! ho! John Holm !" John, however, held
stiffly to his opinions, and the defence of Washington ; and
some of the graver town's-men, chafed by his pertinacity, were
ill-natured enough to say that he was little better than Washington himself. Curious as it may appear, he was, notwithstanding the modern tone of his politics, a rare and singular
piece of antiquity ;—one of that extinct class of mechanics described by Coleridge, " to whom every trade was an allegory,
and had its own guardian saint." He was a connecting link
between two different worlds—the worlds of popular opinion
and of popular mystery ; and, strange as it may seem, both a
herald of the Reform Bill, and a last relic of the age " in which "
(to use the language of the writer just quoted) " the detail of
each art was ennobled in the eyes of its professors, by being

spiritually improved into symbols and mementos of all doctrines and all duties." John had, besides, a strong turn for military architecture, and used to draw plans and construct models. He was one evening descanting to an old campaigner on the admirable works at Fort George (a very recent erection at that time), and illustrating his descriptions with his stick on a hearth-stone strewed over with ashes, when by came the cat, and with one sweep of her tail demolished the entire plan. " Och, Donald !" said John, " it's all in vain ;" a remark which, simple as it may seem, passed into a proverb. When an adventure proved unsuccessful, or an effort unavailing, it was said to be " All in vain, like John Holm's plan of the fort." But John's day was at hand.—We, the people, are excellent fellows in our way, but I must confess not very consistent. I have seen the principles which we would hang a man for entertaining at the beginning of one year, becoming quite our own before the end of the next.

The American War was followed by the French Revolution, and the crash of a falling throne awakened opinion all over Europe. The young inquired whether men are not born equal ; the old shook their heads, and asked what was to come next ? There were gentlemen of the place who began to remark that the tradesfolk no longer doffed to them their bonnets, and tradesfolk that the gentlemen no longer sent them their newspapers. But the people got newspapers for themselves ;— these, too, of a very different stamp from the ones they had been accustomed to ; and a crop of young Whigs began to shoot up all over the place, like nettles in spring. They could not break into the meanings of all the new, hard-shelled words they were meeting with—words ending in *acy* and *archy ;* but no people could understand better that a king is only a kind of justice of the peace, who may be cashiered for misconduct just like any other magistrate ; that all men are naturally equal ; and that one whose grandfather had mended shoes, was every whit as well-born as one whose grandfather was the bastard of an emperor. And seldom were there people more zealous or less selfish in their devotion than the new-made politicians of Cromarty. Their own concerns gave place, as they ought, to the more important business of the state; and they actually hurt their own heads, and sometimes, when the ale was bad, their own bellies, in drinking healths to the French. Light after light gleamed upon them, like star after star in a frosty evening. First of all, Paine's Rights of Man shone upon them through the medium of the newspapers, with the glitter of fifty constellations; then the Resolutions of the Liberty and Equality clubs of the south looked down upon them with the effulgence of fifty more; at length, up rose the scheme for the division of property, like the moon at full, and, flaring with portentous splendour, cast all the others into comparative obscurity. The people looked round them at the parks which the modern scheme of agriculture had so conveniently fenced in with dikes and hedges; and spoke of the high price of potato-land and the coming Revolution.—A countryman went into one of the shops about this time, craving change for a pound-note. "A pound-note!" exclaimed the shopkeeper, snapping his fingers; —"a pound-note!—Man, I widna gie you tippence for't."

There was a young man of the place, the son of a shop-keeper, who had been marked from his earliest boyhood by a smart precocity of intellect, and the boldness of his opinions; his name (for I must not forget that, to borrow one of Johnson's figures, I am walking over ashes the fires of which are not yet extinguished) I shall conceal. He was one of those persons who, like the stormy petrel of the tropics, come abroad when the seas begin to rise, and the heavens to darken; and who find their proper element in a wild mixture of all the four elements jumbled into one. He read the newspapers, and, it was said, wrote for them; he corresponded, too, with the Jacobin clubs of the south, and strove to form similar clubs at home; but the people were not yet sufficiently ripe. No one could say that he was disobliging or ill-tempered; on the contrary, he was a favourite with, at least, his humbler town's-men for being much the reverse of both; but he was poor and clever, and alike impatient of poverty, and of seeing the wealth of the country in the hands of duller men than himself; and so the man who was unfortunate enough to be born to a thousand pounds a year had little chance of finding him either well-tempered or obliging. He had stept into the ferry-boat one morning, and the ferrymen had set themselves to their oars, when a neighbouring proprietor came down to the beach, and called on them to return and take him aboard. "Get on!" shouted the Democrat, "and let the fellow wait;—'tis I who have hired you this time." "O Sir! it's a shentleman," said one of the ferrymen, propelling the boat sternwards, as he spoke, by a back stroke of the oar. "Gentleman!" exclaimed the Democrat, seizing the boat-hook and pushing lustily in a contrary direction—"Gentleman truly!—we are all gentlemen, or shall be so very soon." The proprietor, meanwhile, made a dash at the rudder, and held fast, but with such good-will did the other ply the boat-hook, that ere he had made good a lodgment he was drenched to the armpits. "Nothing like being accustomed to hardship in time," muttered the Democrat, as, glancing his eye contemptuously on the dripping vestments of the proprietor, he laid down the pole and quietly resumed his seat.

There were about a dozen young men in the place who were so excited by the newspaper accounts of the superb processions of their south-country friends, that they resolved on having a procession of their own. They procured a long pole with a Kilmarnock cap fixed to the one end of it, which they termed the cap of liberty, and a large square of cotton, striped blue, white, and red, which they called the tricolor of independence. In the middle stripe there were inscribed in huge Roman capitals, the words Liberty and Equality; and a stuffed cormorant, intended

to represent an eagle, was perched on the top of the staff. They
got a shipmaster of the place prevailed on to join with them.
He was a frank, hearty sailor, who saw nothing unfair in the
anticipated division of property, and hated a pressgang as he
hated the devil. "But how," said he, "will we manage, after
all hands have been served out, should a few of us take a *bouse*
and melt our portions? just divide again, I suppose?" "Highly
probable," replied the revolutionists; "but we have not yet
fully determined on that." "I see, I see," rejoined the sailor;
"everything can't be done at once." On the day of the procession he brought with him his crew attired in their best, and
with all the ship flags mounted on poles. The revolutionists
demurred. "To be sure," they said, "nothing could be finer;
but then the flags were British flags." "And —— it," said
the master, "would you have me bring French flags?" It was
no time, however, to dispute the point; and the procession
moved on, followed by all the children of the place. It reached
an eminence directly above the links; and drawing up beside an
immense pile of brushwood, and a few empty tar-barrels, its
leader planted the tree of liberty amid shouts, and music, and
the shooting of muskets, on the very spot on which the town
gallows had been planted about two centuries before. No one,
however, so much as thought of the circumstance; for people
were too thoroughly excited to employ themselves with anything
but the future; besides, a very little ingenuity could have made
it serve the purpose of either party. After planting the tree,
the brushwood was fired, and a cask of whisky produced, out of
which the republicans drank healths to liberty and the French.
"The French! the French!" exclaimed the shipmaster.
"Well, —— them, I don't care though I do; here's health
to the French; may they and I live long enough to speak
to one another through twelve-pounders!" All the boys and
all the sailors huzza'd; the republicans said nothing, but
thought they had got rather a queer ally. The evening,

however, passed off in capital style; and, ere the crowd dispersed, they had burnt two fishing-boats, a salmon coble, and almost all the paling of the neighbouring fields and gardens.

The day of the procession was also that of a Redcastle market; at that time one of the chief cattle fairs of the north. It was largely attended on this occasion by Highlanders from the neighbouring straths, many of whom had fought for the Prince, and remembered the atrocities of Cumberland; it was attended, too, by parties of drovers from England and the southern counties of Scotland, all of them brimful of the modern doctrines, and scarcely more loyal than the Highlanders themselves; it was attended, besides, by a Cromarty salmon-fisher, George Hossack, a man of immense personal strength and high spirit, now a little past his prime perhaps, but so much a politician of the old school, that he would have willingly fought for his namesake the King with any two men at the fair. But he was no match for everybody, and everybody to-day seemed to hold but one opinion. "Awfu' expensive government this of ours," said an East-Lothian drover; "we maun just try whether we canna manage it mair cheaply for ourselves." "Ay, and what a blockhead of a king have we got!" said an Englishman; "not fit, as Tom Paine says, for a country constable; but, poor wretch, we must turn him about his business, and see whether he can't work like ourselves." "Och, but he's a limmer anyhow, and a creat plack whig!" remarked an old Highlander, "and has nae right till ta crown. Na, na, Charlie my king!" Poor George was almost broken-hearted by the abuse poured out against his sovereign on every side of him; but what could he do? He would look first at one speaker, then at another, and repress his rising wrath by the consideration, that there was little wit in being angry with about three thousand people at once. He had driven a bargain with two Englishmen, and on going in to drink with them, according to custom, was shown

2 H

into a room which chanced at the time, unlike every other room in the house, to be unoccupied. The Englishmen seated themselves at the table; George cautiously fastened the door, and took his place fronting them. "Now, gentlemen," said he, filling the glasses, "permit me to propose a toast :—Health and prosperity to George the Third." He drank off his glass, and set it down before him. One of the Englishmen, a bit of a wag in his way, looked at him with a droll, quizzical expression, and took up his. "Health and prosperity," he said, "to George the *herd*."—"Well, young man," remarked George, "he is, as you say, a herd, and a very excellent one ;—allow me, however, to wish him a less unruly charge." "Health and prosperity," shouted out the other, "to George the ———." This was unbearable : George sprang from his seat, and repaid the insult with a blow on the ear, which drove both man and glass to the floor. Up rose the other Englishman—up rose, too, the fallen one, and fell together upon George ; but the cause of the king was never yet better supported. Down they both went, the one over the other, and down they went a second, and a third, and a fourth time ; till at length, convinced that nothing could be more imprudent than their attempts to rise, they lay just where they fell. George departed, after discharging the reckoning, leaving them to congratulate one another on their liberalism and their wit ; and reached Cromarty as the last gleam of the Jacobin bonfire was dancing on the chimney-tops, to learn that there was scarcely more loyalty among his town's-men than at the market, and that his favourite salmon-coble had perished among the flames about two hours before. I remember George —a shrewd, clear-headed man of eighty-two, full of anecdote and remark ; and I have derived not a few of my best traditions from him. But he is gone, and well-nigh forgotten ; and when the sexton of some future age shall shovel up his huge bones, the men who come to gaze on them may descant, as they turn them over, on modern degeneracy and the might of their fathers, but who among them all will know that they belonged to the last of the loyalists!

The day after the procession came on, pregnant with mystery and conjecture. Rider after rider entered the town, and assembled in front of the council-house;—the town's officer was sent for, together with the sergeant of a small recruiting party that barracked in one of the neighbouring lanes; they then entered the hall, and made fast the doors. The country gentlemen, it was said, had come in to put down the revolutionists. Shortly after, two of the soldiers and a constable glided into the house of the young democrat, and producing a warrant for his apprehension, and the seizure of all his papers, hurried him away to the hall—the soldiers, with their bayonets fixed, guarding him on either side, and the constable, laden with a hamper of books and papers, bringing up the rear. In they all went, and the door closed as before. The curiosity of the town's-people was now awakened in right earnest, and an immense crowd gathered in front of the council-house; but they could see or hear nothing. At length the door opened, and the sergeant came out; he looked round about him, and beckoned on George Hossack. " George," he said, " one of the London smacks has just entered the bay; you must board her and seize on all the parcels addressed to * * * * the Jacobin merchant; there is an information lodged that he is getting a supply of pikes from London for arming the town's-people. Take the customhouse boatmen with you; and bring whatever you find to the hall. And, hark ye, we must see and get up an effigy of the blackguard Tom Paine; —try and procure some oakum and train-oil, and I'll furnish powder enough to blow him to Paris." Away went George, delighted with the commission, and returned in about an hour after, accompanied by some boatmen bearing two boxes large enough to contain pistols and pike-heads for all the men of the place. They were admitted into the hall, where they found the bench occupied by the town and county gentlemen—the soldiers ranged in the area in front, and the Republican, nothing abashed, standing at the bar. He had baffled all his judges, and had given them so much more wit and argument than they wanted, that they had ceased questioning him, and were now employed in turning over his papers. A letter written in cipher had been found on his person, and a gentleman, somewhat skilled in such matters, was examining it with much interest, while his more immediate neighbours were looking over his shoulders. " Bring forward the boxes, George," said one of the gentlemen. George placed them both on the large table fronting the bench, and proceeded to uncord them. The first he opened was filled with gingerbread, the other with girls' dolls and boys' whistles, and an endless variety of trinkets and toys of a similar class. Some of the elderly gentlemen took snuff and looked at one another ;—the younger laughed outright. " Have you deciphered that scrawl, Pointzfield ?" inquired one of the more serious, with a view of restoring the court to its gravity. " Yes," said Pointzfield dryly enough, " I rather think I have."—" Treasonable of course," remarked the other. " No, not quite that now," rejoined the other, " whatever it might have been fifty years ago. It is merely a copy in shorthand of the old Jacobitical ballad, the Sow's Tail to Geordie." A titter ran along the bench as before, and the court broke up after determining that the Democrat should be sent to the jail of Tain to abide further trial, and that Paine should be burnt in effigy at the expense of the county. Paine was accordingly burnt ; and all the children were gratified with a second procession and a second bonfire, quite as showy in their way as those of the preceding evening. The prisoner was escorted to Tain by a party of soldiers ; and on his release, which took place shortly after, he quitted the country for London, where he became the editor of a newspaper on the popular side, which he conducted for many years with much spirit and some ability. Meanwhile the revolutionary cause languished for lack of a leader ; and, on the declaration of war with France, sunk entirely amid the stormy ebullitions of a feeling still more popular than the Republican one.

There are some passions and employments of the human mind which give it a sceptical bias, and others, apparently of a very similar nature, which incline it to credulity. So long as the revolutionary spirit stalked abroad, it seemed as if every other spirit stayed at home. The spectre slept quietly in its churchyard, and the wraith in its pool ; the dead-light was hooded by an extinguisher, and the witch minded her own business without interfering with that of her neighbours. On the breaking out of the war, however, there came on a season of omens and prodigies, and the whole supernatural world seemed starting into as full activity as the fears and hopes of the community. Armies were seen fighting in the air, amid the waving of banners and the frequent flashing of cannon ; and the whole northern sky appeared for three nights together as if deluged with blood. In the vicinity of Inverness, shadowy bands of armed men were descried at twilight marching across the fields—at times half enveloped in smoke, at times levelling their arms as if for the charge. There was an ominously warlike spirit, too, among the children, which the elderly people did not at all like ;— they went about, just as before the American war, with their mimic drums and fifes, and their muskets and halberts of elder, disturbing the whole country with uncouth music, and their zeal against the French. Then came the tug of war ; trade sank ; and many of the mechanics of the place flung aside their tools and entered either the army or navy. Party spirit died ; the Whigs forgot everything but that they were Britons ; and when orders came that such of the males of the place as volunteered their services should be embodied into a kind of domestic militia, old men of seventy and upwards, some of whom had fought at Culloden, and striplings of fifteen, who had not yet left school, came to the house of their future colonel, begging

to be enrolled and furnished with arms.   In less than two days
every man in the town and parish was a soldier.   Then came
the stories of our great sea victories : the glare of illuminations
and bonfires ; the general anxiety when the intelligence first
arrived that a battle had been fought, and the general sadness
when it was ascertained that a town's-man had fallen.   When
the news of Duncan's victory came to the town, a little girl,
who had a brother a sailor, ran more than three miles into the
country, to a field in which her mother was employed in digging
potatoes, and falling down at her feet, had just breath enough
left to say, " Mither, mither, the Dutch are beaten, and Sandy's
safe."   The report of a threatened invasion knit the people still
more firmly together, and they began to hate the French, not
merely as national, but also as personal enemies.   And thus
they continued to feel, till at length the battle of Waterloo, by
terminating the war, reduced them to the necessity of seeking,
as before, their enemies at home.

For more than twenty years the words Whig and Tory had
well-nigh gone out ; and the younger town's-men were for some
time rather doubtful about their meaning.   At length, however,
they learned that *the Whigs* meant the people, and *the Tories*
those who wished to live by them, and yet call them names.
The town's-people, therefore, became Whigs to a man, execrated
the Holy Alliance and the massacre at Manchester, drank
healths to Queen Caroline and Henry Brougham ; and though
they petitioned against Catholic emancipation—for, like most
Scotch folks, they had too thorough a respect for their grandfathers to be wholly consistent—they were yet shrewd enough
to inquire whether any one had ever boasted of his country
because the great statesmen opposed to that measure were his
countrymen.   The Reform Bill, however, set them all right
again, by turning them full in the wake of their old leaders ;
and yet, no sooner was Whiggism intrusted with the keys of
office, than they began to make discoveries which had the effect

of considerably modifying the tone of their politics. They began to discover—will it be believed?—that all men are not born equal, and that there exists an aristocracy in the very economy of nature. It was not merely the choice of his countrymen that made Washington a great general, or Franklin a profound statesman. They have also begun to discover, that a good Whig may be a bad man; nay, that one may be at once Whig and Tory—a Tory to his servants and dependants, a Whig to his superiors and his country. For my own part, I am a Whig —a born Whig; but no similarity of political principle will ever lead me to put any confidence in the man to whom I could not intrust my private concerns; and as for the Whiggism that horsewhipped the poor woman who was picking a few withered sticks out of its hedge, it may wear the laurel leaf and the blue ribbon in any way it pleases, but I assure it—it won't be of my party.

Sanson & Co., Printers, Edinburgh.

Printed in the United States
76097LV00004B/46

9 78076